Experimental Manipulation of Gene Expression

Experimental Manipulation of Gene Expression

EDITED BY

Masayori Inouye

Department of Biochemistry
State University of New York at Stony Brook
Stony Brook, New York

1983

ACADEMIC PRESS

A Subsidiary of Harcourt Brace Jovanovich, Publishers
New York London
Paris San Diego San Francisco São Paulo Sydney Tokyo Toronto

ACADEMIC PRESS, INC.
111 Fifth Avenue, New York, New York 10003

United Kingdom Edition published by
ACADEMIC PRESS, INC. (LONDON) LTD.
24/28 Oval Road, London NW1 7DX

Library of Congress Cataloging in Publication Data

Main entry under title:

Experimental manipulation of gene expression.

Includes index.
1. Genetic engineering--Addresses, essays, lectures.
2. Gene expression--Addresses, essays, lectures.
3. Cloning--Addresses, essays, lectures.
I. Inouye, Masayori. [DNLM: 1. Cloning, Molecular. 2. DNA, Recombinant. 3. Genetic intervention. QH 442 E96]
QH442.E96 1983 575.1 83-2530
ISBN 0-12-372380-9

PRINTED IN THE UNITED STATES OF AMERICA

83 84 85 86 9 8 7 6 5 4 3 2 1

Contents

Appendix 1. Two-Dimensional DNA Electrophoretic Methods Utilizing *in Situ* Enzymatic Digestions

THOMAS YEE AND MASAYORI INOUYE

Appendix 2. Site-Specific Mutagenesis Using Synthetic Oligodeoxyribonucleotides as Mutagens

GEORGE P. VLASUK AND SUMIKO INOUYE

Contributors

Numbers in parentheses indicate the pages on which the authors' contributions begin.

MERVYN J. BIBB (53), John Innes Institute, Norwich NR4 7UH, England

JAMES R. BROACH (83), Department of Microbiology, State University of New York at Stony Brook, Stony Brook, New York 11794

J. PAUL BURNETT (259), Molecular and Cell Biology Research, Lilly Research Laboratories, Indianapolis, Indiana 46285

KEITH F. CHATER (53), John Innes Institute, Norwich NR4 7UH, England

JACK COLEMAN (15), Department of Biochemistry, State University of New York at Stony Brook, Stony Brook, New York 11794

RIK DERYNCK (247), Department of Molecular Biology, Genentech, Inc., South San Francisco, California 94080

DAVID DUBNAU (33), Department of Microbiology, The Public Health Research Institute of the City of New York, Inc., New York, New York 10016

MARY-JANE GETHING (225), Cold Spring Harbor Laboratory, Cold Spring Harbor, New York 11724

ELI GILBOA (175), Department of Biochemical Sciences, Princeton University, Princeton, New Jersey 08544

DAVID V. GOEDDEL (247), Department of Molecuar Biology, Genentech, Inc., South San Francisco, California 94080

MAX E. GOTTESMAN (137), Laboratory of Molecular Biology, Division of Cancer Biology and Diagnosis, National Cancer Institute, National Institutes of Health, Bethesda, Maryland 20205

PATRICK W. GRAY (247), Department of Molecular Biology, Genentech, Inc., South San Francisco, California 94080

DEAN H. HAMER (191), Laboratory of Biochemistry, National Cancer Institute, National Institues of Health, Bethesda, Maryland 20205

PATRICK HEARING (211), Department of Microbiology, Health Sciences Center, State University of New York at Stony Brook, Stony Brook, New York 11794

RONALD A HITZEMAN (247), Department of Molecular Biology, Genentech, Inc., South San Francisco, California 94080

YEN-SEN HO[1] (1), Laboratory of Biochemistry, National Cancer Institute, National Institutes of Health, Bethesda, Maryland 20205

DAVID A. HOPWOOD (53), John Innes Institute, Norwich NR4 7UH, England

BRUCE H. HOWARD (137), Laboratory of Molecular Biology, Division of Cancer Biology and Diagnosis, National Cancer Institute, National Institutes of Health, Bethesda, Maryland 20205

MASAYORI INOUYE (15, 279), Department of Biochemistry, State University of New York at Stony Brook, Stony Brook, New York 11794

SUMIKO INOUYE (291), Department of Biochemistry, State University of New York at Stony Brook, Stony Brook, New York 11794

MAKKUNI JAYARAM (83), Department of Microbiology, State University of New York at Stony Brook, Stony Brook, New York 11794

JOHN D. KEMP (119), Agrigenetics Advanced Research Laboratory, Madison, Wisconsin 53716

YU-YANG LI (83), Department of Microbiology, State University of New York at Stony Brook, Stony Brook, New York 11794

YOSHIHIRO MASUI[2] (15), Department of Biochemistry, State University of New York at Stony Brook, Stony Brook, New York 11794

RICHARD C. MULLIGAN (155), Center for Cancer Research and Department of Biology, Massachusetts Institute of Technology, Cambridge, Massachusetts 02139

MARTIN ROSENBERG[1] (1), Laboratory of Biochemistry, National Cancer Institute, National Institutes of Health, Bethesda, Maryland 20205

JOE SAMBROOK (225), Cold Spring Harbor Laboratory, Cold Spring Harbor, New York 11724

ALLAN SHATZMAN[1] (1), Laboratory of Biochemistry, National Cancer Institute, National Institutes of Health, Bethesda, Maryland 20205

THOMAS SHENK (211), Department of Microbiology, Health Sciences Center, State University of New York at Stony Brook, Stony Brook, New York 11794

GEORGE P. VLASUK (291), Department of Biochemistry, State University of New York at Stony Brook, Stony Brook, New York 11794

LING-CHUAN CHEN WU (83), Department of Microbiology, State University of New York at Stony Brook, Stony Brook, New York 11794

THOMAS YEE (279), Department of Biochemistry, State University of New York at Stony Brook, Stony Brook, New York 11794

[1]Present address: Smith Kline & French Laboratories, Philadelphia, Pennsylvania 19101.
[2]Present address: Peptide Institute, Protein Research Foundation, Osaka 562, Japan.

Preface

Cloning a foreign gene or a gene of interest for expression in a particular host system is a major task in recombinant DNA technology. The manipulations require extensive knowledge of the regulatory mechanisms of gene expression and basic understanding of vector–host relationships as well as skills in recombinant DNA techniques.

It is now possible to clone and express a gene of interest in widely different host systems, from bacteria to yeast, plant cells, and mammalian cells. In this book, we have attempted to cover all of these systems so that the reader may quickly learn the versatility of the systems and obtain an overview of the technology involved in the manipulation of gene expression. Furthermore, I hope that the reader will learn enough from the various approaches to be able to develop systems and to arrange for a gene of particular interest to express in a particular system. The appendix contains two useful state-of-the-art techniques.

The original idea of putting together a book under the present title came from a symposium with the same title that was held at the State University of New York at Stony Brook in May 1982. Although several of the contributors to this volume were, in fact, speakers at that symposium, this volume has gone far beyond the symposium in the breadth and depth of the material covered. We are indeed fortunate that people who are in the forefront of the field have contributed chapters to this book.

I wish to thank my secretary, Janet Koenig, for her invaluable liaison work between editor and contributors.

CHAPTER 1

Use of Phage λ Regulatory Signals to Obtain Efficient Expression of Genes in *Escherichia coli*

ALLAN SHATZMAN*
YEN-SEN HO*
MARTIN ROSENBERG*

Laboratory of Biochemistry
National Cancer Institute
National Institutes of Health
Bethesda, Maryland

*Present address: Molecular Genetics, Smith Kline & French Laboratories, Philadelphia, Pennsylvania.

Experimental Manipulation
of Gene Expression

ISBN 0-12-372380-9

I. Introduction

There are many gene products of biological interest that cannot be studied simply because they cannot be obtained in quantities sufficient for biochemical analysis. Recombinant technology now provides us with some new approaches to the solution of this problem. This chapter describes the design and construction of a plasmid vector system used to achieve high-level expression of a particular phage regulatory protein normally found in only minute amounts in a phage-infected bacterial cell. We also describe the subsequent development of this plasmid into a vector system that offers the potential for efficiently expressing essentially any gene product in *Escherichia coli.*

II. Expression of Prokaryotic Gene Products

A. Overproduction of the Phage λ Regulatory Protein *cII*

The *cII* protein of bacteriophage λ is a transcriptional activator that positively regulates expression from two different phage promoters (Herskowitz and Hage, 1980; Shimatake and Rosenberg, 1981). These two promoters are coordinately controlled by this effector, and this control is essential for normal integration of the phage into the host genome (i.e., formation of a bacterial lysogen). In order to study *cII* and its role in transcriptional activation, it was necessary to express the gene product at levels sufficient to allow its purification and biochemical analysis. Because *cII* is made in only small amounts during a normal phage infection and presents the added difficulty of being rapidly turned over by the host, the protein could not be easily detected or obtained.

1. CLONING STRATEGIES

Initially, we attempted to increase the cellular levels of *cII* by cloning the gene onto a multicopy plasmid vector (i.e., a pBR322 derivative) such that it was transcribed from a "strong" bacterial promoter (Fig. 1). A DNA fragment (1.3 kb) carrying the *cII* gene and the N-terminal 80% of the phage O gene (Fig. 1A) was inserted downstream of different promoter signals that had been cloned onto various pBR322 derivatives (Fig. 1B). Stable transformants containing *cII* as part of a constitutive transcription unit could not be obtained (Fig. 1a). Instead, the only stable recombinants isolated were those carrying the gene in orientation

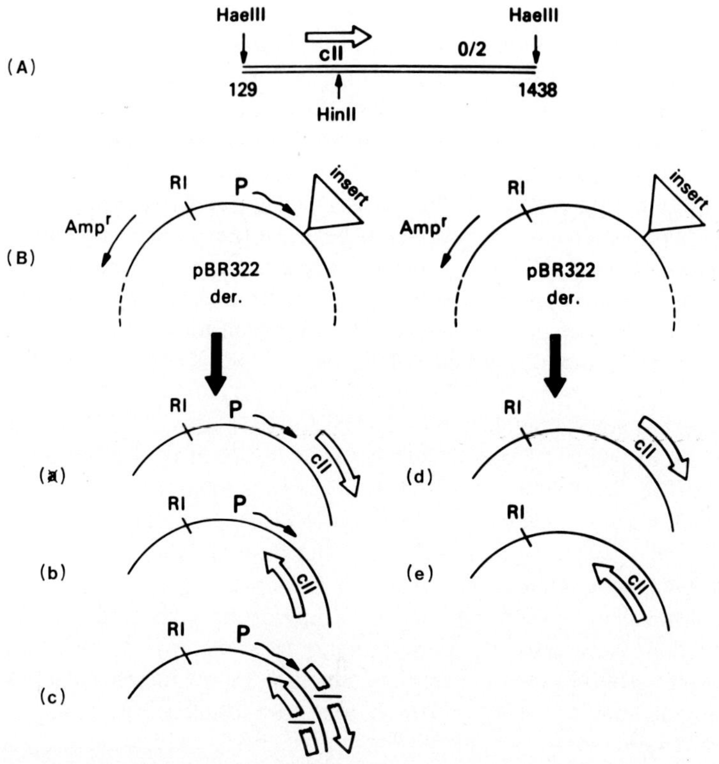

Fig. 1 Scheme of the cloning strategy used initially to overproduce the phage λ *cII* protein. (A) A 1.3-kb *Hae*III restriction fragment from phage λ that carries the entire *cII* gene and part of the O gene (for details see text; Shimatake and Rosenberg, 1981). (B) Cloning scheme for inserting the fragment shown in (A) into pBR322 derivatives that either do or do not carry a promoter signal P upstream of the site of insertion (see text for details).

opposite to the direction of transcription (Fig. 1b). Stable isolates were also obtained using vectors that did not provide for transcription of the *cII* gene (Fig. 1d,e) or by fragmenting the *cII* gene and cloning pieces into a transcription unit (Fig. 1c). These results suggested that increased expression of *cII* is an event lethal to the bacterial cell. We now know that this is a common phenomenon when gene products are overproduced in *E. coli* using recombinant procedures. Clearly, *cII* provides an excellent model system for attempting to obtain efficient expression of a lethal gene product that is relatively unstable in the bacterial cell, besides being an interesting transcriptional regulatory molecule.

In order to clone the *cII* gene onto a multicopy plasmid vector as part of an efficient transcriptional unit, we reasoned that the promoter signal used would have to be regulated. This would enable us to obtain the desired recombinant under conditions in which the *cII* gene is not expressed. Cells carrying this vector could then be grown to high density and subsequently induced for expression of *cII*. Because the cells were already at high density, the lethality of the product would be of little consequence. Moreover, if the induction and expression were efficient, the loss of product due to turnover would be minimized. For these reasons and others to be discussed later, we placed the *cII* gene under the transcriptional regulation of the bacteriophage promoter signal, P_L. This "strong" promoter has been shown to be 8–10 times more efficient than the well-characterized promoter P_{lac}, which regulates the *E. coli* lactose operon. In fact, using a vector system designed specifically for comparing and characterizing transcriptional regulatory signals, P_L was found to be as or more efficient than all other promoters tested (McKenney *et al.*, 1981; Rosenberg *et al.*, 1982b).

The P_L promoter, contained on a 2.4-kb *Hin*dIII–*Bam*HI restriction fragment derived from phage λ, was cloned between the *Hin*dIII and *Bam*HI sites of pBR322 (Fig. 2). The resulting vector pKC30 was found to be highly unstable. In general, plasmids that carry strong promoters are unstable, presumably owing to detrimental effects on replication resulting from high levels of transcription. Our problem of instability was overcome by repressing P_L transcription using bacterial hosts that contain an integrated copy of the λ genome (i.e., bacterial lysogens). In these cells, P_L transcription is controlled by the phage-repressor protein (*cI*), a product synthesized continuously and regulated autogenously in the lysogen (Ptashne *et al.*, 1976). It was demonstrated that certain lysogens synthesize repressor in amounts sufficient to inhibit P_L expression completely on the multicopy vector. Thus the cells can be stably transformed and the vector maintained in these lysogenic hosts. Moreover, by using a lysogen carrying a temperature-sensitive mutation in the *cI* gene (*cI857*; Sussman and Jacob, 1962), P_L-directed transcription can be activated at any time. Induction is accomplished by simply raising the temperature of the cell culture from 32 to 42°C. Thus cells carrying the vector can be grown to high density at 32°C without expression of the cloned gene and subsequently induced at 42°C to synthesize the product.

In addition to providing a strong, regulatable promoter, the system also insures that P_L-directed transcription efficiently traverses the gene insert. This is accomplished by providing both the phage-antitermination function N and a site on the P_L transcription unit necessary for N utilization (*Nut* site). N is provided in single copy from the host lysogen, and its expression is induced by temperature, concomitant with P_L transcription on the vector. N expression removes transcriptional polarity, thereby alleviating termination within the P_L transcription unit. Hence, any transcriptional polarity caused by sequences that occur before or within the gene-coding sequence is eliminated by the N + *Nut* system. This antitermination effect is particularly important for the expression of *cII* because a transcription-termination signal, *tR1* (Fig. 2), is positioned immediately up-

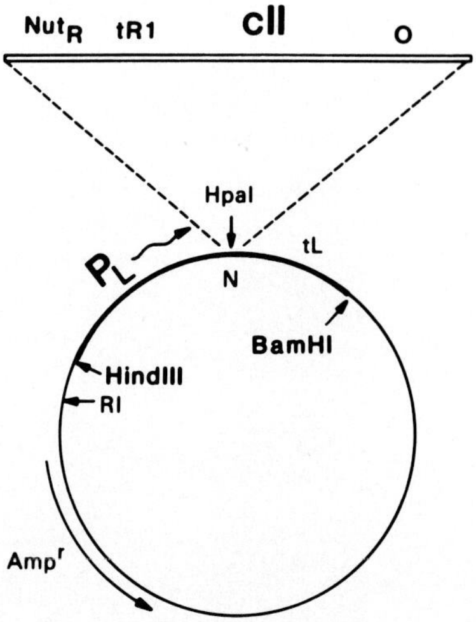

Fig. 2 Schematic diagram of plasmid pKC30 and the insertion of the *Hae*III restriction fragment (see Fig. 1A) carrying the λ *cII* gene into the *Hpa*I restriction site. pKC30 is a derivative of pBR322 and contains a 2.4-kb *Hin*dIII–*Bam*HI restriction fragment derived from phage λ inserted between the *Hin*dIII and *Bam*HI restriction sites within the tetracycline gene of pBR322 (R. N. Rao, unpublished). This insert contains the operator (O_L) and a site for N recognition (*Nut*$_L$) (Franklin and Bennett, 1979). DNA fragments carrying the appropriate gene are cloned into the *Hpa*I restriction site that occurs within the N-gene coding region. The fragment containing the *cII* gene also contains a site for N recognition (*Nut*$_R$), the ρ-dependent transcription termination site *tR1*, and most of the O region (Rosenberg *et al.*, 1978).

stream of the *cII* coding region. Experiments in our laboratory have shown that the single-copy lysogen provides N in amounts sufficient to antiterminate completely all transcription through *tR1*, as well as other terminators positioned on the multicopy plasmid. As we demonstrate later, because the N + *Nut* system generally relieves all transcriptional polarity, it leads to dramatic increases in product yield and allows greater flexibility in inserting gene fragments into the pKC30 vector.

2. EXPRESSION OF THE CLONED *cII* GENE

The *cII* gene was inserted into pKC30 at the unique *Hpa*I restriction site positioned 321 bp downstream from the P_L promoter (Fig. 2). In contrast to our earlier results, stable recombinants were now obtained that carried *cII* in the correct orientation (i.e., pKC30*cII*), and these recombinants could be maintained at 32°C in certain host lyogens. Cells carrying pKC30*cII* were grown at 32°C and induced at 42°C for 40 min, after which total cellular protein was characterized on SDS-polyacrylamide gels and the major accumulated proteins detected by

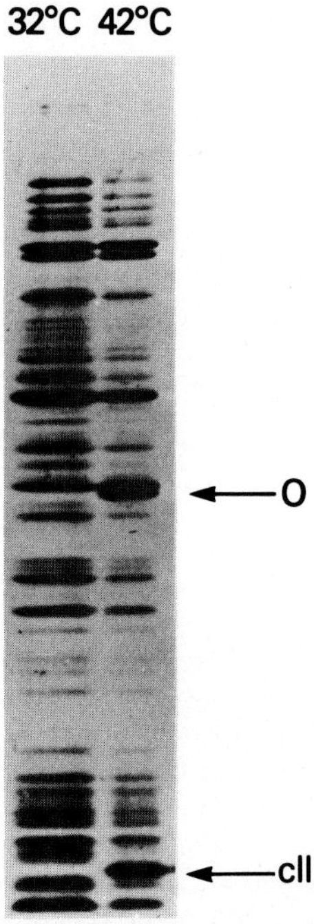

Fig. 3 Coomassie Blue–stained SDS-polyacrylamide gel analysis of total cellular protein made in a λ *cI857* lysogen carrying pKC30*cII* before and after a 40-min heat induction. (For details see Shimatake and Rosenberg, 1981; Ho *et al.*, 1982; and Rosenberg, 1982.) The positions of the *cII* protein and the O fusion product are indicated.

staining (Fig. 3). Two major products accumulated in response to the induction, and these were identified as the *cII* and O proteins. We estimated that the *cII* protein (MW 10,500) represented approximately 5% of total cellular protein, and even higher yields were obtained for O protein. A simple purification procedure was developed (Ho *et al.*, 1982) that allows isolation of pure *cII* at yields approaching 2 mgm/gm wet weight of cell culture. For the first time, *cII* protein was produced and obtained in quantities sufficient for its detailed physical and biochemical analysis (Ho *et al.*, 1982).

B. Expression of Other Genes in pKC30

The pKC30 vector systems has been used to express efficiently a variety of other phage and bacterial gene products including *nusA* (P. Olins, B. Ericksen, and R. Burgess, personal communication), the σ subunit of RNA polymerase

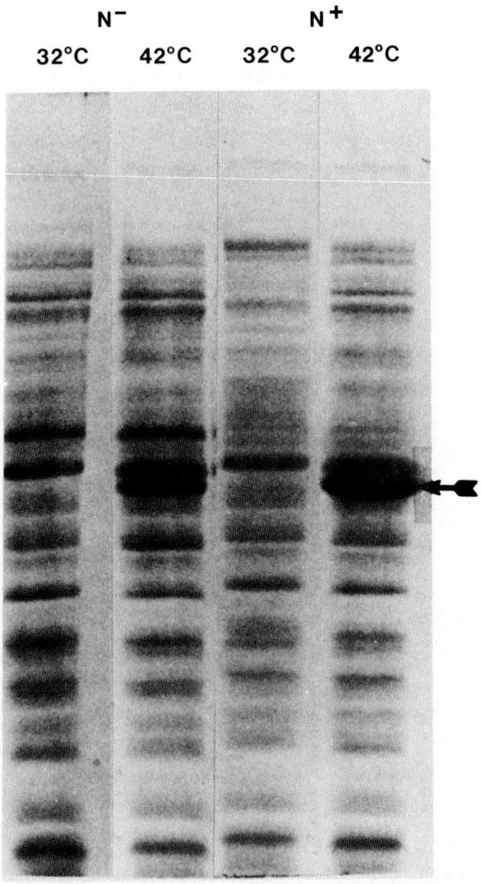

Fig. 4 Coomassie Blue–stained SDS-polyacrylamide gel analysis of total cellular protein obtained from an N^+- and an N^--defective lysogen carrying the plasmid pKC30*galK*. Protein samples were prepared from cells before (at 32°C) and 60 min after thermal induction (at 42°C). The position of the galactokinase gene product is indicated by an arrow.

(M. Gribskow and R. Burgess, personal communication), and *uvrA* (Yoakum *et al.*, 1982) from *E. coli* as well as certain *Bacillus subtilis* phage φ29 proteins (Garcia *et al.*, 1982). In fact, the potential exists for any gene to be expressed that is preceded by a ribosome-binding signal recognized in *E. coli*. The gene is simply inserted on a blunt-ended DNA fragment into the blunt-ended *Hpa*I site on pKC30. The appropriate DNA fragment can be generated (1) directly by restriction, (2) subsequent to the removal of single-strand overhanging ends by the action of S_1 or mung bean nuclease, or (3) subsequent to ''fill-in'' of single-strand overhanging ends by the use of DNA polymerase (Klenow fragment). The *Hpa*I site of pKC30 has also been adapted for the insertion of other DNA fragments by first introducing a synthetic linker that contains multiple unique restriction sites (B. Ferguson, A. Shatzman, and M. Rosenberg, unpublished). In addition, any of these sites can be used in combination with other unique restriction sites positioned downstream from the *Hpa*I site of pKC30. Synthetic linkers carrying additional cloning sites have also been introduced into several of these downstream positions, thereby maximizing the versatility of the system for insertion of DNA fragments.

We noted previously the importance of the N + *Nut* antitermination system for achieving high-level expression of the *cII* protein. Production of *cII* was found to be 8–10 times higher in lysogens that provided N as opposed to those that do not (H. Shimatake and M. Rosenberg, unpublished). More recent experiments indicate that the N + *Nut* system leads to increased expression of other genes cloned into pKC30 that do not have terminator signals preceding them. For example, the *E. coli* galactokinase gene *galK* was cloned into the *Hpa*I site of pKC30. Expression was monitored before and after induction in both an N^+ and N^- lysogen. After induction, 3–4 times more *galK* was produced in the N^+ cells than in the N^- cells (Fig. 4). No *galK* was detected in either background before induction. The effect of N on *galK* may be related to its ability to relieve certain natural polarity occurring before or within the *galK* coding region. However, it is also possible that N may directly affect the translation of genes that come under its transcriptional control. Preliminary experiments (D. L. Wulff, unpublished; Shatzman *et al.*, 1982) suggest that N may directly increase the translational efficiency of certain genes within an N-dependent transcription unit. Whatever the case, the effect is significant (*galK* is produced as 20% of total cellular protein in the N^+ condition) and eliminates the necessity for manipulating sequences upstream of the gene that might otherwise interfere with its expression.

III. Expression of Eukaryotic Genes

A. Vector Construction

In order to extend the pKC30 system to the expression of genes lacking *E. coli* translational regulatory information, an efficient ribosome-recognition and translation-initiation site was engineered into the P_L transcription unit. The site chosen was that of the efficiently translated λ phage gene *cII*. The entire coding region of this gene was removed, leaving only its initiator *f-met* codon and

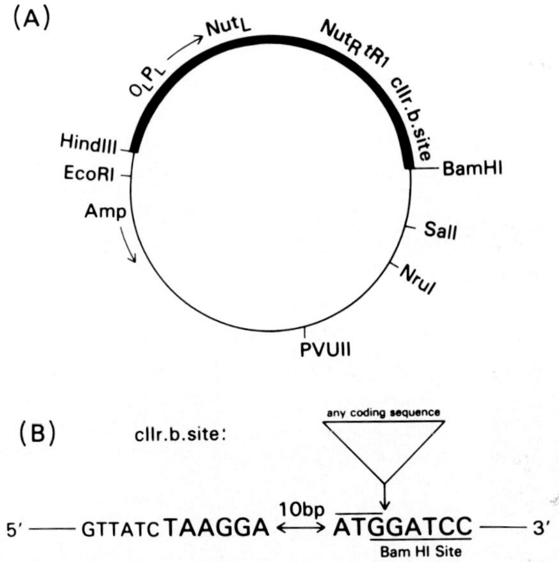

Fig. 5 (A) Scheme of plasmid pAS1 (5878 bp). This vector is a derivative of pKC30 and is made of phage λ sequences (thick segment) inserted between the *Hin*dIII and *Bam*HI restriction sites in pBR322. The region of pKC30 between the *Hpa*I (36060) and *Bam*HI (35301) sites has been deleted, and a portion of the fragment shown in Fig. 1b, extending from the *Hae*III site (38981) to and including the ribosome-binding site (r.b. site) and ATG initiation codon of *cII*, has been inserted (see the text for other details). (B) Region of pAS1 surrounding the *cII* translational regulatory information. The ribosome-binding site and ATG initiation codon (overscored) of *cII* are indicated, as is the unique *Bam*HI site (underscored) providing access to this regulatory information. DNA fragments are inserted into this site (details in the text).

upstream regulatory sequences. Neither the sequence nor the position of any nucleotides in the ribosome-binding region was altered. Instead, a restriction site for insertion of the desired gene was introduced immediately downstream from the ATG initiation codon. This was done by fusing the *Bam*HI site of pBR322 directly to the *cII* ATG codon (Fig. 5A). This fusion retains the *Bam*HI site and positions one side of the staggered cut immediately adjacent to the ATG codon, permitting ready access to the *cII* translational regulatory information (Fig. 5B). The resulting vector pAS1 allows direct fusion of any coding sequence (prokaryotic, eukaryotic, or synthetic) to the *cII* translational regulatory signal. Of course, it is important that all fusions between the gene-coding sequence and the *cII* initiation codon maintain the correct translational reading frame. Essentially, any gene can be adapted for insertion into the pAS1 vector; various examples are described later. Note that expression of genes cloned into pAS1 is controlled by temperature induction, exactly analogous to the pKC30 vector system. Thus

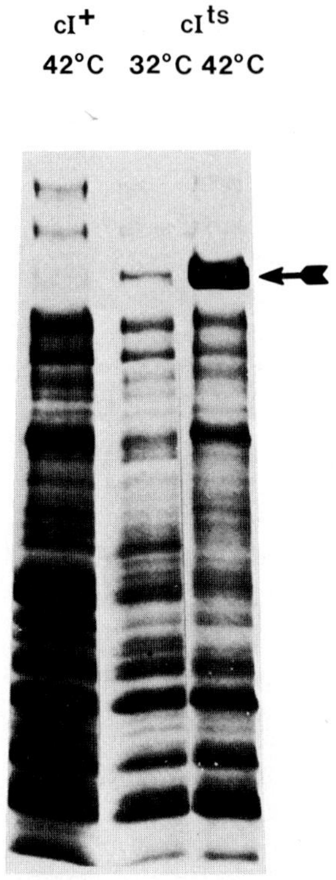

Fig. 6 Coomassie Blue–stained SDS-polyacrylamide gel analysis of total cellular protein obtained from a cI^+ and a cI^{ts} host lysogen carrying the plasmid pAS1β-*gal*. The position of the β-galactosidase gene product is indicated by an arrow.

"foreign" gene products that prove lethal to bacteria or that are "turned over" in bacteria potentially can be expressed using this system.

B. Expression of *lacZ* in pAS1

In order to test the ability of the *cII* ribosome-binding site to direct translation of another gene, we initially fused the *E. coli lacZ* gene to the *cII* ATG initiation codon. This was accomplished by using a *lacZ* gene construction into which a unique *Bam*HI restriction site had been engineered near the 5′ end of the gene (a gift from M. Casadaban). Direct ligation of this *Bam*HI site to the *Bam*HI site in pAS1 created the appropriate in-frame fusion of *lacZ* to the *cII* ATG codon. In this vector *lacZ* expression is controlled entirely by the transcriptional and translational signals provided on pAS1. As shown in Fig. 6, the pAS1 *lacZ* construction results in high-level expression of β-galactosidase. After only 1 hr of temperature induction, β-galactosidase accounted for 30–40% of total cellular protein.

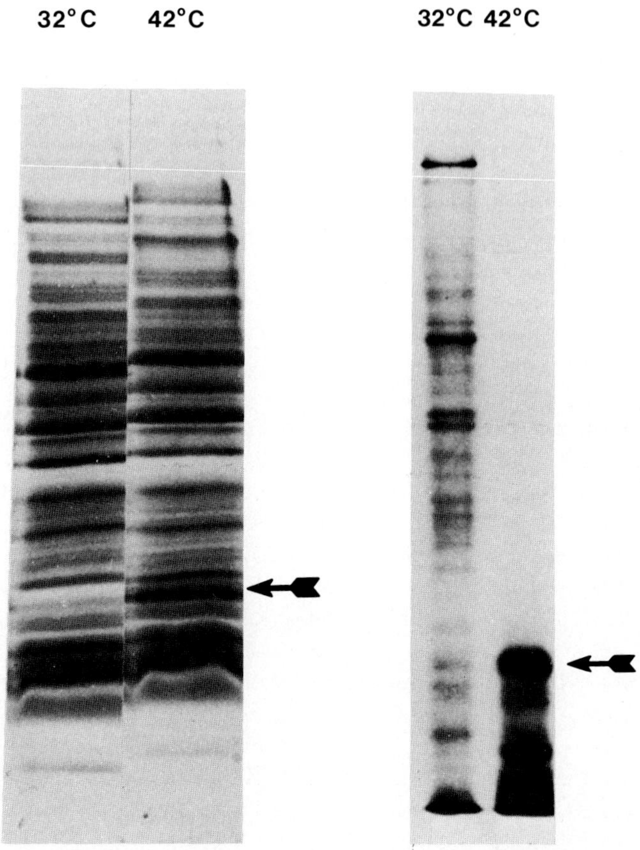

Fig. 7 (Left) Analysis of total cellular protein obtained from a cI^{ts} lysogen carrying the plasmid pAS1t. The position of the SV40 small-t antigen is indicated by an arrow. (Right) Autoradiogram of an SDS-polyacrylamide gel analysis of [^{35}S]methionine-pulse-labeled (45 sec) proteins synthesized in a cI^{ts} lysogen carrying pAS1t before and 60 min after heat induction.

C. Expression of SV40 Small-t Antigen

Unlike the *lacZ* construction, most genes do not contain the restriction information necessary for their direct insertion into the *Bam*HI site of pAS1. Thus it was necessary to provide greater flexibility for inserting DNA fragments into the vector. This was accomplished by converting the *Bam*HI site of pAS1 into a blunt-ended cloning site. The 4-base 5′ overhanging end of the *Bam*HI cleavage site can be removed using mung bean nuclease, thereby creating a blunt-ended cloning site immediately adjacent to the *cII* initiation codon. Any gene containing any restriction site properly positioned at or near its 5′ end can now be inserted into this vehicle. Blunt-ended fragments can be inserted directly, whereas other restriction fragments must first be made blunt-ended. This is accomplished by either removing the 5′ and 3′ overhanging ends with mung bean nuclease (as described previously) or by "filling in" the 5′ overhanging ends with DNA polymerase. Of course, even this procedure still limits the use of pAS1 to those genes that contain appropriate restriction information near their 5′ termini. In order to make the pAS1 system generally applicable to the expression of any gene, a procedure was developed that allows precise placement of a new restriction site at the second codon (or any other codon) of any gene. Creation of this site permits fusion of the gene in frame to the *cII* initiation codon of pAS1. The general procedure has been outlined by Maniatis *et al.* (1982) and described in detail by Rosenberg *et al.* (1982a). This procedure has been used to adapt several genes for insertion and expression into pAS1.

The small-t-antigen gene of SV40 does not contain an appropriate restriction site at its 5′ end. Using *Bal*31 exonucleolytic digestion from an upstream restriction site, the first base (G) of the second codon of the small-t gene (ATG GAT. . .) was fused to an upstream, filled-in *Ava*I restriction site (. . .CCCGA) (C. Queen, personal communication). The fusion, (...CCCGAGAT...) recreates the *Ava*I site precisely at the second codon of the small-t gene. Restriction of this vector with *Ava*I followed by mung bean nuclease digestion produces a blunt end that was fused in frame to the blunt-ended *Bam*HI site of pAS1. The resulting vector pAS1t expresses authentic SV40 small-t antigen entirely from phage-regulatory signals. After only a 60-min induction period, small-t antigen represented some 10% of the total cellular protein (Fig. 7, left). Moreover, ^{35}S-pulse-labeling experiments indicate that small t is the major product synthesized in these bacteria after temperature induction (Fig. 7, right).

D. Expression of Functional Mammalian Metallothioneines

Procedures similar to those just described have also been used to adapt, insert, and express both the mouse and monkey metallothioneine genes in the pAS1 vector system. The mouse gene (a gift from R. Palmiter) contains a unique *Ava*II restriction site at its second codon (5′ ATG GAC CCC 3′). Cleavage with *Ava*II, followed by fill-in of the 3-base 5′ overhanging end with DNA polymerase, creates a blunt end before the first base pair of the second codon. This blunt-ended fragment was inserted into the mung bean nuclease-treated pAS1 vector. The resulting fusion places the second codon of metallothioneine in frame with the initiation codon of *cII*. This construction (pAS1–MTL1) results in high levels of synthesis of the authentic metallothioneine gene product in *E. coli* (Fig. 8). The protein produced in bacteria has been partially purified and shown to have

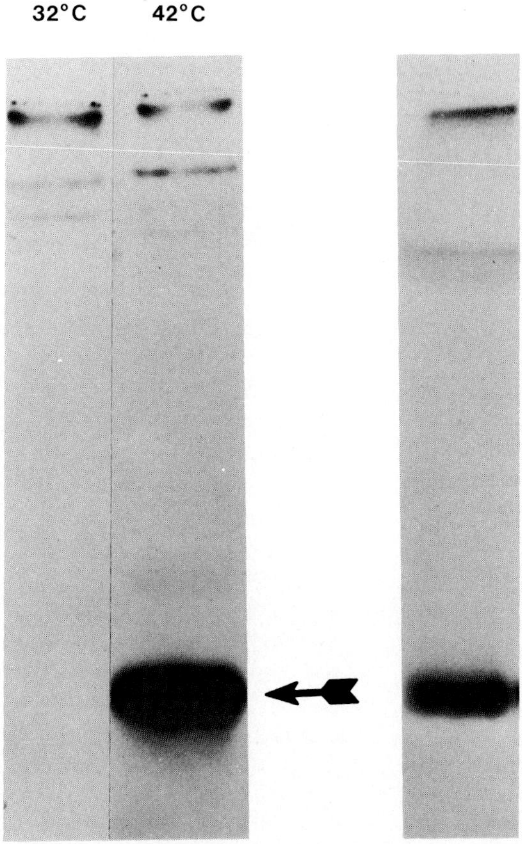

Fig. 8 (Left) Autoradiogram of an SDS-polyacryamide gel analysis of [^{35}S]cysteine-pulse-labeled proteins synthesized in a defective *cI*^{ts} lysogen, carrying pASI–MTL1, before and 20 min after heat induction. (Right) Autoradiogram of [^{35}S]cysteine-pulse-labeled, Cd^{2+}-induced mouse metallothioneine, shown for comparison.

electrophoretic characteristics identical to the protein obtained from Cd^{2+}-induced mouse cells. Similar results were obtained with a monkey metallothioneine gene (a gift from D. Hamer).

More recent experiments indicate that the induction of the metallothioneine gene product confers resistance to heavy-metal toxicity in *E. coli.* Bacteria synthesizing metallothioneine survive and continue growing in media to which heavy metals such as silver or mercury have been added. Cells not carrying the metallothioneine gene, or those carrying the gene but not induced for its expression, are killed readily by the addition of heavy metals. These bacterial cultures, which can be precisely regulated to efficiently produce metallothioneines, have a variety of potentially useful applications. It becomes increasingly clear that the pKC30–pAS1 vector system offers the potential for efficiently expressing functional gene products from essentially any organism within the bacterial cell.

Acknowledgment

We thank Linda Hampton for typing and editing the manuscript.

References

Franklin, N. C., and Bennett, G. N. (1979). *Gene* **8,** 197–205.

Garcia, J., Pastrana, R., Prieto, I., and Salas, M. (1982). *Gene,* in press.

Herskowitz, I., and Hage D. (1980). *Annu. Rev. Genet.* **14,** 309–445.

Ho, Y. S., Lewis, M., and Rosenberg, M. (1982). *J. Biol. Chem.* **257,** 9128–9134.

Maniatis, T., Fritsch, E., and Sambrook, J. (1982). *In* "Molecular Cloning—A Laboratory Manual," pp. 418–421. Cold Spring Harbor Lab., Cold Spring Harbor, New York.

McKenney, K., Shimatake, H., Court, D., Schmeissner, U., Brady, C., and Rosenberg, M. (1981). *In* "Gene Amplification and Analysis" (J. G. Chirikjian, and T. Papas, eds.), Vol. 2, pp. 383–415. Elsevier-North Holland, New York.

Ptashne, M., Backman, K., Humayun, M. Z., Jeffery, A., Maurer, R., Meyer, B., and Sauer, R. T. (1976). *Science (Washington, D.C.)* **194,** 156–161.

Rosenberg, M., Court, D., Shimatake, H., Brady, C., and Wulff, D. (1978). *Nature (London)* **272,** 414–423.

Rosenberg, M., Ho, Y. S., and Shatzman, A. (1982a). *Methods Enzymol,* in press.

Rosenberg, M., McKenney, K., and Schumperli, D. (1982b). *In* "Promoters: Structure and Function" (M. Chamberlin, and R. L. Rodriguez, eds.), pp. 387–406. Praeger, New York.

Shatzman, A., Debouck, C., and Rosenberg, M. (1982). *Cold Spring Harbor Bacteriophage Symp. Abstr.,* p. 83.

Shimatake, H., and Rosenberg, M. (1981). *Nature (London)* **292,** 128–132.

Sussman, R., and Jacob, F. (1962). *C.R. Hebd. Seances Acad. Sci. Ser. A* **254,** 1517–1519.

Yoakum, G., Yeung, A., Matles, W., and Grossman, L. (1982). *Proc. Natl. Acad. Sci. USA* **79,** 1766–1770.

CHAPTER 2

Multipurpose Expression Cloning Vehicles in *Escherichia coli*

YOSHIHIRO MASUI*
JACK COLEMAN
MASAYORI INOUYE

Department of Biochemistry
State University of New York at Stony Brook
Stony Brook, New York

I. Introduction

This chapter describes a series of versatile high-level expression vectors that utilize the very efficient *Escherichia coli* lipoprotein promoter as well as various other portions of the lipoprotein gene *lpp*. Needless to say, *E. coli* provides an

*Present address: Peptide Institute, Protein Research Foundation, Osaka, Japan.

Experimental Manipulation
of Gene Expression

ISBN 0-12-372380-9

ideal system for gene expression because it is one of the most well-studied and understood organisms. There are two major reasons for us to choose the *lpp* gene for development of expression cloning vectors:

1. The lipoprotein is the most abundant protein in *E. coli* in terms of numbers of molecules (for a review see Inouye, 1979). Therefore, it is considered that the *lpp* gene has one of the strongest promoters in *E. coli.*
2. The lipoprotein is an outer-membrane protein, and it is synthesized initially as a secretory precursor, prolipoprotein, which has an N-terminus extension consisting of 20 amino acid residues (signal peptide). The signal peptide plays an important role for protein secretion across the cytoplasmic membrane (Inouye *et al.,* 1977; Halegoua and Inouye, 1979). After the cleavage of the signal peptide, the new N-terminus is extensively modified to form a glycerylcysteine residue, to which a fatty acid residue is linked by an amide linkage and two other fatty acid residues attached by ester linkages (see review by Inouye, 1979). This N-terminal structure is believed to play an important role in the assembly of the lipoprotein into the outer membrane.

Because of these properties, *lpp* provides an ideal system to construct multipurpose expression cloning vectors that would not only allow abundant production of a cloned gene product but that also could guide the final localization of the cloned gene product, depending on where within the *lpp* gene a foreign gene has been inserted. Thus one may be able to localize the cloned gene product into the various compartments of *E. coli* cells (e.g., cytoplasm, cytoplasmic membrane, and the periplasmic space—the space between the cytoplasmic and the outer membranes) and the outer membrane.

In this chapter we describe unique features of various expression vectors developed in our laboratory and focus mainly on how to utilize these vectors. The pIN-I vectors express the inserted gene at high levels constitutively. For proteins that may be lethal to the cell when produced at high levels, we have constructed the pIN-II vectors, which contain the *lacUV5* promoter–operator inserted downstream of the *lpp* promoter. This arrangement represses transcription in the absence of a *lac* inducer, if the host strain can produce enough *lac* repressor molecules. We have also constructed pIN-III vectors by inserting the *lacI* gene into the pIN-II vectors so that the expression of the cloned gene can be regulated by the *lac* repressor produced by the vector itself. Therefore, in the case of pIN-III vectors, any host strain may be used.

The pIN vectors just described use pBR322 as their vehicle. In order to increase the copy number per cell of a vector, we have also constructed pIM vectors that at high temperatures are able to increase their copy numbers to as high as 2000 copies/cell.

We also describe vectors (pIC vectors) that can easily be used to clone and express a portion of a foreign protein to produce a hybrid protein with β-galactosidase. This hybrid protein can then be used for developing antiserum against that foreign protein.

Finally, we describe two other vectors that can be used to clone a promoter and to determine the efficiency of that promoter. The first vector (pKEN005) turns on tetracycline resistance when a DNA fragment carrying a promoter is inserted in the correct orientation. The second promoter-fusion vector (pKM005) produces β-galactosidase when a promoter is inserted in the correct orientation.

One of the unique features of all these plasmids is that they share several restriction sites upstream as well as downstream of the insertion site for a foreign gene. Therefore, once a foreign DNA fragment is cloned in one of the vectors just described, the fragment can be easily transferred to any one of the other vectors, like a cassette. Thus it is possible that the foreign gene product can be expressed in a wide variety of fashions, depending on the requirements.

II. pIN-I Vectors

The *E. coli lpp* gene coding for the prolipoprotein consists of 78 amino acid residues, and its mRNA is only 322 bases long (Nakamura and Inouye, 1980). The two-thirds of the mRNA from the 3′ end is highly enriched with stable stem-and-loop structures, and the 3′ end is typical of a ρ-independent transcription-termination signal. Expression of the *lpp* gene is constitutive, and its promoter region (45 bp) is extremely A-T rich (Nakamura and Inouye, 1979). Furthermore, the spacer region ~160 bp upstream from the promoter region is also A-T rich. Because these structural characteristics of the *E. coli lpp* gene are highly conserved in other bacterial *lpp* genes, these features are thought to play important roles in efficient transcription, translation, and high stability of the *lpp* mRNA (Yamagata *et al.*, 1981). We attempted to use all of these structural characteristics of the *lpp* gene for construction of expression vehicles (Nakamura and Inouye, 1982).

A. A Sites

In order to clone and express a foreign DNA fragment in the vector, unique *Eco*RI, *Hin*dIII, and *Bam*HI sites were created by inserting a DNA sequence of 22 bp immediately after the initiation codon (Fig. 1A; Nakamura and Inouye, 1982). Furthermore, this sequence was inserted in three different reading phases so that any DNA fragment can be expressed by choosing one of the three cloning vectors (A1, A2, A3; see Fig. 1A).

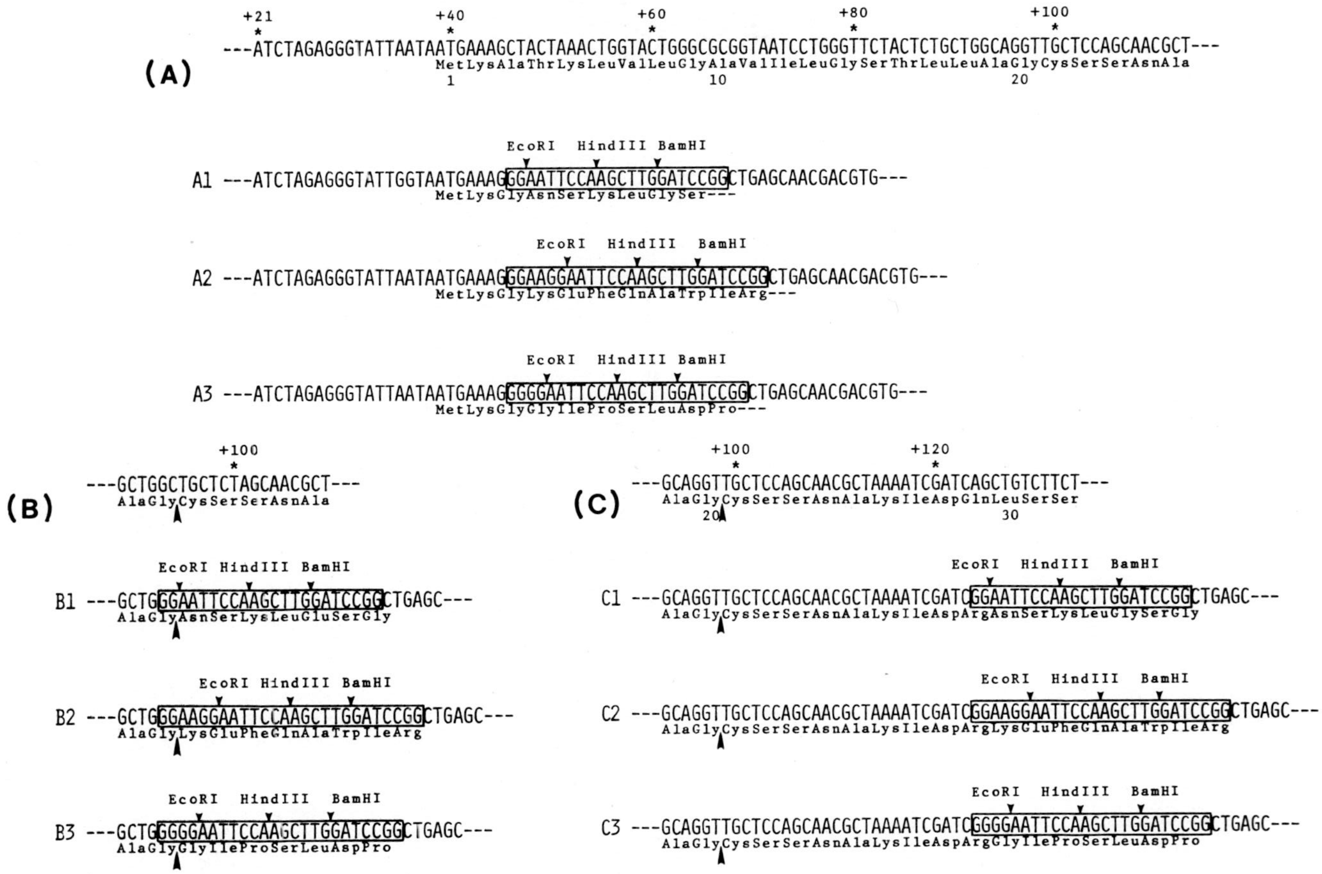

Fig. 1 DNA sequence showing the three reading frames in the (A) pIN-IA site, (B) pIN-IB site, and (C) pIN-IC site. Large arrows indicate the sites of cleavage of the signal peptide. The inserted linker DNA is boxed in. Numbers refer to the distance in base pairs from the transcription-initiation site (Nakamura and Inouye, 1979).

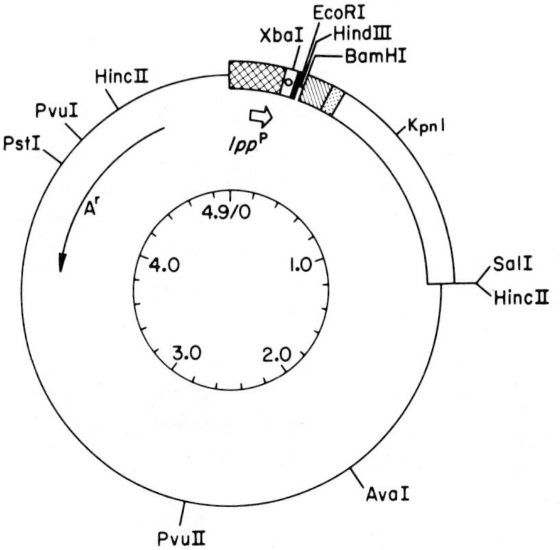

Fig. 2 Restriction map of pIN-I. The numbers indicate size in kilobase pairs.

The restriction map of pIN-IA vectors is shown in Fig. 2. It should be noted that downstream of the unique restriction sites for cloning, there are stop codons in all three reading frames to prevent unnecessary read-through of the inserted foreign gene if it does not have its own termination codon. Furthermore, the normal lipoprotein ρ-independent transcription-termination site is also retained to facilitate efficient transcription termination of the cloned foreign gene. This expression unit is in the place of the tetracycline-resistance gene on the multicopy plasmid pBR322, retaining the ampicillin-resistance marker. Construction of these vectors has been described elsewhere (Nakamura and Inouye, 1982; Inouye *et al.*, 1983).

B. B Sites

We have also constructed expression vectors (pIN-IB) with the same restriction sites (*Eco*RI, *Hin*dIII, and *Bam*HI) as those just described in three different reading frames (Fig. 1B) inserted immediately after the coding region of the lipoprotein signal peptide. With these vectors, the foreign gene product is produced as a hybrid with the lipoprotein signal peptide, so it should be translocated across the cytoplasmic membrane where the protein may be less prone to proteolytic attack than it would be inside the cytoplasm (Inouye *et al.*, 1983). If the protein is located in the periplasmic space, the purification of the protein is considered to be easier, because there are fewer proteins there than in the cytoplasm.

C. C Sites

We have constructed a third type of expression vector (pIN-IC) that has the unique restriction sites *Eco*RI, *Hin*dIII, and *Bam*HI, again in three different reading frames (C1, C2, and C3; Fig. 1C), inserted into the coding region of the mature lipoprotein (Inouye *et al.*, 1983). Regardless of the different reading frames, all of the cloned gene products have identical signal peptides plus eight amino acid residues at their N-termini. Therefore, the hybrid protein is secreted across the cytoplasmic membrane, and subsequently, after the signal peptide is cleaved off, the resultant protein is a hybrid between eight amino acid residues of the N-terminus of the lipoprotein and the protein coded by the cloned foreign gene. The N-terminal cysteine residue is modified as in the case of the lipoprotein; this is considered to play an important role for localization of the lipoprotein in the outer membrane. Therefore, the hybrid protein may be inserted into the outer membrane.

D. Cloning Strategies

It should be noted in Fig. 2 that the pIN vectors also have a unique *Xba*I site immediately upstream of the Shine–Dalgarno sequence for ribosome binding derived from the *lpp* gene (Nakamura and Inouye, 1979). This *Xba*I site enables the insertion of a gene if the gene contains its own ribosome-binding site.

To use these vectors, a gene to be cloned must have an *Eco*RI, *Hin*dIII, or *Bam*HI restriction site at or near the N-terminus of the gene to be expressed. (Note that *Bcl*I, *Bgl*II, or *Sau*3A sticky ends will fit into the *Bam*HI site without further modification of the DNA.) The order of the unique restriction-enzyme sites in the vector is *Eco*RI, *Hin*dIII, and *Bam*HI (Figs. 1 and 2). Therefore, an *Eco*RI–*Eco*RI, *Eco*RI–*Hin*dIII, *Eco*RI–*Bam*HI, *Hin*dIII–*Hin*dIII, *Hin*dIII–*Bam*HI, or *Bam*HI–*Bam*HI fragment can be inserted into the vector. It is convenient if the downstream restriction-enzyme site used is different from the restriction enzyme used at the N-terminus because the correct orientation of the fragment of interest can be assured. If these restriction-enzyme sites are not available, one can still insert a foreign gene at a convenient restriction-enzyme site, using an appropriate linker DNA or by blunt-end ligation. If an appropriate restriction-enzyme cleavage site is not available within the coding region of the foreign gene to be cloned, the DNA may be cleaved at a site upstream of the gene, then digested back with the double-stranded exonuclease *Bal*31 to an appropriate position. For blunt-end ligation, the *Hin*dIII site can be used by converting the sticky ends into blunt ends by filling in the sticky ends with the Klenow enzyme or by removing the sticky ends with S1 nuclease digestion. *Eco*RI or *Bam*HI sites can be used for the same purpose. However, the *Hin*dIII site would be preferable, because the inserted foreign gene can be easily removed

```
AluI              BstNI    RNA Polymerase            Pribnow
                           Recognition Site            Box
AGCTCACTCATTAGGCACCCCAGGCTTTACACTTTATGCTTTCCGGCTCGTATAATGTGTGGAA

                                   AluI                    HaeIII
TTGTGAGCGGATAACAATTTCACACAGGAAACAGCTATGACCATGATTACGGATTCACTGGCC
```

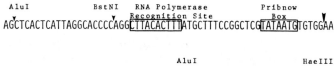

Fig. 3 DNA sequence of the *lac* promoter–operator region (Reznikoff and Abelson, 1978). The *Alu*I, *Bst*NI, and *Hae*III restriction sites are marked. The Shine–Dalgarno sequence and the initiating ATG are underscored. The RNA polymerase recognition site (−35 region) and the Pribnow box (−10 region) are boxed in. The large arrow indicates the transcription-initiation site.

by *Eco*RI and *Bam*HI digestion. It should be noted that no matter which sites are used, one can mix vectors of three different reading frames (A1 + A2 + A3, B1 + B2 + B3, or C1 + C2 + C3) for cloning if the exact reading frame of the foreign gene is not known.

It is important to notice that the products from these vectors always have a few extra amino acid residues at their N-termini because of the junction sequence. Therefore, if the expression using these vectors is successful, the next step, if necessary, is to correct the N-terminal sequence to the sequence of the natural product. This can be achieved by replacing the DNA fragment between the unique *Xba*I site (14 bp upstream of the initiation codon) and the restriction site used for cloning with a synthetic DNA fragment so that one can recreate the identical sequence from the *Xba*I site to the initiation codon plus the DNA sequence that corresponds to the N-terminus of the natural product.

pIN-I vectors have been used successfully for expressing prokaryotic as well as eukaryotic proteins; for example, when the gene for human growth hormone is inserted into the pIN-I vector, 10–15% of the total cellular protein of cells carrying this plasmid is the human growth hormone (unpublished results). The rat sarcoma virus oncogene (v-Ha-*ras*) has been inserted into the *Hin*dIII site of pIN-IA to successfully produce v-Ha-*ras* gene product p21 in *E. coli* (E. Scolnick, personal communication).

III. pIN-II Vectors

The pIN-I vectors (Section II) express the inserted foreign gene product constitutively. This creates a problem if the gene product is harmful to the cell. For this reason, we have constructed an inducible expression vector in which the foreign gene product is expressed only in the presence of an inducer, allowing the cells to grow to a high density before the gene is turned on.

For this purpose, a 95-bp *Alu*I DNA fragment (Fig. 3) carrying the *lacUV5* promoter–operator region was inserted into the unique *Xba*I site of the pIN-I

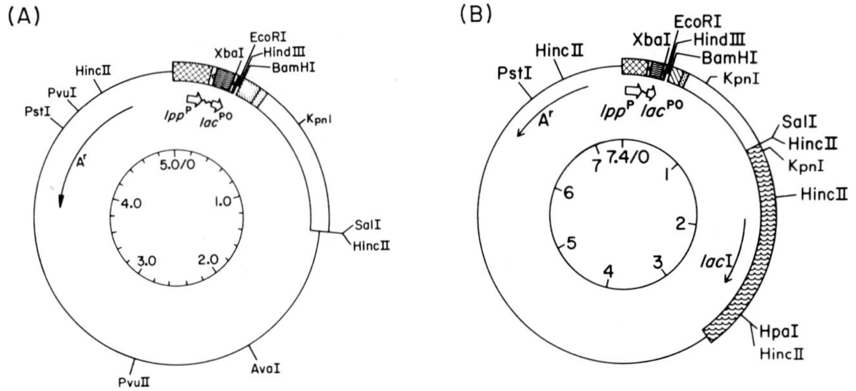

Fig. 4 (A) Restriction map of pIN-II. (B) Restriction map of pIN-III. Numbers indicate size in kilobase pairs.

vectors just upstream of the lipoprotein ribosome-binding site producing the pIN-II vectors (Fig. 4A; Nakamura and Inouye, 1982). These vectors express the inserted foreign gene product only in the presence of a *lac* inducer such as isopropyl-β-D-thiogalactopyranoside (IPTG). The pIN-II vectors must be used in a host strain that is able to overproduce the *lac* repressor, the *lacI* gene product. We used the strain JA221 (*hsdR, ΔtrpE5, leuB6, lacY, recA, thi*) (Clarke and Carbon, 1978) containing F′ ($lacI^{q}$, $lacZ^{+}$, $lacY^{+}$, $lacA^{+}$, $proA^{+}$, $proB^{+}$) (Inouye *et al.*, 1982) as a recipient, because it carries $lacI^{q}$, which is able to overproduce the lactose repressor. The F′ containing the $lacI^{q}$ gene can be transferred into other strains by conjugation, which can then be used as a recipient for the pIN-II vectors.

We have also constructed pIN-II vectors with a larger *lac* promoter–operator fragment (113 bp) from the *Bst*NI-to-*Hae*III site (Fig. 3). These pIN-II vectors are able to produce the cloned gene product at a rate a fewfold higher than the pIN-II vectors with the shorter *lac* promoter–operator fragment (95 bp) when fully induced (Y. Masui and M. Inouye, unpublished). The exact reason for the higher production with pIN-II (lac^{po}: 113 bp) than with pIN-II (lac^{po}: 95 bp) is unknown at present. At any rate, for higher production of a cloned gene product, pIN-II (lac^{po}: 113 bp) would be recommended, although one might get higher background production in the absence of IPTG.

The pIN-II vectors have been used for various proteins. For example, Ohta *et al.* (1982) inserted the hook-protein gene of *Caulobacter cresentus* into the pIN-IIA vector and successfully induced the production of the hook protein upon addition of IPTG.

The pIN-II vectors are as convenient to use as the previously mentioned pIN-I vectors, yet they also offer the ability to express proteins that are harmful to *E. coli*. It is important to note that even normal *E. coli* proteins can be harmful to *E. coli* if they are overproduced, as has been shown in the case of the lipoprotein

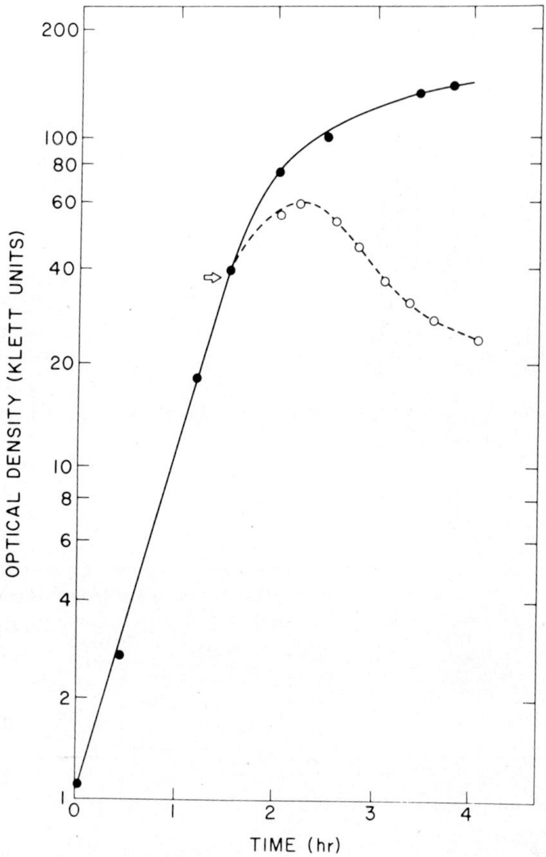

Fig. 5 Growth curve of cells harboring the pIN-II vector containing the MS2 lysis gene (pJDC030) with (○) and without (●) IPTG. The arrow indicates addition of IPTG (Coleman *et al.*, 1983).

(Lee *et al.*, 1981). Figure 5 shows an example of a gene cloned in pIN-II that is lethal to the cell. As can be seen, when IPTG is added to the culture medium, cells soon start to lyse because of the lytic enzyme (MS2 phage lysis protein) cloned in the plasmid (Coleman *et al.*, 1983).

IV. pIN-III Vectors

The major drawback with the pIN-II vectors is that the host cell must be able to overproduce the *lac* repressor. To overcome this problem, we have inserted the *lacI* gene into pIN-II to construct pIN-III (Fig. 4B; Y. Masui and M. Inouye,

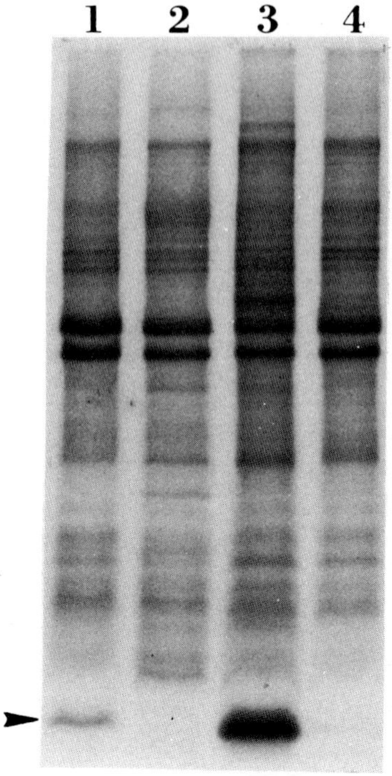

Fig. 6 Production of the lipoprotein with the use of the pIN-III vector. *Escherichia coli* JE5505 (*lpp*$^-$) carrying pIN-III*lpp* was grown in L-broth medium (Miller, 1972) in the presence or absence of 2 m*M* IPTG. The membrane fractions were prepared from the cells at the stationary phase and subjected to sodium dodecyl sulfate (SDS)-polyacrylamide gel electrophoresis, with the use of 17.5% acrylamide gel. The gel was stained with Coomassie Brilliant Blue. The pIN-III vector used in this experiment carries a 540-bp DNA fragment containing the *lac* promoter–operator region (Nakamura *et al.*, 1982). The arrow indicates the position of the lipoprotein. (Lane 1) Membrane fraction from wild-type *lpp*$^+$ *E. coli* (JE5506). (Lane 2) From isogenic *lpp*$^-$ *E. coli* (JE5505). (Lane 3) From JE5505 containing pIN-III*lpp* induced overnight with IPTG. (Lane 4) Same as lane 3, but grown without IPTG.

unpublished). Besides all the good assets of the inducible pIN-II vectors, pIN-III vectors have the additional advantage that they can be used in any host strain capable of replicating the vectors.

We have inserted the gene for human leukocyte interferon into one of the pIN-III vectors. Strains carrying this plasmid produce a high level of biologically active interferon when they are fully induced with IPTG. Figure 6 shows another example of a pIN-III vector used for the *lpp* gene. As can be clearly seen, there was no detectable amount of the lipoprotein in the absence of IPTG (lane 4), whereas on addition of IPTG a large quantity of the lipoprotein was produced (lane 3) in comparison with the amount of the lipoprotein of the wild-type *lpp*$^+$ strain (lane 1).

V. pIM Vectors: High-Copy-Number Vectors

Because the pIN vectors just described are derived from pBR322, which is itself derived from the colicin E1 plasmid, they occur at the frequency of ~20 copies/cell. In general, protein production is proportional to the number of the

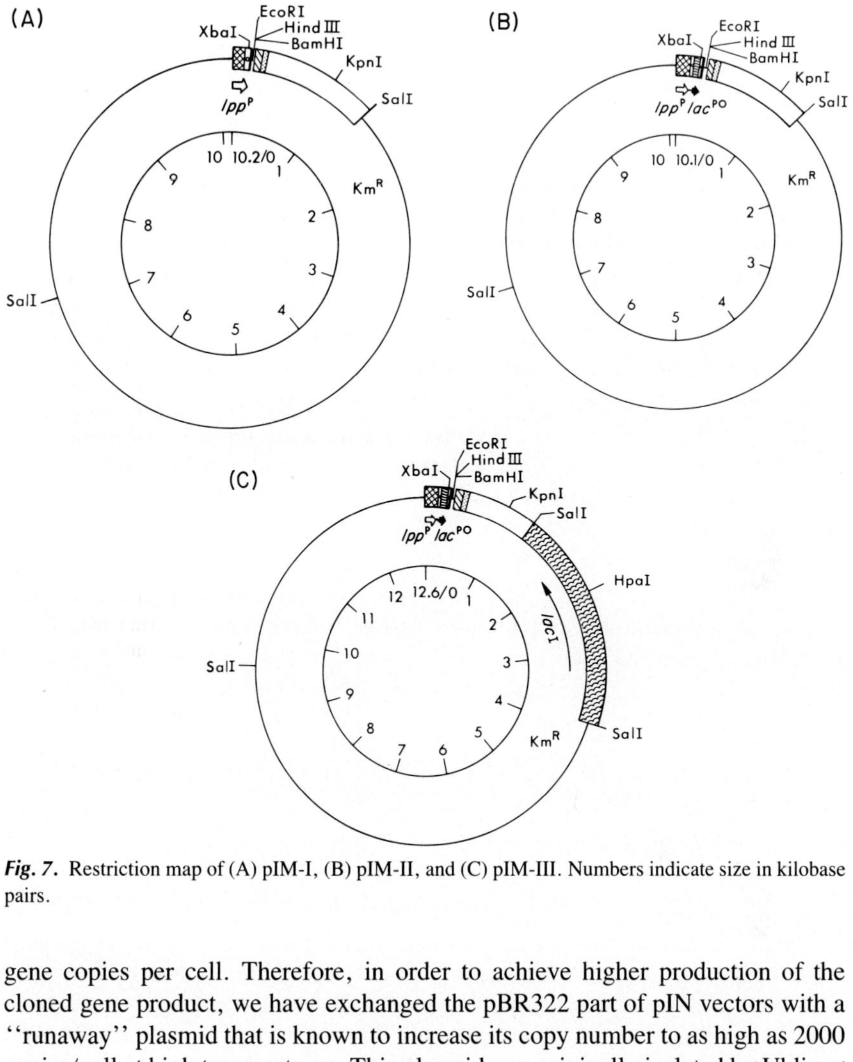

Fig. 7. Restriction map of (A) pIM-I, (B) pIM-II, and (C) pIM-III. Numbers indicate size in kilobase pairs.

gene copies per cell. Therefore, in order to achieve higher production of the cloned gene product, we have exchanged the pBR322 part of pIN vectors with a "runaway" plasmid that is known to increase its copy number to as high as 2000 copies/cell at high temperatures. This plasmid was originally isolated by Uhlin *et al.* (1979) from plasmid RI. The mutant plasmid occurs at a frequency of ~15 copies/cell at 30°C. However, when the temperature is raised above 35°C, the plasmid replicates without copy-number control for approximately five doublings of the bacteria; during this time the plasmid copy number increases to as high as 2000/cell, and the cells eventually die because of overproduction of the plasmid. Protein synthesis continues normally until cell death, allowing synthesis of a cloned protein to be greatly amplified at higher temperatures.

Utilizing a kanamycin-resistant derivative of this temperature-sensitive replication mutant (pSY343; S. Yasuda, personal communication; Arai *et al.*, 1981) we have, by inserting the expression region of the pIN-I, -II, and -III vectors constructed, respectively, pIM-I, -II, and -III (Fig. 7). These vectors act as the pIN vectors at 30°C, but above 35°C and in the presence of 2 m*M* IPTG for pIM-II and pIM-III, the plasmid replicates in an uncontrolled manner, and the cloned gene's product is made in very large quantities. For example, we have inserted the gene for *envZ* (T. Mizuno, unpublished) into pIM-I. The *envZ* protein is normally present at ~10 copies/cell; but using the pIM-I vector, the production of the *envZ* protein becomes as high as 5×10^6 molecules/cell.

The pIM vectors have all the unique restriction-enzyme sites for expression as well as all the other features used for pIN vectors, except for the portion derived from pBR322. We have constructed only A-site vectors for pIM because for the expression of a foreign gene, pIN vectors should be tried first, and, if the attempt is successful, the foreign gene should then be transferred to one of the pIM vectors. Because one can use the unique *Xba*I site at the Shine–Dalgarno sequence for the upstream restriction-cleavage site for this purpose, it is not necessary to construct B- and C-site vectors. More precise cloning strategies are discussed later.

There are a few features concerning the pIM vectors that should be kept in mind when they are used. First, the pIM vectors have a kanamycin-resistance marker, in contrast to the ampicillin-resistance marker found in the pIN vectors. Bacteria carrying the pIM vector are resistant to 20 μgm/ml of kanamycin. Second, cells carrying the pIM vectors are temperature sensitive when grown above 35°C, so the transformations should be carried out at 30°C.

VI. pIC Vectors: Hybrid Expression Vectors

The pIN and pIM vectors just described are designed to express a foreign gene when it is rather well characterized in terms of its DNA sequence, restriction map, and position of the initiation codon. However, in many cases a gene of interest may not be characterized well enough to be cloned in pIN or pIM vectors. Yet one may want to express it to obtain antiserum against the gene product. For this purpose, special cloning vectors have been developed that enable one to detect easily the expression of the gene of interest (the entire gene or a portion of the gene) and to produce a protein that is a hybrid between the gene product and *E. coli* β-galactosidase (Gray *et al.*, 1982). A similar yeast vector is also described in Chapter 5 by Broach *et al.*

We have developed similar vectors (pIC vectors) using the pIN vectors. The pIC-III vector was derived from pIN-III *lacZ* in which the *lacZ* gene (the gene for β-galactosidase) was inserted into the *Bam*HI–*Sal*I sites of pIN-IIIA3. The *Hin*

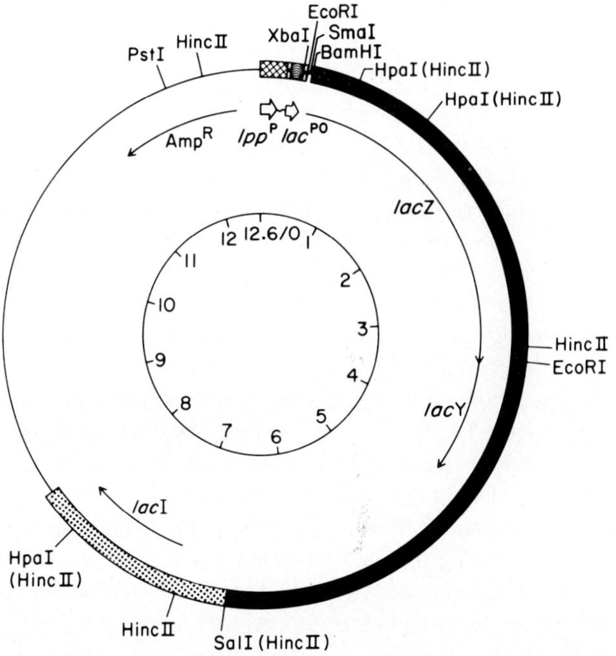

Fig. 8. Restriction map of pIC-III. Numbers indicate size in kilobase pairs.

dIII site sandwiched by the *Eco*RI and the *Bam*HI sites was then converted to a *Sma*I site (CCCGGG) in such a way that the reading frame of the *lacZ* gene was no longer the same as that of the initiation codon (J. Coleman and M. Inouye, unpublished). This was achieved by cleaving pIN-III *lacZ* with *Hin*dIII, followed by ligation with a oligodeoxynucleotide (5′ AGCTCCCGGG 3′). When the plasmid is ligated with the oligodeoxynucleotide, the *Hin*dIII site is converted to a *Sma*I site (Fig. 8), and the newly formed plasmid becomes phenotypically $LacZ^-$ because the reading frame of the *lacZ* gene is no longer the same as that of the initiation codon. However, when a DNA fragment from a gene of interest is inserted in the *Sma*I site by blunt-end ligation, the plasmid may become $LacZ^+$ again, if the DNA fragment has an open reading frame in the same phase as the initiation codon and its length is $3n + 2$. The gene product is thus a hybrid between the peptide coded by the DNA fragment and β-galactosidase. β-Galactosidase is known to remain active even if it is fused to a number of peptides at its N-terminus (e.g., see Bassford *et al.*, 1978).

Because the hybrid protein could be harmful to the cell, pIC-III is an ideal vector, with which the expression of the hybrid gene can be controlled by the *lac* repressor. One of the advantages of using the fusion with the *lacZ* gene is that because of the extremely large size of the hybrid protein (>116 kdaltons), the purification of the hybrid protein is relatively easy and the hybrid protein may become more stable. However, the expression of the DNA fragment alone can be attempted by removing it with the upstream *Xba*I or *Eco*RI and the downstream *Bam*HI sites (Fig. 8) and transferring it to the appropriate pIN or pIM vectors. It is also recommended to use pIC-I or pIC-II vectors for higher expression. This can be easily achieved by simply transferring the same DNA fragment obtained previously to pIN-I*lacZ* or pIN-II*lacZ,* both of which are available and have the same *Xba*I, *Eco*RI, *Sal*I, and *Bam*HI sites as pIN-III*lacZ.*

The pIC-III vector can also be used to shotgun-clone DNA fragments into the vector either to find the coding region of a specific gene within a fragment or to create an open-reading-frame library of some specific DNA, cDNA, or genomic DNA. To do the shotgun-cloning, the DNA of interest is fragmented with a deoxyribonuclease I limited digest or with a restriction enzyme. It is important that before cloning the DNA fragments into the *Sma*I site of a pIC vector, the length of the fragment be randomized by a limited digestion with the exonuclease *Bal*31. Also, fragments shorter than about 200 bp should be removed by gel filtration or polyacrylamide gel electrophoresis to lower the chance of a random open reading frame getting in. The fragments are then blunt-end ligated to the *Sma*I site of the pIC-III vector. After transformation, the $lacZ^{+}$ transformants are candidates for an open-reading-frame library.

VII. Promoter-Proving Vectors

Here we describe other vectors that can be used to clone a DNA fragment carrying a promoter and to examine promoter efficiency. The first promoter-proving vector is pKEN005 (Fig. 9A; Nakamura and Inouye, 1982). This vector was constructed from pBR322 by removing the promoter for the tetracycline-resistance gene, leaving a unique *Eco*RI site upstream of the translational start of the tetracycline-resistance gene (Fig. 9A). This vector will only confer the tetracycline-resistance phenotype if a promoter is inserted in the correct orientation into the *Eco*RI site. This vector is useful for determining which part of a particular DNA fragment is a promoter. This is done by fragmenting the DNA piece of interest with a restriction enzyme or by deoxyribonuclease I partial digestion. The ends of the fragments thus generated should be changed to blunt ends by filling in the sticky ends with DNA polymerase I Klenow fragment or by removing the sticky ends with S1 nuclease. Next, *Eco*RI linkers are added to the ends of the DNA fragments and ligated into the *Eco*RI site of pKEN005. This chimera can now be used to transform any tetracycline-sensitive *E. coli* capable of replicating the plasmid. These transformed cells should be plated onto plates

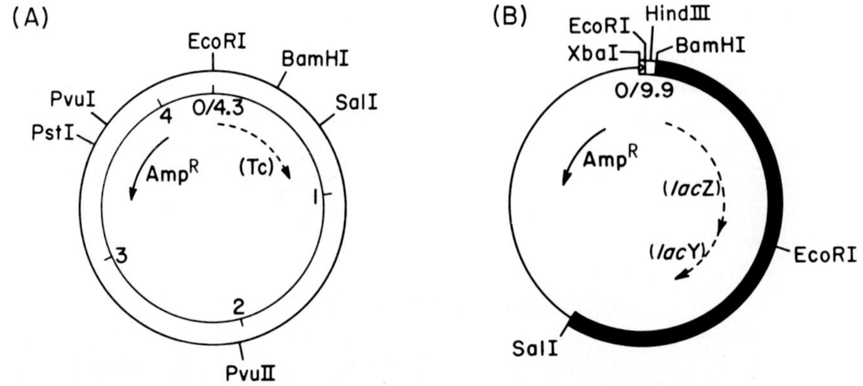

Fig. 9. Restriction map of pKEN005 (A) and pKM005 (B). Numbers indicate size in kilobase pairs.

containing 8 μgm/ml of tetracycline (more tetracycline if one has a strong promoter) and 50 μgm/ml of ampicillin. Any transformant growing on these plates should contain a promoter. The resulting chimeric plasmid may be used to study not only the efficiency of the promoter cloned by measuring the degree of tetracycline resistance but also the regulatory mechanism of the promoter by growing the strain under various growth conditions.

The second promoter-proving vector is pKM005 (Fig. 9B; Y. Masui, H. Kimura, and M. Inouye, unpublished). This vector was constructed from pIN-I A3*lacZ* by removing its *lpp* promoter region. The *lacZ* gene was inserted between the *Bam*HI site and the *Sal*I site of pIN-IA3. The *Bam*HI site is at the eighth amino acid of β-galactosidase, putting the β-galactosidase in phase with the pIN-I lipoprotein ATG initiating codon (Fig. 1). A DNA fragment upstream of the unique *Xba*I site, which carries the *lpp* promoter, was then removed, leaving the *Xba*I site (Fig. 9B). Therefore, the only way to recover β-galactosidase activity is to reinsert a promoter in the correct orientation into the unique *Xba*I site. To use this vector, a DNA fragment carrying a promoter should be religated in the *Xba*I site with use of *Xba*I linkers, and transformation should be carried out with a *lac* deletion strain of *E. coli* such as SB4288 [F^-, *recA thi-1 rel*A *mal-24 spc-12 supE50* DE5 (Δ*proB lac*)]. Cells carrying plasmids with a new promoter will then produce red colonies on MacConkey plates containing 50 μgm/ml of ampicillin.

Because the vector pKM005 contains the entire ribosome-binding-site sequence for β-galactosidase production, the only factor contributing to the difference (between promoters) in activity of β-galactosidase in the cell is the activity of the inserted promoter. Therefore, this vector is particularly suitable for examining the promoter efficiency of any gene of interest. β-Galactosidase activity can be easily and accurately measured using the substrate *o*-nitrophenyl-β-D-galactoside (ONPG), as described by Miller (1972).

It should be noted that both vectors described here are superior to other promoter-proving vectors using protein fusion because the *lacZ* gene to be expressed is already equipped with its own ribosome-binding site and the initiation codon, so adjustment of the reading frame is not required (in contrast to protein-fusion vectors). Also, because the ribosome-binding site remains constant, the rate of translation initiation does not contribute to β-galactosidase production.

VIII. General Cloning Strategy

We have described in this chapter several vectors for the high-level expression of a foreign gene. To use these vectors most efficiently, we recommend first to clone a foreign gene of interest into one of the pIN-IIIA vectors. Once the gene is expressed properly in pIN-IIIA, the gene can be easily moved from pIN-III to any one of the other pIN vectors, including pIN-IIIB, pIN-IIIC, pIN-IIA, pIN-IIB, pIN-IIC, pIN-IA, pIN-IB, and pIN-IC. This can be easily achieved by using the same restriction-enzyme site as used for constructing the pIN-III clone or by transferring the entire coding region from the unique *Xba*I site to the unique *Kpn*I site (like a cassette) to the same unique sites of pIN-I or pIN-II, resulting in pIN-IA or pIN-IIA configurations, respectively.

If higher gene expression of the gene is wanted, one can transfer the same coding cassette (from *Xba*I-to-*Kpn*I sites) into appropriate vectors so that one can easily construct pIM-IA, pIM-IB, pIM-IC, pIM-IIA, pIM-IIB, pIM-IIC, pIM-IIIA, pIM-IIIB, and pIM-IIIC.

In the case that the gene has not been well mapped, the pIC-III vector should be used as described in the text for preliminary characterization of the gene and for the development of antiserum against the gene product.

It is important to note that the *Xba*I site immediately upstream of the Shine–Dalgarno sequence is important in our vector systems not only because it can be used for transferring the cloned gene from one vector to another, but also because it provides the most convenient site for sequencing the junction of the cloned gene. A general strategy is first to cleave a vector DNA carrying a foreign DNA with *Xba*I, followed by filling in the cleaved sites with [α-^{32}P]+XTP with use of the Klenow enzyme (labeled at the 3′ ends). Subsequently, the [^{32}P]DNA is digested with a restriction enzyme to generate a DNA fragment of an appropriate size for DNA sequencing.

We have also described two vectors for cloning fragments of DNA containing a promoter or promoter–operator. These vectors are useful for (1) finding the fragment of DNA that has promoter activity and (2) characterizing the property and the function of the promoter of interest.

IX. Summary

We have constructed several high-level expression vectors using the *Escherichia coli* lipoprotein promoter and part of the protein. These vectors are designed to produce a large amount of an inserted gene product and to localize it to a specific compartment of the *E. coli* cell (cytoplasm, cytoplasmic membrane, periplasm, or outer membrane). These vectors have three restriction enzyme sites (*Eco*RI, *Hin*dIII, and *Bam*HI) in each of the three reading frames at various positions along the lipoprotein gene. The pIN-I vectors constitutively produce a large amount of the inserted gene product. The pIN-II and pIN-III vectors produce the foreign gene product only in the presence of an inducer. The pIM vectors contain a temperature-sensitive replication control, so at elevated temperatures the plasmid replicates to a very high copy number, producing a large amount of the inserted gene product.

Once a gene has been inserted properly into one of these vectors it can easily be moved to the other vectors, like a cassette, with the use of several possible unique restriction-enzyme sites on either side of the gene.

We also have described another type of vector, pIC vectors, that can be used to clone and express in the presence of an inducer a portion of a foreign protein in a hybrid with active β-galactosidase. This hybrid protein can easily be purified for developing antiserum against the foreign protein.

Last, we have described two promoter cloning vectors that produce a protein (tetracycline resistance or β-galactosidase) when a promoter is inserted.

References

Arai, K., Yasuda, S., and Kornberg, A. (1981). *J. Biol. Chem.* **256,** 5247–5252.

Bassford, P., Beckwith, J., Berman, M., Brickman, E., Casadaban, M., Guarte, L., Saint-Girous, I., Sarthy, A., Schwartz, M., and Silhavy, T. (1978). *In* "The Operon" (J. H. Miller, and W. S. Reznikoff, eds.), pp. 245–262. Cold Spring Harbor Lab., Cold Spring Harbor, New York.

Clarke, L., and Carbon, J. (1978). *J. Mol. Biol.* **120,** 517–534.

Coleman, J., Inouye, M., and Atkins, J. (1983). *J. Bacteriol.* **153,** 1098–1100.

Gray, M. R., Colot, H. V., Guarente, L., and Risbash, M. (1982). *Proc. Natl. Acad. Sci. USA.* **79,** 6598–6602.

Halegoua, S., and Inouye, M. (1979). *In* "Bacterial Outer Membranes: Biogenesis and Function" (M. Inouye, ed.), pp. 67–113. Wiley, New York.

Inouye, M. (1979). *Biomembranes* **10,** 141–208.

Inouye, M., Nakamura, K., Inouye, S., and Masui, Y. (1983). *In* "The Future in Nucleic Acid Research" (I. Watanabe, ed.), in press. Academic Press Japan, Tokyo.

Inouye, S., Wang, S. S., Sekizawa, J., Halegoua, S., and Inouye, M. (1977). *Proc. Natl. Acad. Sci. USA* **74,** 1004–1008.

Inouye, S., Soberon, X., Franceschini, T., Nakamura, K., Itakura, K., and Inouye, M. (1982). *Proc. Natl. Acad. Sci. USA* **79,** 3438–3441.

Lee, N., Nakamura, K., and Inouye, M. (1981). *J. Bacteriol.* **146,** 861–866.

Miller, J. H. (ed.) (1972). "Experiments in Molecular Genetics." Cold Spring Harbor Lab., Cold Spring Harbor, New York.

Nakamura, K., and Inouye, M. (1979). *Cell* **18,** 1109–1117.

Nakamura, K., and Inouye, M. (1980). *Proc. Natl. Acad. Sci. USA* **77,** 1369–1373.

Nakamura, K., and Inouye, M. (1982). *EMBO J.* **1,** 771–775.

Nakamura, K., Masui, Y., and Inouye, M. (1982). *J. Mol. Appl. Genet.* **1,** 289–299.

Ohta, N., Chen, L., and Newton, A. (1982). *Proc. Natl. Acad. Sci. USA* **79,** 4863–4867.

Reznikoff, W. S., and Abelson, J. (1978). *In* "The Operon" (J. H. Miller, and W. S. Reznikoff, eds.), pp. 221–243. Cold Spring Harbor Lab., Cold Spring Harbor, New York.

Uhlin, B. E., Molin, S., Gustafsson, P., and Nordstrom, K. (1979). *Gene* **6,** 91–106.

Yamagata, H., Nakamura, K., and Inouye, M. (1981). *J. Biol. Chem.* **256,** 2194–2198.

CHAPTER 3

Molecular Cloning in *Bacillus subtilis*

DAVID DUBNAU

Department of Microbiology,
The Public Health Research Institute of the City of New York, Inc.
New York, New York

I. Introduction

After *Escherichia coli*, the most intensively studied prokaryote is *Bacillus subtilis*. Considerable effort has consequently been devoted to the development of recombinant DNA methods in this organism, for use as research tools. These efforts have been further motivated by the considerable industrial potential of *B.*

Experimental Manipulation
of Gene Expression

ISBN 0-12-372380-9

subtilis. The bacilli are already widely used by the fermentation industry, and a great deal of practical know-how has accrued (reviewed by Debabov, 1982). Many bacilli can be grown to very high yield on inexpensive media. They are largely innocuous to humans and to plants and animals of commerical importance. The bacilli (*B. subtilis* in particular) do not produce endotoxin, and, therefore, the purification of pharmaceuticals produced by *B. subtilis* should present minimal problems. The potentially most important property of the bacilli for commercial exploitation is their ability to secrete protein products into the culture medium. These products often accumulate to very high yield and can be recovered simply and inexpensively.

The development of gene manipulation methodologies in *B. subtilis* has not been straightforward. This is owing to certain inherent peculiarities of the transformation system that have been circumvented by the use of various, unusual stratagems. We therefore begin this review with a brief discussion of transformation in *B. subtilis*. We then describe the available vectors and stratagems for cloning and conclude by briefly outlining the current methods for studying the expression of cloned genes. Various aspects of molecular cloning in *B. subtilis* have been reviewed by Dubnau *et al.* (1980b), Lovett (1981), Ehrlich *et al.* (1982), Gryczan (1982), Hoch *et al.* (1982), and Kreft and Hughes (1982).

II. Plasmid Transformation

Bacillus subtilis has been one of the principal organisms used for the investigation of genetic transformation. A great deal is known, therefore, about this process (for a review see Dubnau, 1982). Because the transformation of competent cells by plasmid DNA continues to be the most widely used means of introducing recombinant DNA, we shall discuss this process in some detail later. However, Chang and Cohen (1979) have described a highly efficient ($\sim 10^7$ transformants/μgm plasmid DNA) protoplast-transformation system that does not require the use of competent cells. This method is certainly useful for some purposes. Unlike competent cells (see later), protoplasts are transformable by plasmid monomers, and by nicked, gapped, and linear plasmid molecules (Chang and Cohen, 1979; Gryczan *et al.*, 1980a; Scherzinger *et al.*, 1980). However, protoplast transformation is limited in its usefulness by the following properties:

1. It requires the use of complex media for protoplast regeneration and therefore does not permit the direct selection of nutritional markers.
2. The efficiency of protoplast transformation has a very steep inverse dependence on DNA molecular weight; ligated (hence concatameric) plas-

mid DNA transforms poorly (Chang and Cohen, 1979; Mottes *et al.*, 1979).

3. The procedure is fairly laborious, and protoplasts, unlike competent cells, cannot be stored.

The transformation of competent cells (which can be maintained frozen for several years) by plasmid DNA occurs at high frequency ($\sim 10^6$ transformants/μgm DNA) but low efficiency. About 10^3–10^4 plasmid molecules are taken up per transformant (Contente and Dubnau, 1979a). Linear and open circular plasmid molecules do not transform (Contente and Dubnau, 1979a). Remarkably, the transforming activity of native plasmid preparations is entirely owing to the presence of oligormeric forms (Canosi *et al.*, 1978). In fact, covalently closed circular (CCC) plasmid monomer is completely inactive in transformation, even at very high input concentrations (R. Villafane and D. Dubnau, unpublished). Various models have been presented to explain these and other properties of plasmid transformation (Dubnau *et al.*, 1980a; Canosi *et al.*, 1981; Haykinson *et al.*, 1982). These models are beyond the scope of this chapter. However, the properties of plasmid transformation just discussed place a severe limitation on the efficiency of shotgun-cloning in *B. subtilis*. The requirement for CCC DNA prevents use of techniques that leave gaps or nicks in ligated recombinant DNA (e.g., homopolymer tailing or use of alkaline phosphatase to enrich for recombinant molecules). This is not a severe limitation, because a new vector permits direct selection for recombinant molecules (see Section III,D). A more serious limitation in practice is imposed by the requirement for plasmid oligomers. In a complex ligation mixture, the probability of forming a head-to-tail vector oligomer carrying a desired "foreign" fragment is low. Although shotgun cloning of fragments from simple mixtures (e.g., restriction digests or plasmid or bacteriophage DNA) is readily accomplished, the straightforward cloning of chromosomal fragments has proven very difficult. This problem has been surmounted using stratagems based on certain further properties of the plasmid transformation system. If the competent culture carries a plasmid that is homologous to the vector, then plasmid monomer can transform. The properties of this "helper" transformation system have been explored in some detail (Contente and Dubnau, 1979b), and they form the basis of a successful cloning stratagem (Gryczan *et al.*, 1980a). Another form of homology-facilitated transformation also permits the use of plasmid monomers and is potentially useful for cloning (Bensi *et al.*, 1981; Lopez *et al.*, 1982). Michel *et al.* (1981) and Haykinson *et al.* (1982) have reported that plasmid monomers carrying short direct or indirect repeats can transform with high efficiency. The latter authors have adapted this observation to the construction of several potentially useful plasmid vectors. We discuss these various cloning stratagems in more detail in Section IV.

III. Plasmid Vectors

For a time, only cryptic plasmids native to the bacilli were identified. Now, however, several vectors based on indigenous *Bacillus* plasmids have been constructed (reviewed by Kreft and Hughes, 1982). The indigenous *Bacillus* plasmids will not be discussed further in this chapter.

A. Vectors Derived from *Staphylococcus aureus* Plasmids

Ehrlich (1977) reported that several resistance plasmids isolated from *Staphylococcus aureus* could be introduced into *Bacillus subtilis* by transformation and that they are capable of expression and stable replication in the latter organism. This observation made available a large number of plasmids carrying a variety of useful selective markers and restriction sites. Shortly after the transfer of these plasmids into *B. subtilis,* they were successfully used for several *in vitro* molecular cloning experiments (Ehrlich, 1978; Gryczan and Dubnau, 1978; Keggins *et al.*, 1978). Most of the plasmid vectors currently employed in *B. subtilis* are derived from the resistance plasmids listed in Table I. These plasmids generally replicate quite stably in *B. subtilis* and confer useful levels of antibiotic resistance. Interestingly, the tetracycline- (Tc)-resistance plasmid pBC16 (Bernhard *et al.*, 1978), which is indigenous to *B. cereus,* has a replication system that is identical or nearly identical to that of pUB110, originally isolated from *S. aureus* (Polak and Novick, 1982). It is thus not clear that pUB110, at least, should properly be considered as an *S. aureus* rather than a *Bacillus* plasmid.

These plasmids, as well as those listed in Table II, are multicopy plasmids. They range in copy number from ~10 (pE194; Weisblum *et al.*, 1979) to ~50 (pUB110; Shivakumar and Dubnau, 1978). Mutants of pE194 with elevated copy number have been isolated (Weisblum *et al.*, 1979). These, and the wild-type pE194, display temperature-sensitive replication properties (Scheer-Abramowitz *et al.*, 1981). Above 44°C, pE194 fails to replicate, whereas *B. subtilis* grows well up to 52°C. This is a useful property, facilitating the curing of strains carrying pE194 and its derivatives. pE194 is probably the best characterized of the plasmids listed in Table I; its replication properties have been explored (Scheer-Abramowitz *et al.*, 1981; Gryczan *et al.*, 1982). The polypeptide products of pE194 have been identified in minicells (Shivakumar *et al.*, 1979), and the product of the erythromycin- (Em)-reistance gene has been purified and characterized (Shivakumar and Dubnau, 1981). The determinants for the known

Table I *Staphylococcus* Resistance Plasmids Transferred to *Bacillus subtilis*

Plasmid	Resistance markers[a]	Molecular weight ($\times 10^6$)	Reference[b]
pC194	Cm^R	2.0	Ehrlich (1977)
pC221	Cm^R	2.9	Ehrlich (1977)
pC223	Cm^R	2.9	Ehrlich (1977)
pUB112	Cm^R	2.7	Ehrlich (1977)
pT127	Tc^R	2.9	Ehrlich (1977)
pUB110	Km^R	3.0	Gryczan *et al.* (1978)
pSA0503	Sm^R	2.8	Gryczan *et al.* (1978)
pSA2100	Cm^R, Sm^R	4.8	Gryczan *et al.* (1978)
pE194	Em^R	2.3	Weisblum *et al.* (1979)

[a]Cm, Chloramphenicol; Em, erythromycin; Km, kanamycin; Sm, streptomycin; Tc, tetracycline.
[b]These are references to the introduction of the plasmids into *B. subtilis*.

polypeptide products have been located on the physical map of pE194, and the positions of the RNA polymerase binding sites determined (Shivakumar *et al.*, 1980a). Finally, the regulation of synthesis of the Em-resistance gene product (an rRNA methylase) has been intensively studied (Gryczan *et al.*, 1980b; Horinouchi and Weisblum, 1980, 1981; D. Dubnau *et al.*, 1981; Hahn *et al.*, 1982). The pE194 genome has been completely sequenced (Gryczan *et al.*, 1980b; Horinouchi and Weisblum, 1980, 1982). pUB110 has also been characterized fairly extensively; restriction-site maps have been published (Gryczan *et al.*, 1978; Jalanko *et al.*, 1981), and the map position of the replication origin and the direction of replication of pUB110 determined (Scheer-Abramowitz *et al.*, 1981). Plasmids replicating by means of the pUB110 replicon can be amplified, using certain dna^{ts} mutants (Shivakumar and Dubnau, 1978). Incubation of these strains (carrying pUB110) at a nonpermissive temperature causes a halt in chromosome but not pUB110 replication. Copy numbers of nearly 1000 can be obtained in this way. The major polypeptide products of pUB110 have been identified in minicells (Shivakumar *et al.*, 1979), and the product of the Km^R determinant (a kanamycin nucleotidyl transferase) has been purified (Sadaie *et al.*, 1980). The replication origin region of pUB110 has been sequenced (T. Tanaka, T. McKenzie, and N. Sueoka, personal communication). pC194 has also been entirely sequenced (Horinouchi and Weisblum, 1982), and its protein products examined in minicells (Shivakumar *et al.*, 1979).

Many chimeric derivatives of these and other plasmids have been constructed that are useful or potentially useful as cloning vectors. Some representative plasmids of this sort are described in Table II. This is not an exhaustive list, and many other such vectors, published and unpublished, have been constructed. The

Table II Chimeric Cloning Vectors

Plasmid	Molecular weight ($\times 10^6$)	Parental genomes	Insertional inactivation: Marker[a]	Insertional inactivation: Sites	Noninactivating unique sites	References[b]
pBD6	5.8	pSA0501, pUB110	Km^R	*Bgl*II	*Bam*HI, *Tac*I	1,2,3
			Sm^R	*Eco*RI, *Hin*dIII		
pBD8	6.0	pSA2100, pUB110	Km^R	*Bgl*II	*Bam*HI, *Xba*I	1,2,4
			Sm^R	*Eco*RI, *Hin*dIII		
			Cm^R	*Stu*I		
pBD9	5.4	pE194, pUB110	Km^R	*Bgl*II	*Bam*HI, *Eco*RI, *Pst*I, *Tac*I, *Cla*I, *Ava*I, *Acc*I	1,2,4,6
			Em^R	*Sst*I, *Hpa*I, *Bcl*I		
pBD10	4.4	pBD8, pE194	Km^R	*Bgl*II	*Bam*HI, *Xba*I	1,2,4,5
			Cm^R	*Stu*I		
			Em^R	*Sst*I, *Hpa*I, *Bcl*I		
pBD11	4.4	pBD8, pE194	Km^R	*Bgl*II	*Bam*HI, *Xba*I	1,2,5
			Em^R	*Sst*I, *Hpa*I, *Bcl*I		

pBD12	4.5	pUB110, pC194	Km^R Cm^R	*Bgl*II *Stu*I	*Bam*HI, *Xba*I, *Eco*RI, *Hin*dIII	1,2,4
pBD64	3.2	pUB110, pC194	Km^R Cm^R	*Bgl*II *Stu*I	*Bam*HI, *Eco*RI, *Xba*I, *Ava*I	2,4
pHV11	3.5	pC194, pT127	Cm^R Tc^R	*Stu*I *Kpn*I	*Hpa*I, *Xba*I	4,7,8
pCW48	4.0	pC221, pCW3	Cm^R Tc^R	*Bst*EII *Kpn*I	—	4,9
pCW59	3.5	pCW7, pCW3	Cm^R Tc^R	*Bst*EII, *Bgl*II *Kpn*I	—	4,9
pSL103	5.0	pUB110, *Bacillus pumilus* chromosome	Km^R *trpC*	*Bgl*II *Hin*dIII	—	10,11
pSL105	5.4	pUB110, *B. licheniformis* chromosome	Km^R *trpC*	*Bgl*II *Hin*dIII	—	10,11

[a]See Table I for resistance marker abbreviations.

[b]1, Gryczan and Dubnau (1978); 2, Gryczan *et al.* (1980c); 3, Löfdahl *et al.* (1978); 4, D. Dubnau, unpublished; 5, Gryczan *et al.* (1980b); 6, Horinouchi and Weisblum (1982); 7, Ehrlich (1978); 8, Michel *et al.* (1980); 9, Wilson *et al.* (1981); 10, Keggins *et al.* (1978); 11, Keggins *et al.* (1979).

original references listed provide more information, including restriction-site maps.

B. Shuttle Vectors

A commonly used cloning stratagem involves the use of *E. coli* as an intermediate host for the isolation of *Bacillus* fragments (Section IV,C). For this purpose, shuttle vectors capable of replication in both hosts and carrying selectable markers that express in these hosts are useful. These have been constructed in several laboratories, and several are listed and described in Table III. As discussed in Section V, *E. coli* genes generally do not express in *B. subtilis,* whereas *B. subtilis* and *S. aureus* genes do express in *E. coli.* Thus the ampicillin- (Ap)- and Tc-resistance genes of pBR322 do not express in *B. subtilis,* whereas the chloramphenicol- (Cm)- and kanamycin- (Km)-resistance genes of pC194 and pUB110 express in *E. coli* (Ehrlich, 1978; Kreft *et al.,* 1978; Iomantas *et al.,* 1979; Table III).

C. Promoter Cloning Vectors

Williams *et al.* (1981a) and Lovett *et al.* (1982) have constructed a vector that permits selection of fratments promoting expression of a Cm^R gene. A 1.44-Mdalton *Eco*RI fragment derived from the *B. pumilus* chromosome and carrying a chloramphenicol acetyl transferase gene was cloned into pUB110. The resulting plasmid (pPL531) conferred Km^R and high-lcvcl Cm^R. The plasmid specified an inducible chloramphenicol acetyl transferase. A 0.55-Mdalton *Pst*I fragment was deleted from pPL531, resulting in a derivative that expressed only a low-level Cm^R. When the cloned *Eco*RI fragment of this derivative was recloned in the reverse orientation, high-level Cm^R was restored. The authors concluded that the structural portion of the Cm^R gene was therefore intact and that removal of the *Pst*I fragment must have removed an essential expression signal (probably transcriptional). Further engineering generated pPL603 (3.1 Mdaltons), which contained a single *Eco*RI site situated near the deleted *Pst*I fragment. Shotgun cloning of *Eco*RI and *Eco*RI* fragments from a variety of chromosomal, plasmid, and phage DNA samples into this *Eco*RI site yielded recombinant plasmids with restored high-level Cm^R. This vector appears to permit the selection of promoter-bearing fragments (Lovett *et al.,* 1982). However, in the absence of transcription-mapping and sequence data it is difficult to interpret these results definitively. For instance, deletion of the *Pst*I fragment may have removed essential translation signals. This interesting system deserves further investigation.

Table III Shuttle Vectors for *Escherichia coli* and *Bacillus subtilis*

Plasmid	Molecular weight ($\times 10^6$)	Parental plasmids	Markers[a]		Reference
			E. coli	*B. subtilis*	
pHV14	4.6	pC194, pBR322	Ap, Cm	Cm	Ehrlich (1978)
pHV33	4.6	pC194, pBR322	Ap, Tc, Cm	Cm	Primrose and Ehrlich (1981)
pHV23	6.1	pC194, pT127, pBR322	Ap, Cm	Tc, Cm	Michel *et al.* (1980)
pJJ1	5.6	pBR322, pUB110	Ap, Km	Km	Iomantas *et al.* (1979)
pIS3	5.8	pBD64, pBR322	Ap, Tc, Cm, Km	Km, Cm	I. Smith (personal communication)

[a]Ap, Ampicillin; see Table I for other abbreviations.

McLaughlin *et al.* (1982) have described another promoter-fishing vector, pSYC423. This plasmid was constructed from a chimera (pSYC310-2), which was derived from pBR322 and the *B. subtilis* plasmid pOG1196. pSYC310-2, which replicates in both *E. coli* and *B. subtilis,* contains an intact *B. licheniformis* β-lactamase gene (*penP*) and a chloramphenicol acetyltransferase gene derived from pOG1196. These determinants express in both hosts. The Ap^R function of pBR322 was inactivated by introduction of a small deletion. A further deletion was introduced, which removed the *penP* promoter, leaving the coding sequence and ribosome-binding site intact. This deleted sequence was replaced by a 28-bp fragment, carrying *Eco*RI, *Sma*I, and *Mbo*I sites, to yield the promoter-probing vector pSYC423. When the *E. coli lacUV5* promoter fragment was inserted into the *Eco*RI site on pSYC423, good activation of the *penP* gene was observed in *E. coli,* but very poor activation in *B. subtilis* (Section V). When an *Eco*RI–*Rsa*I fragment, which comprises a portion of the deleted *penP* promoter-bearing fragment was reintroduced into pSYC423, which had been cleaved with *Eco*RI and *Sma*I, full *penP* activity was restored in both hosts. This shuttle vector may prove useful, especially for comparative studies of expression in gram-negative and -positive hosts. It has the considerable advantage that the sequence of the *penP* gene is known (Neugebauer *et al.,* 1981), and the transcriptional and translational start sites have been determined (McLaughlin *et al.,* 1982).

We have used the *ermC* gene of plasmid pE194 to construct a similar vector (J. Hahn, T. J. Gryczan, and D. Dubnau, unpublished). This gene confers high-level Em^R, it has been sequenced, and both its transcriptional and its translational regulation have been intensively studied. The promoter of *ermC* has been inferred from sequence data (Gryczan *et al.,* 1980b; Horinouchi and Weisblum, 1980) and by S1 nuclease mapping (Gryczan *et al.,* 1980b). We have sequenced the 5′ terminus of *in vitro* RNA polymerase runoff products and thus confirmed the inferred transcription-start site (T. J. Gryczan, M. Reches, and D. Dubnau, unpublished). Two deletants of pE194, isolated after treatment of the DNA with the nuclease *Bal*31, were shown to lack the *ermC* promoter but to contain the entire structural gene and most of the translational and regulatory signals intact (Hahn *et al.,* 1982). Both of these derivatives were constructed in such a way that the deleted segment was replaced by *Hin*dIII linker. Cm^R fragments derived from pC194 were introduced elsewhere on these plasmids to provide a selective marker, and upstream promoters that provided read-through transcription of *ermC* were removed. The resulting Cm^R Em^S plasmids were used as vectors for the shotgun cloning of *Hin*dIII fragments derived from *B. licheniformis* chromosomal DNA, with selection for Em^R colonies (T. J. Gryczan and D. Dubnau, unpublished). These were obtained with high efficiency, and all contained derivatives of the vectors carrying inserts. The expression of Em^R by these plasmids was inducible. This was expected, because *ermC* induction is mediated on the level of translation (Shivakumar *et al.*, 1980b). It is likely that these vectors will prove useful for the selection of promoter-bearing fragments.

D. Positive Selection of Recombinant Plasmids

A plasmid vector has been constructed that permits direct selection for recombinant plasmids (Gryczan and Dubnau, 1982). A *thy* gene carried by *B. subtilis* phage β22 (Yehle and Doi, 1967) was cloned (on a *Bgl*II fragment) into vector pBD64 (Table II). This *thy* gene is not homologous to the *thy* locus of *B. subtilis* (Duncan *et al.*, 1978). The cloned *thy* gene contains inactivating sites for *Eco* RV, *Bcl*I, *Eco*RI, and *Pvu*II. The *Eco*RV and *Bcl*I sites were unique; *Eco*RI and *Pvu*II sites on the vector were removed, so all four sites were unique inactivating sites within the *thy* determinant. The resulting plasmid (pBD214) has a molecular weight of about 3.8 Mdaltons. In *B. subtilis,* as in *E. coli,* Thy^- strains are resistant to trimethoprim (TMP), and Thy^+ strains are TMP^S. pBD214 complements a chromosomal *thy* (TMP^R) mutant of *B. subtilis,* conferring a Thy^+ (TMP^S) phenotype. pBD214 also carries a Cm^R determinant derived from the pBD64 vector. Insertion of foreign DNA into any of the four inactivating sites, followed by transformation and simultaneous selection for Cm^R and TMP^R yields colonies ($\sim 1 \times 10^5/\mu gm$ foreign DNA), all of which contain plasmids with inserts. This selective system has been used successfully to clone fragments of the replicative form of Pf3, a filamentous *Pseudomonas* phage (D. Gluck Putterman, T. J. Gryczan, D. Dubnau, and L. Day, unpublished), to isolate sporulation genes from *B. subtilis* (E. Dubnau and I. Smith, personal communication), to isolate α-amylase genes from *B. licheniformis* (S. Ortlepp, personal communication), to clone a β-lactamase-bearing fragment from *B. cereus* (A. Sloma and M. Gross, personal communication), and to clone a fragment carrying the insertion site for the *S. aureus* transposon Tn554 (E. Murphy, personal communication).

This system, to a certain extent, overcomes the consequences of the inability of nicked and gapped plasmid DNA to transform in *B. subtilis.* This inability precludes the use of homopolymer tailing and alkaline phosphatase treatment to enrich for recombinant plasmid molecules. The TMP system directly selects for such molecules; it is also useful for the construction of gene libraries. Following transformation with the ligated mixture, selection is applied in liquid culture for Cm^R TMP^R transformants. When the culture has grown, plasmid DNA can be isolated and stored. These samples can be used as gene libraries to isolate desired clones by transformation into secondary recipients with simultaneous selection for Cm^R and a desired marker, or for Cm^R alone (which yields only plasmids with inserts), followed by screening for inserts for which no direct selection is available.

Modification of the TMP-selection system is desirable and under way, to permit the use of helper cloning or one of the other stratagems that facilitate shotgun cloning in *B. subtilis* (Section IV).

E. Bacteriophage Vectors

Some progress has been made in the search for useful bacteriophage cloning systems for use in *B. subtilis*. Several phages have been proposed as cloning vectors, including SPO2 (Graham *et al.*, 1979), ρ14 (Kroyer *et al.*, 1980), φ1 (Kawamura *et al.*, 1980), and SPPlV (Heilmann and Reeve, 1982). Only the latter system has been used to isolate specific fragments, although some recombinant phages were obtained using φ1. SPP1 is a well-characterized virulent bacteriophage of *B. subtilis*, which has a molecular weight of 27.4×10^6, contains a terminal redundancy consisting of 1.2 Mdaltons and transfects efficiently. Viable deletion mutants have been described that have lost up to 10% of the SPP1 genome (Amann *et al.*, 1981). Heilmann and Reeve (1982) constructed a derivative of SPP1 (SPP1V) that contains a deletion of a nonessential region of the phage genome and a unique *Bam*HI site adjacent to the deletion. *Bam*HI treatment of SPP1V DNA reduces the transfection frequency (ordinarily $\sim 10^5/\mu$gm DNA) by a factor of 100–1000. Cloning of *Bgl*II, *Bcl*I, or *Mbo*I fragments into the *Bam*HI site destroys the site. Thus treatment with *Bam*HI after ligation and before transfection selects for recombinant phage molecules, as long as the cloned insert lacks *Bam*HI sites. The latter problem is obviated by treatment of the DNA to be cloned with the enzyme *Bst*15031. This modification methylase specifically protects *Bam*HI sites (Levy and Welker, 1981). In some cases a cloned fragment can be excised using *Xma*III sites that flank the *Bam*HI site on SPP1V. A present limitation of this promising system is the maximum size of clonable fragments (~4 Mdaltons). However, Heilmann and Reeve (1982) have outlined a plausible approach to removing this limitation.

Another type of bacteriophage-based cloning system has been described (Kawamura *et al.*, 1979; Yoneda *et al.*, 1979; Iijima *et al.*, 1980) using the temperate phages ρ11, φ3T, and φ105. We shall illustrate the use of these systems by describing the one based on ρ11 (MW $\sim 80 \times 10^6$). Kawamura *et al.* (1979) digested ρ11 DNA with *Eco*RI (to yield more than 30 fragments) and ligated this mixture to *Eco*RI-digested *B. subtilis* chromosomal DNA. The ligation mixture was used to transform a ρ11 lysogen of *B. subtilis*. A fragment of chromosomal DNA in the ligation mixture could, presumably, integrate at its site of homology in the chromosome, or, if ligated to flanking ρ11 fragments in the correct orientation, at the locus of the ρ11 prophage. In these experiments a *spoA lys hisA leu* lysogen was used, and selection was for Lys^+, His^+, or Leu^+ transformants. Transformants of each type were pooled and induced with mitomycin C to produce phage. The resulting lysates were then used to transduce a nonlysogenic recipient, and Lys^+ and His^+ transductants were obtained. Several of these contained either plaque-forming or defective high-frequency transducing phages. In addition to His^+ and Lys^+ phages, the *spoF* gene was isolated in this way (Kawamura *et al.*, 1981).

IV. Cloning Stratagems

As described previously, the major obstacle to shotgun cloning in *B. subtilis* seems to derive from the failure of plasmid monomers to transform. The frequency of formation of head-to-tail oligomers, carrying a given chromosomal fragment, is obviously quite low. Several systems will now be described that obviate this problem. One of them (helper cloning) has been used fairly widely. The others are potentially useful.

A. Homology-Facilitated Cloning: The "Helper" System

Monomer plasmid DNA can transform with high efficiency when the competent culture carries an homologous or partially homologous "helper" plasmid (Contente and Dubnau, 1979b). pBD64 (Km^R Cm^R) is an *in vitro*–derived chimera, constructed from pUB110 and pC194 (Table II). When used to transform cultures carrying pUB110 (Km^R) with selection for Cm^R, CCC monomers, linearized plasmid DNA, and open circular (OC) molecules transform readily (10^5–10^6 transformants/μgm DNA). This vector–helper pair has been used in several shotgun cloning experiments (Gryczan *et al.*, 1980a; Iomantas *et al.*, 1980; Docherty *et al.*, 1981; E. Dubnau *et al.*, 1981; Gryczan and Dubnau, 1982). It has also been used to clone α-amylase genes (S. Ortlepp, personal communication). Two disadvantages are inherent in this system. First, the multicopy pUB110 helper plasmid persists, making it difficult to utilize the inactivation of the Km^R gene by *Bgl*II insertions to screen for inserts. This is only important if no selection is available for the desired insert. A second limitation derives from the need to use Rec^+ recipients for helper cloning, because the helper effect is Rec-dependent. This has effectively limited use of the helper system to the cloning of heterologous DNA, because homologous (*B. subtilis*) DNA will simply integrate. Other pairs of vector and helper plasmids are also available (Tables I–III) and may be more suitable for particular purposes. For instance, the pE194 incompatibility system is "strong" (Gryczan *et al.*, 1982).

When pE194 is used as a helper and selection applied for a donor marker [e.g., using pBD9 as a vector with selection for Km^R (Table II)], the helper is lost rapidly. Thus insertional inactivation of Em^R by cloning into the *Sst*I, *Hpa*I, or *Bcl*I sites is readily detectable. pE194 helpers have been used by E. Dubnau, K. Cabane, and I. Smith to clone the *hisH* gene from *B. licheniformis*. We have used pE194 derivatives as helpers together with the promoter probe vector described previously (Section III,C). It is worth noting that helper cloning has been used in *Streptococcus* (Macrina *et al.*, 1981; Ash Tobian and Macrina, 1982).

B. Vectors with Sequence Repeats

Michel *et al.* (1981) have reported that plasmid monomers carrying direct repeats can transform competent cultures. The efficiency of transformation increases steeply with the length of the repeats. Haykinson *et al.* (1982) have made the same observation and have further established that indirect repeats also permit monomer transformation. These authors have developed a potentially powerful cloning system based on derivatives of the *Streptococcus pyrogenes* plasmid pSM19035, which has been shown to contain a long inverted repeat (Behnke *et al.*, 1979). pSM19035 confers Em^R, can be maintained in *B. subtilis,* and transforms as a monomer due to its inverted repeats. Shotgun cloning with pSM19035 derivatives as vectors presumably can, therefore, be carried out in a straightforward manner. Haykinson *et al.* (1982) have also developed a cloning system using a vector carrying an extensive direct repeat. This vector (pMX20) is derived from the shuttle vector pJJ1 (Table III), but it carries a duplication that includes the Km^R determinant. It can be maintained in *recA E. coli* strains. pMX20 CCC monomer transforms *B. subtilis* strains efficiently, but it rapidly dissociates to yield large and small daughter plasmids, presumably owing to intramolecular recombination. pMX20, isolated from *E. coli recA,* is thus available as a vector for transformation into *B. subtilis*. Owing to the intramolecular recombination event, cloned fragments appear in *B. subtilis* carried on a pMX20 derivative that *lacks* the direct repeat. These systems are both very promising.

C. Use of *Escherichia coli* as an Intermediate Host

Because cloning systems for *E. coli* are relatively advanced, many investigators have chosen to use these systems for the isolation of *Bacillus* genes, followed by recloning into *B. subtilis,* often using shuttle vectors (Table III). The β-lactamase from *B. licheniformis* has been cloned in *E. coli* by several laboratories (Brammar *et al.*, 1980; Gray and Chang, 1981; Imanaka *et al.*, 1981; Neugebauer *et al.*, 1981). This gene was then recloned into *B. subtilis* by the first two groups. Many other *Bacillus* genes have been cloned into *E. coli* (e.g., see Mahler and Halvorson, 1977; Segall and Losick, 1977; Chi *et al.*, 1978; Duncan *et al.*, 1978; Nagahari and Sakaguchi, 1978). In many cases expression of *Bacillus* genes has been obtained in *E. coli;* see Section V). Several laboratories have constructed *B. subtilis* libraries in *E. coli.* Hoch and collaborators have utilized λ Charon phages for this purpose (reviewed in Hoch *et al.*, 1982). To obtain presumably random, partial *Eco*RI digests of *B. subtilis* DNA for cloning, chromosomal DNA was partially methylated using *Eco*RI methylase and then cut to completion with *Eco*RI. DNA extracted from the λ particle transformed *B. subtilis* for several markers with high efficiency. Surprisingly, intact λ particles, when added to competent cultures, also yielded transformants with high efficiency. Thus, λ clones carrying large *B. subtilis* fragments were readily isolated. Hoch *et al.* (1982) estimated that "10% of the *B. subtilis* chromosome now exists as identified and isolated fragments in λ [p. 165]." These can now presumably be recoloned into suitable *B. subtilis* vectors.

Hutchison and Halvorson (1980) constructed a bank of randomly sheared *B. subtilis* DNA fragments in the *E. coli* vector pMB9 using the dA-dT tailing method. The average size of the cloned fragments obtained was 7 kb. Three clones were identified that contained fragments of ϕ105 prophage DNA. Bott *et al.* (1981) have used this bank successfully to probe for clones containing rDNA genes. Another bank was constructed by Rapoport *et al.* (1979), also using dA-dT tailing, with the shuttle vector pHV33 (Table III). Thirty clones were isolated that hybridized specifically to *B. subtilis* rRNA. Other plasmids were isolated from this collection that complemented *B. subtilis thr, leuA, hisA, glyB,* and *purB* mutants; these plasmids were unstable in Rec^+ *B. subtilis* hosts, but their stability improved in a *recE4* background.

V. Expression of Cloned Genes

In addition to the direct detection of cloned genes that confer phenotypic changes on the host organism, two systems are currently available in *B. subtilis* for the analysis of cloned genetic determinants. The first utilizes the minicell system isolated and developed by Reeve *et al.* (1973). This system is straightforward and simple, and examples of its use to study the expression of plasmid genes have been provided by Shivakumar *et al.* (1979, 1980a,b), Mahler and Halvorson (1980), Docherty *et al.* (1981), E. Dubnau *et al.* (1981), and Hahn *et al.* (1982). A second system that is potentially powerful is the *B. subtilis in vitro* coupled transcription–translation system (Leventhal and Chambliss, 1979; McLaughlin *et al.*, 1981a).

Bacillus subtilis does not express most genes from *E. coli* and other gram-negative organisms. The reverse does not seem to be true; many *B. subtilis* genes express readily in *E. coli*. McLaughlin *et al.* (1981b) and Murray and Rabinowitz (1982) have presented evidence suggesting that this asymmetric barrier is owing, at least in part, to a requirement in *B. subtilis* for a strong complementarity between the mRNA ribosome-binding site and the 3′ end of 16-S rRNA. In addition, the recognition requirement for transcriptional initiation signals may be more stringent for the *B. subtilis* than for the *E. coli* RNA polymerase (Moran *et al.*, 1982; Murray and Rabinowitz, 1982). It is not clear whether this requirement simply involves a better fit with the promoter consensus sequence in the "−10" and "−35" regions, or whether additional recognition elements are involved. In any event, it will be useful to design expression vectors for use in *B. subtilis* that contain cloning sites located downstream from appropriate transcriptional and translational signals. Several such constructions have been reported; for examples, see Goldfarb *et al.* (1981), Hardy *et al.* (1981), Williams *et al.* (1981b), and Lovett *et al.* (1982).

VI. Conclusions

Considerable progress has been made in developing *Bacillus subtilis* as a cloning host. The chief obstacle, which derives from the requirement for oligomers in plasmid transformation, has been largely overcome by the use of homology-facilitated cloning. The development of vectors carrying direct or indirect repeats is particularly promising. Although a large catalog of plasmid vectors is now available, emphasis in the future will be placed on the development of *expression vectors* and on those that permit the secretion of foreign proteins. Finally, the further development of bacteriophage vectors is desirable and almost certain to be successful.

Acknowledgments

The author thanks T. J. Gryczan, E. Dubnau, and I. Smith for useful discussions, and A. Howard for expert secretarial assistance.

During the preparation of this review the author's laboratory was supported by grants from the National Institutes of Health (AI-10311 and AI-17472) and the American Cancer Society (MV99B) and by funds from the Italian Hydrocarbon Authority (ENI).

References

Amann, E. P., Reeve, J. N., Morelli, G., Behrens, B., and Trautner, T. A. (1981). *Mol. Gen. Genet.* **182,** 292–298.

Ash Tobian, J., and Macrina, F. (1982). *J. Bacteriol.* **152,** 215–222.

Behnke, D., Golubkov, V. I., Malke, H., Boitsov, A. S., and Totolian, A. A. (1979). *FEMS Microbiol. Lett.* **6,** 5–9.

Bensi, G., Iglesias, A., Canosi, U., and Trautner, T. A. (1981). *Mol. Gen. Genet.* **184,** 400–404.

Bernhard, K., Schrempf, H., and Goebel, W. (1978). *J. Bacteriol.* **133,** 897–903.

Bott, K. F., Wilson, F. E., and Stewart, G. C. (1981). *In* "Sporulation and Germination" (H. S. Levinson, A. L. Sonenshein, and D. J. Tipper, eds.), pp. 119–122. Amer. Soc. Microbiology, Washington, D.C.

Brammar, W. J., Muir, S., and McMorris, A. (1980). *Mol. Gen. Genet.* **178,** 217–224.

Canosi, U., Morelli, G., and Trautner, T. A. (1978). *Mol. Gen. Genet.* **166,** 259–267.

Canosi, U., Iglesias, A., and Trautner, T. A. (1981). *Mol. Gen. Genet.* **181,** 434–440.

Chang, S., and Cohen, S. N. (1979). *Mol. Gen. Genet.* **168,** 111–115.

Chi, N.-Y. W., Ehrlich, S. D., and Lederberg, J. (1978). *J. Bacteriol.* **133,** 816–821.

Contente, S., and Dubnau, D. (1979a). *Mol. Gen. Genet.* **167,** 251–258.

Contente, S., and Dubnau, D. (1979b). *Plasmid* **2,** 555–571.

Debabov, V. G. (1982). *In* "The Molecular Biology of the Bacilli" (D. Dubnau, ed.), Vol. 1, pp. 331–370. Academic Press, New York.

Docherty, A., Grandi, G., Grandi, R., Gryczan, T. J., Shivakumar, A. G., and Dubnau, D. (1981). *J. Bacteriol.* **145,** 129–137.

Dubnau, D. (1982). *In* "The Molecular Biology of the Bacilli" (D. Dubnau, ed.), Vol. 1, pp. 147–178. Academic Press, New York.

Dubnau, D., Contente, S., and Gryczan, T. J. (1980a). *In* "DNA-Recombination Interactions and Repair" (S. Zadrazil, and J. Sponar, eds.), pp. 365–386. Pergamon, Oxford and New York.

Dubnau, D., Gryczan, T., Contente, S., and Shivakumar, A. G. (1980b). *Genet. Eng.* **2,** 115–132.

Dubnau, D., Grandi, G., Grandi, R., Gryczan, T. J., Hahn, J., Kozloff, Y., and Shivakumar, A. G. (1981). *In* "Molecular Biology, Pathogenicity, and Ecology of Bacterial Plasmids" (S. B. Levy, R. C. Clowes, and E. L. Koenig, eds.), pp. 157–167. Plenum, New York.

Dubnau, E., Ramakrishna, N., Cabane, K., and Smith, I. (1981). *J. Bacteriol.* **147,** 622–632.

Duncan, C. H., Wilson, G. A., and Young, F. E. (1978). *Proc. Natl. Acad. Sci. USA* **75,** 3664–3668.

Ehrlich, S. D. (1977). *Proc. Natl. Acad. Sci. USA* **74,** 1680–1682.

Ehrlich, S. D. (1978). *Proc. Natl. Acad. Sci. USA* **75,** 1433–1436.

Ehrlich, S. D., Niaudet, B., and Michel, B. (1982). *Curr. Top. Microbiol. Immunol.* **96,** 19–29.

Goldfarb, D. S., Doi, R. H., and Rodriguez, R. L. (1981). *Nature (London)* **293,** 309–311.

Graham, S., Yoneda, Y., and Young, F. E. (1979). *Gene* **7,** 69–77.

Gray, O., and Chang, S. (1981). *J. Bacteriol.* **145,** 422–428.

Gryczan, T. J. (1982). *In* "The Molecular Biology of the Bacilli" (D. Dubnau, ed.), Vol. 1, pp. 307–329. Academic Press, New York.

Gryczan, T. J., and Dubnau, D. (1978). *Proc. Natl. Acad. Sci. USA* **75,** 1428–1432.

Gryczan, T. J., and Dubnau, D. (1982). Gene **20,** 459–469.

Gryczan, T. J., Contente, S., and Dubnau, D. (1978). *J. Bacteriol.* **134,** 318–329.

Gryczan, T. J., Contente, S., and Dubnau, D. (1980a). *Mol. Gen. Genet.* **177,** 459–467.

Gryczan, T. J., Grandi, G., Hahn, J., Grandi, R., and Dubnau, D. (1980b). *Nucleic Acids Res.* **8,** 6081–6097.

Gryczan, T., Shivakumar, A. G., and Dubnau, D. (1980c). *J. Bacteriol.* **141,** 246–253.

Gryczan, T. J., Hahn, J., Contente, S., and Dubnau, D. (1982). *J. Bacteriol.* **152,** 722–735.

Hahn, J., Grandi, G., Gryczan, T. J., and Dubnau, D. (1982). *Mol. Gen. Genet.* **186,** 204–216.

Hardy, K., Stahl, S., and Küpper, H. (1981). *Nature (London)* **293,** 481–483.

Haykinson, M., Rabinovitch, P., and Stepanov, A. (1982). *Dokl. Akad. Nauk. SSSR* **265,** 975–978.

Heilmann, H., and Reeve, J. N. (1982). *Gene* **17,** 91–100.

Hoch, J. A., Nguyen, A., and Ferrari, E. (1982). *Basic Life Sci.* **19,** 163–173.

Horinouchi, S., and Weisblum, B. (1980). *Proc. Natl. Acad. Sci. USA* **77,** 7079–7083.

Horinouchi, S., and Weisblum, B. (1981). *Mol. Gen. Genet.* **182,** 341–348.

Horinouchi, S., and Weisblum, B. (1982). *J. Bacteriol.* **150,** 804–814.

Hutchison, K. W., and Halvorson, H. O. (1980). *Gene* **8,** 267–278.

Iijima, T., Kawamura, F., Saito, H., and Ikeda, Y. (1980). *Gene* **9,** 115–126.

Imanaka, T., Tanaka, T., Tsunekawa, H., and Aiba, S. (1981). *J. Bacteriol.* **147,** 776–786.

Iomantas, Iu. V., Rabinovitch, P. M., Bandrin, S. V., Demyanova, N. G., Kozlov, Y. I., and Stepanov, A. I. (1979). *Dokl. Akad. Nauk SSSR* **244,** 993–996.

Iomantas, Iu. V., Rabinovitch, P. M., Rebentish, B. A., and Stepanov, A. I. (1980). *Dokl. Akad. Nauk SSSR* **254,** 493–495.

Jalanko, A., Palva, I., and Söderlund, H. (1981). *Gene* **14,** 325–328.

Kawamura, F., Saito, H., and Ikeda, Y. (1979). *Gene* **5,** 87–91.

Kawamura, F., Saito, H., and Ikeda, Y. (1980). *Mol. Gen. Genet.* **180,** 259–266.

Kawamura, F., Shimotsu, H., Saito, H., Hirochika, H., and Kobayashi, Y. (1981). *In* "Sporulation and Germination" (H. S. Levinson, A. L. Sonenshein, and D. J. Tipper, eds.), pp. 109–118. Amer. Soc. Microbiology, Washington, D.C.

Keggins, K. M., Lovett, P. S., and Duvall, E. J. (1978). *Proc. Natl. Acad. Sci. USA* **75,** 1423–1427.

Keggins, K. M., Lovett, P. S., Marrero, R., and Hoch, S. O. (1979). *J. Bacteriol.* **139,** 1001–1006.

Kreft, J., and Hughes, C. (1982). *Curr. Top. Microbiol. Immunol.* **96,** 1–17.

Kreft, J., Bernhard, K., and Goebel, W. (1978). *Mol. Gen. Genet.* **162,** 59–67.

Kroyer, J. M., Perkins, J. B., Rudinski, M. S., and Dean, D. H. (1980). *Mol. Gen. Genet.* **177,** 511–517.

Leventhal, J. M., and Chambliss, G. H. (1979). *Biochim. Biophys. Acta* **564,** 162–171.

Levy, W. P., and Welker, N. E. (1981). *Biochemistry* **20,** 1120–1127.

Löfdahl, S., Sjöström, J. E., and Philipson, L. (1978). *Gene* **3,** 161–172.

Lopez, P., Espinosa, M., Stassi, D. L., and Lacks, S. A. (1982). *J. Bacteriol.* **150,** 692–701.

Lovett, P. S. (1981). *In* "Sporulation and Germination" (H. S. Levinson, A. L. Sonenshein, and D. J. Tipper, eds.), pp. 40–47. Amer. Soc. Microbiology, Washington, D.C.

Lovett, P. S., Williams, D. M., and Duvall, E. J. (1982). *Basic Life Sci.* **19,** 51–57.

Macrina, F. L., Jones, K. R., and Welch, R. A. (1981). *J. Bacteriol.* **146,** 826–830.

Mahler, I., and Halvorson, H. O. (1977). *J. Bacteriol.* **131,** 374–377.

Mahler, I., and Halvorson, H. O. (1980). *J. Gen. Microbiol.* **120,** 259–263.

McLaughlin, J. R., Murray, C. L., and Rabinowitz, J. C. (1981a). *J. Biol. Chem.* **256,** 11273–11282.

McLaughlin, J. R., Murray, C. L., and Rabinowitz, J. C. (1981b). *J. Biol. Chem.* **256,** 11283–11291.

McLaughlin, J. R., Chang, S.-Y., and Chang, S. (1982). *Nucleic Acids Res.* **10,** 3905–3919.

Michel, B., Palla, E., Niaudet, B., and Ehrlich, S. D. (1980). *Gene* **12,** 147–154.

Michel, B., Palla, E., and Ehrlich, S. D. (1981). *In* "Transformation—1980" (M. Polsinelli, and G. Mazza, eds.), pp. 189–199. Cotswold Press, Oxford.

Moran, C. P., Jr., Lang, N., LeGrice, S. F. J., Lee, G., Stephens, M., Sonenshein, A. L., Pero, J., and Losick, R. (1982). *Mol. Gen. Genet.* **186,** 339–346.

Mottes, M., Grandi, G., Sgaramella, V., Canosi, U., Morelli, G., and Trautner, T. A. (1979). *Mol. Gen. Genet.* **174,** 281–286.

Murray, C. L., and Rabinowitz, J. C. (1982). *J. Biol. Chem.* **257,** 1053–1062.

Nagahari, K., and Sakaguchi, K. (1978). *Mol. Gen. Genet.* **158,** 263–270.

Neugebauer, K., Sprengel, R., and Schaller, H. (1981). *Nucleic Acids Res.* **9,** 2577–2588.

Polak, J., and Novick, R. P. (1982). *Plasmid* **7,** 152–162.

Primrose, S. B., and Ehrlich, S. D. (1981). *Plasmid* **6,** 193–201.

Rapoport, G., Klier, A., Billault, A., Fargette, F., and Dedonder, R. (1979). *Mol. Gen. Genet.* **176,** 239–245.

Reeve, J. N., Mendelson, N. H., Coyne, S. I., Hallock, L. L., and Cole, R. M. (1973). *J. Bacteriol.* **114,** 860–873.

Sadaie, Y., Burtis, K. C., and Doi, R. H. (1980). *J. Bacteriol.* **141,** 1178–1182.

Scheer-Abramowitz, J., Gryczan, T. J., and Dubnau, D. (1981). *Plasmid* **6,** 67–77.

Scherzinger, E., Grieder-Lauppe, H., Voll, N., and Wanke, M. (1980). *Nucleic Acids Res.* **8,** 1287–1305.

Segall, J., and Losick, R. (1977). *Cell* **11,** 751–761.

Shivakumar, A. G., and Dubnau, D. (1978). *Plasmid* **1,** 405–416.

Shivakumar, A. G., and Dubnau, D. (1981). *Nucleic Acids Res.* **9,** 2549–2562.

Shivakumar, A. G., Hahn, J., and Dubnau, D. (1979). *Plasmid* **2,** 279–289.

Shivakumar, A. G., Gryczan, T. J., Kozlov, Y. I., and Dubnau, D. (1980a). *Mol. Gen. Genet.* **179,** 241–252.

Shivakumar, A. G., Hahn, J., Grandi, G., Kozlov, Y., and Dubnau, D. (1980b). *Proc. Natl. Acad. Sci. USA* **77,** 3903–3907.

Weisblum, B., Graham, M. Y., Gryczan, T., and Dubnau, D. (1979). *J. Bacteriol.* **137,** 635–643.

Williams, D. M., Duvall, E. J., and Lovett, P. S. (1981a). *J. Bacteriol.* **146,** 1162–1165.

Williams, D. M., Schoner, R. G., Duvall, E. J., Preis, L. H., and Lovett, P. S. (1981b). *Gene* **16,** 199–206.

Wilson, C. R., Skinner, S. E., and Shaw, W. V. (1981). *Plasmid* **5,** 245–258.

Yehle, C. O., and Doi, R. H. (1967). *J. Virol.* **1,** 935–947.

Yoneda, Y., Graham, S., and Young, F. E. (1979). *Biochem. Biophys. Res. Commun.* **91,** 1556–1564.

CHAPTER 4

Developments in *Streptomyces* Cloning

MERVYN J. BIBB
KEITH F. CHATER
DAVID A. HOPWOOD

John Innes Institute
Norwich, England

Experimental Manipulation
of Gene Expression

ISBN 0-12-372380-9

I. Introduction

Bacteria of the genus *Streptomyces* possess many unusual features. They are gram-positive organisms characterized by a mycelial form of growth that, after a complex process of differentiation, normally culminates in sporulation. They produce a wide range of antibiotics, numerically unsurpassed by any other genus (Hopwood, 1979), and a large number of exoenzymes, commensurate with their free-living saprophytic life-style as soil bacteria. They have relatively large genomes (10^4 kb, approximately three times that of *Escherichia coli;* Benigni *et al.*, 1975) composed of DNA of extremely high guanosine plus cytosine (G + C) content (up to 73 mol %), close to the upper limit observed in nature (Enquist and Bradley, 1971). These many characteristics and their consequences cannot be studied in other genetically well-defined prokaryotic systems (e.g., *Escherichia* or *Bacillus*).

A wide range of techniques are now available to study streptomycetes. These include many examples of plasmid-mediated conjugation (Hopwood *et al.*, 1973; Bibb and Hopwood, 1981; Bibb *et al.*, 1981; Kieser *et al.*, 1982); phage-mediated transduction (Stuttard, 1979); protoplast fusion induced by polyethylene glycol (PEG), permitting high levels of chromosomal recombination (between strains with DNA of similar sequence; Hopwood *et al.*, 1977; Baltz, 1978; Hopwood, 1981); highly efficient PEG-mediated transformation and transfection of protoplasts by plasmid (Bibb *et al.*, 1978) and phage (Suarez and Chater, 1980a; Krügel *et al.*, 1980; Isogai *et al.*, 1980) DNA, respectively; and PEG-induced transformation of protoplasts by chromosomal DNA enclosed in liposomes (Makins and Holt, 1981). The main purpose of this chapter is to illustrate how DNA cloning systems for these organisms can now be used to analyze many of their interesting biological features. Although cloning of *Streptomyces* DNA into alternative hosts can assist in the manipulation of their genes, the existence of indigenous *Streptomyces* cloning systems is obligatory for the analysis of complex properties such as antibiotic production and differentiation.

II. Vectors

The desirable properties of a plasmid or phage vector depend on its intended application. Therefore, for a given organism a number of possible vectors with different characteristics ought, ideally, to be available. Vectors, preferably of wide host range, should contain a minimal region for stable replication and maintenance, an easily selectable marker (frequently an antibiotic-resistance gene), and, preferably, a method to detect cloned DNA fragments. Unique insertion sites for a variety of restriction endonucleases within nonessential regions of the replicon, where possible in the target gene for insert-recognition itself, should be available. Vectors with a low or single copy number per genome are to be preferred for most physiological studies, whereas those with a high or amplifiable copy number are likely to be valuable for the investigation of certain cloned gene products. Systems should also be available for cloning particularly large fragments of genomic DNA for the construction and functional screening of gene libraries. The following compilation describes currently available *Streptomyces* cloning vectors that possess some of these qualities. Although most of them can be used as generalized cloning vectors in appropriate *Streptomyces* species, some have been constructed for particular purposes. They are categorized by the parental replicon from which they were derived. Although there are now many reports of plasmids and phages with the potential for use as cloning systems in *Streptomyces,* the following overview includes only those already employed as vectors.

A. Plasmid Vectors

In several studies, 15–25% of the *Streptomyces* strains examined have been shown by physical analysis to contain plasmids (Hayakawa *et al.,* 1979; Okanishi *et al.,* 1980; Hopwood *et al.,* 1981). These extrachromosomal elements vary in size, copy number and host range, and a few instances of incompatibility have been recognized. All naturally occurring *Streptomyces* plasmids studied (except some natural deletion derivatives; Kieser *et al.,* 1982) elicit the property of lethal zygosis (Ltz^+) on conjugal transfer into a plasmidless recipient and also make their host resistant to this effect (Ltz^R). The lethal zygosis phenotype has proved to be very useful for the recognition of plasmid-containing individuals after transformation of protoplasts or conjugal transfer. Where they have been amenable to study, most of these plasmids (SLP3 and SLP4 are exceptions) have been shown to be fertility factors of varying efficiency for promoting generalized chromosomal recombination (Hopwood *et al.,* 1983).

1. SLP1.2 Derivatives

SLP1.2, the largest of a family of plasmids isolated from *S. lividans* 66, has a size estimated at 12.2 (Bibb *et al.,* 1981) or 14.5 kb (Thompson *et al.,* 1982c). SLP1 sequences originate in the chromosome of *S. coelicolor* A3(2) (mapping close to the *strA* locus), but on interspecific mating with *S. lividans* they may take on an autonomous form in the recipient and replicate independently of the chromosome. In such a form, SLP1.2 has a copy number of 4–5 per chromo-

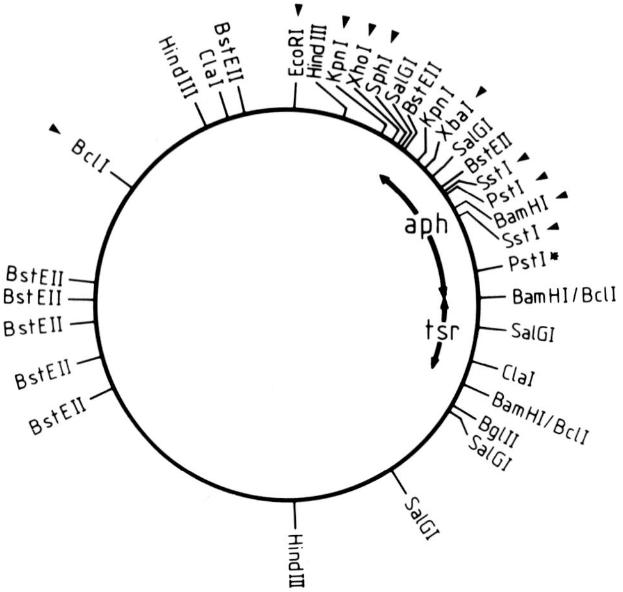

Fig. 1 Restriction map of pIJ41 (and pIJ61) (15.8 kb). *aph,* Aminoglycoside phosphotransferase gene (neomycin resistance); *tsr,* thiostrepton-resistance gene. The *Pst*I site marked with an asterisk is absent from pIJ61. Arrowheads indicate cloning sites referred to in the text. From Thompson *et al.* (1982c).

some and promotes chromosomal recombination at a frequency of about 10^{-4} when present in one or both parents in a cross. In the absence of selection, SLP1.2 derivatives are somewhat unstable, a property that has proved useful for plasmid curing. SLP1.2 derivatives appear to have a narrow host range, currently known to include *S. lividans, S. reticuli, S. clavuligerus,* and *S. vinaceus.* SLP1.2 was used in *S. lividans* to clone genes containing a neomycin (aminoglycoside) phosphotransferase (*aph*) and a neomycin acetyltransferase (*aac*) from *S. fradiae* (Thompson *et al.,* 1980, 1982c), a viomycin (Vio) phosphotransferase (*vph*) from *S. vinaceus* (Thompson *et al.,* 1982a,c), ribosomal RNA methylases conferring resistance to thiostrepton (*tsr*) and erythromycin (*mls*) from *S. azureus* (Thompson *et al.,* 1980, 1982c) and *S. erythreus* (Thompson *et al.,* 1980; 1982b,c), respectively, and a gene(s) (*mmr*) conferring resistance to methylenomycin (mechanism unknown) from the *S. coelicolor* plasmid SCP1 (Bibb *et al.,* 1980). Since then, various derivatives of SLP1.2 have been constructed *in vitro* using these antibiotic resistance genes to provide more useful cloning vectors.

pIJ41 and pIJ61 (Fig. 1) are derivatives of SLP1.2 that retain the replication, transfer, incompatibility, and lethal-zygosis properties of the parental plasmid but in addition possess the *aph* gene of *S. fradiae* and the *tsr* gene of *S. azureus.*

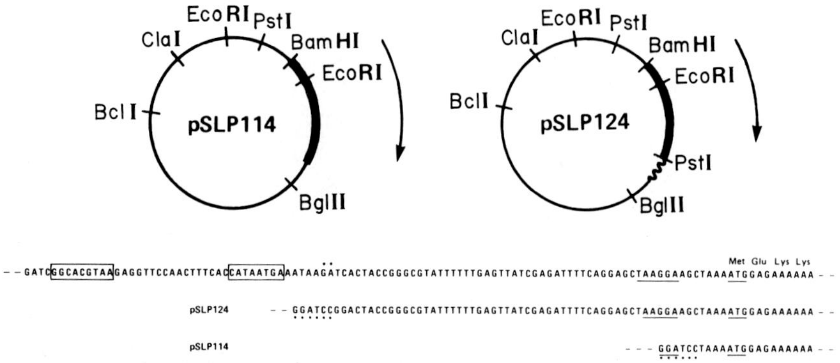

Fig. 2 pSLP114 and pSLP124. (Above) Restriction maps of pSLP114 and pSLP124. The thick segment represents the modified *Escherichia coli* CAT-gene fragment and the wavy segment of pSLP124 represents residual pACYC177 sequences. Arrows show the direction of transcription of the CAT gene. (Below) DNA sequences at the 5′ end of the CAT gene before manipulation (top) and after insertion into the SLP1.2 derivative pSLP101 to yield the promoter-probe plasmids. The site of initiation of transcription in *E. coli* is denoted by asterisks, and postulated RNA polymerase recognition and binding sites are boxed. ATG translation-start codons and potential ribosome-binding sites that show complementarity to the 3′ end of the 16-S rRNA of *E. coli* (top) or *Streptomyces lividans* (pSLP124 and pSLP114) are underscored. The *Bam*HI sites used for the insertion of fragments with potential transcriptional activity into the promoter probes are indicated by dots. From Bibb and Cohen (1982).

Both markers can be used independently to select for the vector and the occurrence of a unique *Bam*HI site, and in pIJ61 a unique *Pst*I site in the *aph* gene allows for the possibility of insertional inactivation. Cloning into pIJ41 and pIJ61 is also possible with *Bcl*I, *Eco*RI, *Kpn*I, *Sph*I, *Sst*I, *Xba*I, and *Xho*I.

All of the transcription-initiation signals (promoters) of prokaryotic and eukaryotic organisms that have been identified contain sequences rich in adenine plus thymine (A + T) (Rosenberg and Court, 1979). It will therefore be of interest, in addition to elucidating the precise nature of *Streptomyces* promoters, to determine if and how these transcriptional control sequences have changed during a course of evolution that has resulted in such an extreme genome base composition as 73 mol % G + C. pSLP114 and pSLP124 (Fig. 2) are vectors constructed for the isolation and examination of DNA sequences having promoter activity in *S. lividans* (Bibb and Cohen, 1982). DNA fragments that contained the coding sequences of the *E. coli* chloramphenicol acetyl transferase (CAT) gene but that lacked the promoter region normally recognized by the RNA polymerase of *E. coli* for initiation of transcription were inserted into a derivative of SLP1.2. The two resulting plasmids contain a unique *Bam*HI site upstream of the CAT-coding sequences and differ from each other substantially only in the position of this site with respect to the ribosome binding site of the chloramphenicol- (Cm)-resistance gene. Insertion of a variety of restriction-endonuclease-generated fragments (e.g., from DNA cleaved with *Bam*HI, *Bcl*I, *Bgl*II, or *Sau*3A/*Mbo*I) into the *Bam*HI site of either vector and selection or screening for Cm resistance after transformation of a Cm-sensitive recipient have made possible the isolation of *Streptomyces* promoter sequences and the study of some of the parameters involved in heterospecific gene expression. Analysis of such vegetatively expressed promoters should also be of value in determining how these organisms control the temporal expression of genes involved in antibiotic synthesis and differentiation.

2. SCP2* Derivatives

SCP2* (~31 kb) is a naturally occurring high-fertility variant of SCP2 with enhanced expression of lethal zygosis; it probably represents a mutant of SCP2 derepressed for transfer (Bibb *et al.*, 1977; Bibb and Hopwood, 1981). It is present in low copy number (1–5) per chromosome and promotes chromosomal recombination at a frequency of about 10^{-3} in *S. coelicolor* A3(2), its natural host (Bibb and Hopwood, 1981), and at $\sim 10^{-4}$ in *S. lividans* 66 (Hopwood *et al.*, 1983). It can also replicate stably in *S. parvulus* ATCC 12434.

Smaller, potentially more useful derivatives consist of either the largest *Pst*I or the largest *Bam*HI fragment of SCP2* (Bibb *et al.*, 1980); both derivatives are frequently lost during nonselective growth, but stability can be restored by the inclusion of DNA sequences of the second largest *Pst*I fragment of SCP2* (e.g., as in pSCP103; Bibb *et al.*, 1980). Stable derivatives of SCP2* containing selectable antibiotic-resistance genes have been constructed (D. J. Lydiate, personal communication). For example, pIJ913 contains the thiostrepton-resistance gene *tsr* from *S. azureus* inserted in place of a nonessential *Bam*HI segment of the SCP2* derivative pSCP103 (Bibb *et al.*, 1980). Sites definitely available for cloning include those for *Eco*RI, *Pst*I, and *Bgl*II (resulting in the loss of dispensable segments from the vector, which may lead to instability) and a unique *Eco*RV site the location of which within the *tsr* gene allows for insertional inactivation. Other sites probably available for insertion include those for *Xho*I and *Xba*I. A similar derivative (pIJ915) contains a *Bam*HI fragment carrying the Vio phosphotransferase gene *vph*. Sites available for cloning include those for *Bgl*II (with loss of dispensable segments), *Eco*RI, and probably, *Xho*I, *Xba*I, and *Cla*I.

3. pIJ101 Derivatives

pIJ101 (8.9 kb) is the largest of a series of four plasmids initially found as a mixture in *S. lividans* ISP 5434 (isolated independently of *S. lividans* 66; Kieser

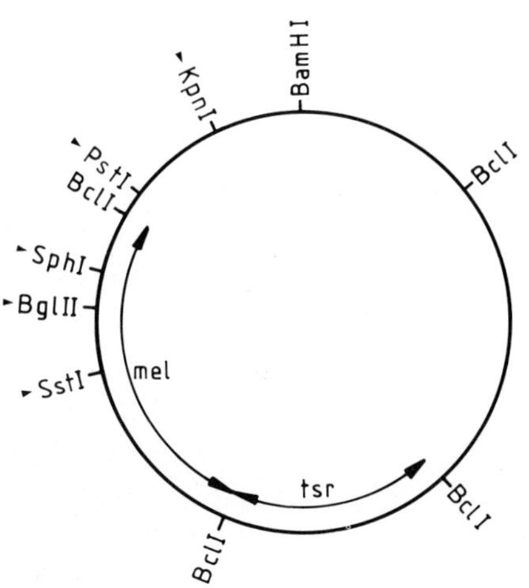

Fig. 3 Simplified restriction map of pIJ702 (5.7 kb). *mel*, Tyrosinase (melanin-production) gene; *tsr*, thiostrepton-resistance gene. Arrowheads indicate unique cloning sites. Data of Katz *et al.* (1983).

et al., 1982). The smaller members of the series (pIJ102–4) appear to be naturally occurring deletion derivatives of pIJ101. pIJ101 is a transmissible Ltz$^+$ plasmid and promotes chromosomal recombination at a very high frequency (10^{-2}–10^{-3}) in matings. pIJ101 and its derivatives have copy numbers varying between 40 and 300 per chromosomal equivalent, the highest values occurring predominantly in the stationary phase of growth. pIJ101 derivatives have a wide host range in *Streptomyces*. Transformation experiments with recombinant plasmids containing either the *aph* or *tsr* genes indicated replication and maintenance in 15 of 18 wild-type strains originally tested, including *S. coelicolor* and many unrelated species (Kieser *et al.*, 1982).

Some deletion derivatives of pIJ101 constructed *in vitro* are Ltz$^-$ and non-transmissible (a useful property for the isolation of primary transformants), though they can be mobilized by several conjugative plasmids, including SCP2* and SLP1.2. Analysis of various deleted forms indicates that a region of less than 2.1 kb is all that is required for plasmid replication. pIJ702 (Fig. 3) is a derivative of the Ltz$^-$ pIJ102 that contains the *tsr* gene from *S. azureus* and the tyrosinase gene *mel* from *S. antibioticus* (Section IV,F). Transformants can be detected either by selecting for resistance to thiostrepton or by screening for the production of black melanin pigment on tyrosine-containing agar. Unique sites for *Bgl*II, *Sph*I, and *Sst*I occur within the tyrosinase gene, permitting recognition of cloned fragments by insertional inactivation of *mel,* leading to the inability of colonies to form melanin on agar containing tyrosine (Fig. 4).

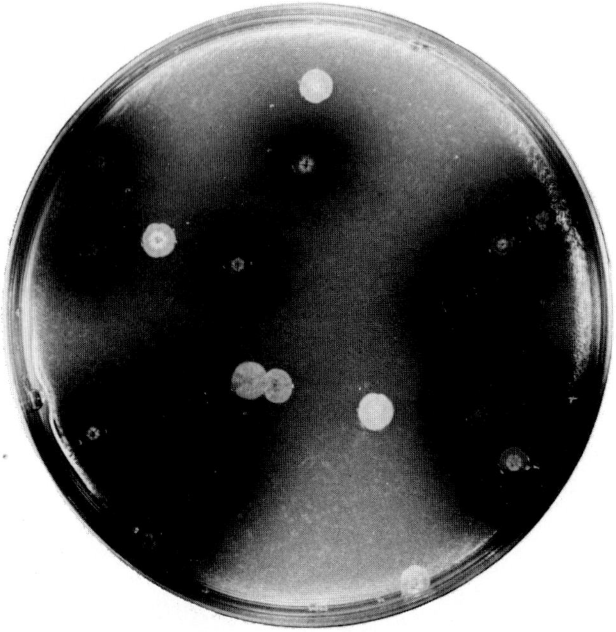

Fig. 4 Colonies of *Streptomyces lividans*, mostly containing the *mel* (tyrosinase) gene cloned on pIJ702; they produce dark melanin pigment. A few (white) colonies contain pIJ702 with foreign DNA inserted at the *Bgl*II site, thereby inactivating the *mel* gene. From Katz *et al.* (1983).

B. Phage Vectors

1. φC31

The temperate phage φC31 (41.2 kb) is the most extensively characterized *Streptomyces* phage and the first to be used for cloning (Suarez and Chater, 1980b). Reviews have covered its genetics and development as a cloning vector (Lomovskaya *et al.*, 1980; Chater *et al.*, 1981; Chater, 1983). φC31 and its derivatives have a wide host range within *Streptomyces* and infect over 70% of wild-type strains tested. For lysogeny to occur in *S. coelicolor*, φC31 normally has to integrate into the host chromosome through the specific *att* sites of both chromosome and phage to yield a prophage with a single copy per genome. The positions of the *attP* site (Chater *et al.*, 1982b) and the repressor gene *c* (Sladkova *et al.*, 1980) have been mapped physically, permitting a rational approach to the construction of derivatives that are more suitable for use as cloning vectors. φC31 DNA has cohesive ends, and from 37.5 to at least 42.4 kb of DNA can be packaged to produce a viable phage particle, although only 32 kb or less is absolutely required for plaque formation (Chater *et al.*, 1981).

Initially, φC31 vectors were constructed using a heat-inducible mutant (*c*ts1) to allow for control of induction. Difficulty in obtaining stable lysogens with this mutant has sometimes necessitated the construction of c^+ derivatives by natural recombination. These early derivatives were obtained as chelating-agent-resistant deletions of a hybrid replicon [φC31 *c*ts1ΔMΔ23::pBR322 (KC100)] that consists of pBR322 cloned into one of the *Eco*RI sites of a deleted form of φC31. This "ancestral" molecule has single *Bam*HI and *Pst*I sites in the nonessential segment of pBR322 that in principle allow for cloning in both *E. coli* and *Streptomyces* (Suarez and Chater, 1980b).

a. φC31 c$^+$*Δ23::pBR322ΔW12 (KC112).* Originally constructed in a *c*ts1 form (and converted to c^+ by natural recombination), this vector arose by a single large spontaneous deletion (ΔW12) from φC31 KC100 with the loss of the β-lactamase gene and *Pst*I site of the *E. coli* replicon (and probably sequences essential for replication in *E. coli;* Chater *et al.*, 1981). Up to at least 4.8 kb of DNA can be inserted into the unique *Bam*HI site of the vector, and retention of the phage *attP* site (Chater *et al.*, 1982b) enables normal integration into the host chromosome to occur upon lysogenization.

b. φC31 cts1*::pBR322ΔW17 (KC117).* This shuttle vector originated by deletion of φC31-specific DNA sequences from φC31 KC100 and retained the replication and selection properties of the vector in *E. coli*. The deletion is of a region that includes the *attP* site (as well as the ΔM and Δ23 deletions) and precludes lysogenization by the normal route (Chater *et al.*, 1982b). The phage has been used as a replacement vector in *Streptomyces*. Cleavage of the *E. coli*-grown plasmid form (pIJ502) with *Bam*HI and *Pst*I and separation of the larger fragment containing the φC31 component of the vector permits insertion of fragments of up to 8 kb having TGCA 3′ and 5′ GATC single-strand extensions (e.g., produced by digestion of DNA samples with *Pst*I and *Mbo*I). In principle, fragments of a similar size having 5′ CG (*Cla*I or *Taq*I) and TGCA 3′ (*Pst*I) extensions can be inserted into the larger fragment produced by digestion of the vector with *Cla*I and *Pst*I (although the vector contains two *Cla*I sites, only that contained within φC31 DNA appears to be protected by methylation during growth in a *dam*$^+$ methylase *E. coli,* effectively producing a vector with only one cleavable *Cla*I site (Hopwood and Chater, 1982; Harris *et al.*, 1983).

The absence of the *attP* site prevents normal lysogeny from occurring and would normally preclude the opportunity for the phenotypic detection of expression of any gene cloned into the vector. However, reconstruction experiments have indicated that transfection into a lysogenized indicator strain allows for such expression through (presumably host-mediated) homologous recombination of the vector into the resident prophage (Chater *et al.*, 1982b). Problems caused by infrequent recombination preventing efficient screening of plaques by

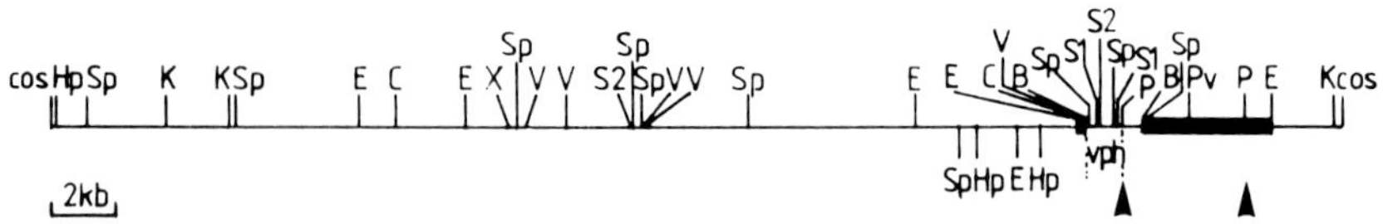

Fig. 5 Restriction map of ϕC31 KC400. *vph,* Viomycin phosphotransferase gene (viomycin resistance); cos, ϕC31 DNA cohesive ends. Arrowheads indicate *Pst*I sites used in cloning experiments (see text and Fig. 6); thick segments indicate pBR322 DNA. Restriction-enzyme target sites: B, *Bam*HI; C, *Cla*I; E, *Eco*RI; H, *Hpa*I; K, *Kpn*I; P, *Pst*I; Pv, *Pvu*II; S1, *Sst*I; S2, *Sst*II; Sp, *Sph*I; V, *Eco*RV; X, *Xba*I. After Harris *et al.* (1983).

replication have been overcome by the addition of an equal amount of nonlysogenic phage-sensitive host to the indicator mixture, allowing amplification of the phages *in situ.* The occurrence of the resulting tandemly repeated integrated ϕC31 DNA sequences in the double lysogens does not lead to gross instability of the selected phenotype; 85% of colonies retained both prophages after a round of confluent growth and sporulation on nonselective medium (Chater *et al.*, 1982b).

c. ϕC31 c$^+$*::pBR322ΔW17::*vph *(KC400).* This vector (Figs. 5 and 6) is a c^+ *vph*-containing derivative of ϕC31 KC117 produced by natural recombination, with the *Bam*HI fragment containing the *vph* gene inserted into the unique *Bam*HI site of the pBR322 portion of the vector (Chater, 1983; Chater *et al.*, 1982b). It has most of the properties of its parental vector but in addition allows for the selection of lysogens by expression of Vio resistance (this resistance can be used in *E. coli*). A *Pst*I fragment of 3.9 kb containing most of pBR322 and sequences from the *S. vinaceus* DNA fragment inessential for expression of Vio resistance can be replaced, permitting the insertion of fragments of up to at least 6.2 kb in length. Because the vector also lacks the *attP* site of ϕC31, stable lysogens expressing Vio resistance can normally only result from homologous recombination of the vector into the host genome (Fig. 6). With a lysogenic indicator strain this may occur through a resident prophage, as described previously; in the case of a nonlysogenic indicator strain, stable lysogens can be produced only by recombination between the cloned fragment in the vector and homologous sequences in the chromosome (or, conceivably, present on an extrachromosomal element; Chater, 1983). Used in the latter mode this vector allows for the possibility of gene disruption and mutational cloning (Fig. 7). If the cloned fragment is internal to a transcription unit and lacks both the promoter region and the translation-stop codon (or, more precisely, any sequences coding for the C-end of the protein that are essential for its function), then homologous recombination into genomic sequences at any point within the cloned fragment will prevent the synthesis of a molecule of mRNA that can code for a functional protein, and a corresponding mutant phenotype will result. Alternatively, if the cloned fragment contains either, then homologous recombination will permit

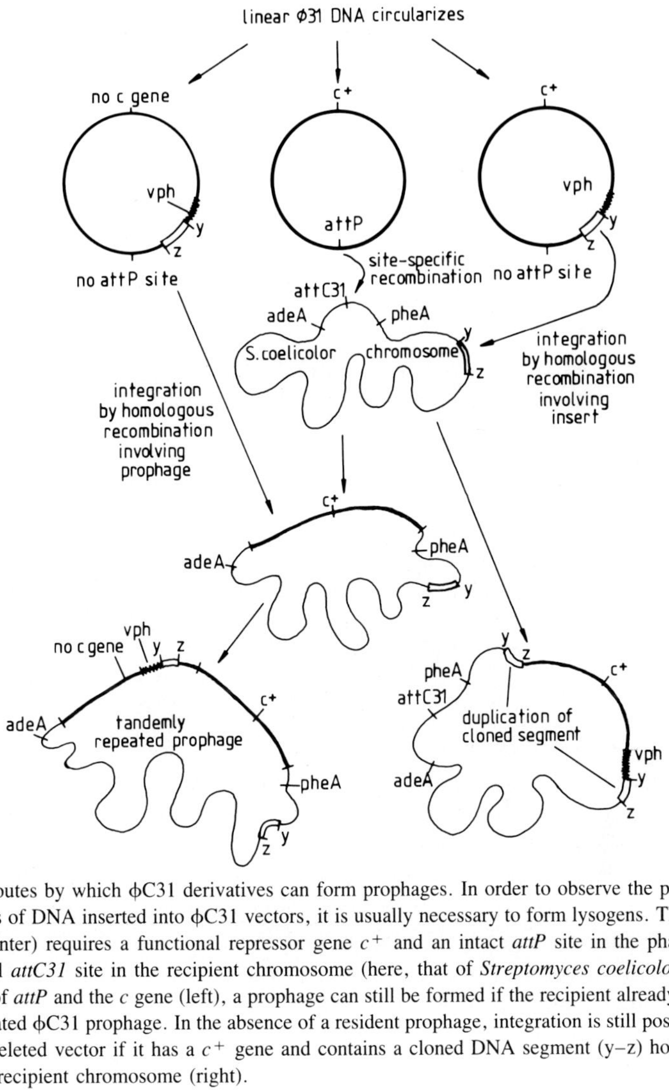

Fig. 6 Routes by which ϕC31 derivatives can form prophages. In order to observe the phenotypic properties of DNA inserted into ϕC31 vectors, it is usually necessary to form lysogens. The natural mode (center) requires a functional repressor gene c^+ and an intact *attP* site in the phage and a functional *attC31* site in the recipient chromosome (here, that of *Streptomyces coelicolor*). In the absence of *attP* and the *c* gene (left), a prophage can still be formed if the recipient already contains an integrated ϕC31 prophage. In the absence of a resident prophage, integration is still possible with an *attP*-deleted vector if it has a c^+ gene and contains a cloned DNA segment (y–z) homologous with the recipient chromosome (right).

synthesis of mRNA capable of producing a functional protein and a mutant phenotype will not arise. Mutational cloning should allow for the isolation of any gene, including its transcription-initiation and -termination signals, for which there is a detectable phenotype, because DNA fragments obtained by mutational cloning can be used as probes for screening gene libraries or Southern blots. In principle, the technique can be applied to genetically uncharacterized strains for

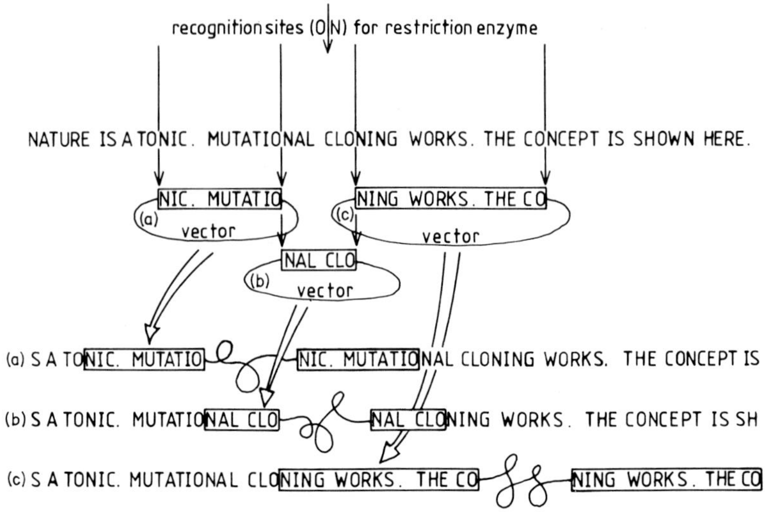

Fig. 7 How mutational cloning works. When, as in (b) the cloned information is entirely within a sentence (in molecular biology, the sentence is equivalent to a promoter and those of its subordinate coding sequences that are essential for expression of a wild phenotype), integration of the vector—through the homologous sequences shared by the cloned segment and the recipient—disrupts the message. When, as in (a) or (c) the cloned segment includes the beginning or end of a sentence (i.e., the promoter or the last translation-stop codon), an intact version of the message remains after integration. From Chater and Bruton (1983).

which transformation and transfection systems are not available, provided, of course, that they are susceptible to ϕC31 infection and lysogenization. The initial cloning could be carried out in a suitable host (e.g., *S. lividans*), and then the cloned fragments could be introduced into the strain of interest by the normal mechanism of phage infection. Viomycin-resistant clones of the uncharacterized strain could then be screened for the desired phenotype.

Clearly, it would be useful to be able to use enzymes other than *Pst*I and resistance genes other than *vph* (Vio is not widely available) in vectors for mutational cloning. To meet these needs, vectors containing *tsr* and suitable for use with *Bam*HI and *Sst*I have been constructed, and ϕC31 KC400 derivatives with sites for *Xho*I and *Bgl*II have been fortuitously obtained during shotgun-cloning experiments (M. R. Rodicio, K. F. Chater, and C. J. Bruton, unpublished).

*d. ϕC31ΔC3::pBR322ΔW17::*vph *(KC401).* This derivative was obtained after a 1.8-kb deletion of ϕC31 KC400 that included the repressor gene *c* (Harris *et al.*, 1983). Consequently, 1.8–8 kb of DNA can be inserted into the *Pst*I-cleaved vector without affecting expression of Vio resistance. Alternatively, up to 4.8 or up to 6.3 kb of DNA can be inserted into *Sst*I- or *Bam*HI-cleaved ϕC31

KC401, respectively, but in both cases the Vio-resistance marker is lost. Use of the *E. coli* plasmid form also permits insertion of fragments of up to 4.3 kb into the unmethylated *Cla*I site of the vector (Section II,B,1,*b*). Lack of the *c* gene requires use of a lysogenized recipient and prevents the easy use of ϕC31 KC401 for mutational cloning. (Recombination of the vector into the chromosome of a lysogenized host is presumably most likely to occur between the longer homologous phage sequences than between the cloned insert and its corresponding chromosomal sequence, and the use of ϕC31 KC400 is to be preferred for mutational cloning.)

2. OTHER PHAGES

At least in order to extend phage cloning to the 30% or so of streptomycetes excluded from the host range of ϕC31 and doubtless with other benefits, vectors based on other phages will be needed. Foor and Morin (1982) have developed phage C20 (TG1), a temperate phage with many similarities to ϕC31, but with an as yet undisclosed host range, along lines parallel to those used in developing ϕC31. Another vector system is provided by phage R4, a wide-host-range temperate phage isolated from soil by Chater and Carter (1979) and possibly identical to the independently isolated phage Bα (Nakano *et al.*, 1981). The R4 genome (53 kb) has been reduced in size by selecting chelating-agent-resistant mutants (Takahashi *et al.*, 1983; S. G. Foster and K. F. Chater, unpublished) and the unique *Pvu*II site used to introduce either *Sma*I-cut *Streptomyces* DNA or a synthetic linker containing a *Bam*HI site (Takahashi *et al.*, 1983). A third phage (SH10) has also been extensively studied as a potential vector, but apparently, it has not yet been used as a receptacle for foreign DNA (Walter *et al.*, 1981).

C. Shuttle Vectors

Several shuttle replicons capable of stable maintenance in both *E. coli* and *Streptomyces* have been derived from various streptomycete plasmids and from ϕC31.

SLP1.2-derived shuttle vectors are generally easily transformed between *E. coli* and *S. lividans*. Of the earlier recombinant plasmids, pSLP120 consists of the *E. coli* replicon pACYC 184 joined at its unique *Bam*HI site to pSLP111 [a derivative of SLP1.2 containing *mmr*, a gene(s) conferring resistance to methylenomycin (Mm), allowing selection for Cm resistance in *E. coli* and Cm and Mm resistance in appropriate derivatives of *S. lividans* (Schottel *et al.*, 1981)]. pSLP125 contains the *E. coli* replicon pACYC177 joined at its unique *Pst*I site to pSLP111, allowing selection for kanamycin (Km) resistance in *E. coli* and Km and Mm resistance in *S. lividans* (Schottel *et al.*, 1981). pIJ28 comprises an *aph-*

containing derivative (pIJ2) of SLP1.2 inserted by its *Bcl*I site into the *Bam*HI site of pBR322, allowing selection for ampicillin (Ap) resistance in *E. coli* and neomycin (Nm) resistance in *S. lividans* (Thompson *et al.*, 1982c).

Derivatives of SCP2* capable of replication in both *E. coli* and *Streptomyces* include pIJ903 (D. J. Lydiate, personal communication). This vector contains most of the sequences of pBR327 (making possible replication and expression of Ap resistance in *E. coli*) fused to a deleted form of pSCP103 into which had been inserted the *tsr* gene, providing selection for thiostrepton resistance in *Streptomyces*. Sites available for cloning in both potential hosts include those for *Eco*RI, *Hin*dIII, *Bam*HI, and, probably, *Xba*I.

Shuttle vectors derived from the pIJ101 series of plasmids include pIJ361. This plasmid contains the entire sequence of pBR322 (permitting replication and expression of Ap resistance in *E. coli*) joined to a derivative of pIJ102 that includes the *tsr* and *vph* genes and that provides for replication and selection in *Streptomyces*. The *vph* gene of pIJ361 is also expressed in *E. coli* [presumably from the tetracycline- (Tc)-resistance gene promoter]. Single sites available for cloning in both hosts include those for *Eco*RI and *Hin*dIII.

Shuttle vectors derived from φC31 include φC31 KC100, φC31 KC117, φC31 KC400, and φC31 KC401 (Section II,B,1). Each can replicate in *E. coli* (as a multicopy plasmid) and in *Streptomyces* (as a phage). Uptake of the covalently closed circular form by *Streptomyces* protoplasts after isolation from *E. coli* is only approximately 10 times less efficient than transfection with linear phage DNA, and no rearrangements have been detected after transfer between the different hosts.

Apart from studies of heterospecific gene expression and use in the construction of modified *Streptomyces* replicons, these early examples of shuttle vectors have been put to only limited use; it is anticipated that such vectors will be of value in several further areas, for example, in cloning the DNA sequences lethal to one or another of the alternative hosts, in the application of well-developed *E. coli* technology (such as transposon and directed site-specific mutagenesis, or suppressor systems) to *Streptomyces* DNA, and in the use of host-determined modification systems for protection of DNA from restriction-enzyme cleavage.

III. Use of Tn5 in Relation to Streptomyces DNA

In the absence of well-defined endogenous transposon systems in *Streptomyces*, the possibility of using the *E. coli* transposon Tn5 (Kleckner, 1981) in *Streptomyces* has been investigated. In a first stage, the ability of Tn5 to integrate into *Streptomyces* DNA present in shuttle vectors in *E. coli* was tested; potential transpositions from a chromosomal location into the high-copy-number replicons were identified by selection for increased Km resistance, and plasmid-linked resistance was shown by cotransformation for Km resistance and a vector-associated resistance. In the case of pIJ502 (φC31::pBR322ΔW17), Tn5 insertions were often into the φC31-specific portion of the molecule, at various locations (S. G. Foster, personal communication). However, because φC31 has a G + C content of 63% (Chater, 1980), it is not necessarily typical of *Streptomyces* DNA. Two other experiments with *Streptomyces* DNA have given conflicting results. When pIJ361 (containing sequences from pBR322, pIJ101, and the *S. vinaceus vph* region) was used, Tn5 insertions into the *vph* gene were detected initially by finding Km-resistant, Vio-sensitive colonies and then were confirmed by physical analysis (S. G. Foster, personal communication). In similar experiments with pIJ818 and pIJ819 (containing pIJ41, pBR322, and *S. griseus pab* sequences that were expressed in *E. coli;* see Section IV,G), however, the Km-resistant Pab^- colonies obtained proved to have Tn5 insertions exclusively in the region of pBR322 from which transcription of the *pab* gene was initiated (J. A. Gil, personal communication). Thus it seems that Tn5 can insert at reasonable frequencies into certain *Streptomyces* sequences of high G + C content (*vph* contains 73% G + C; see Section IV,C) but not others.

The second stage of these investigations was to reintroduce the vector::Tn5 molecules into *S. lividans* and examine Tn5 expression in *Streptomyces*. This was possible only with the pIJ361, pIJ818, and pIJ819 derivatives; the pIJ502 derivatives contained too much DNA to allow packaging into virions. With each of the former three derivatives, expression of Km resistance was detectable in *S. lividans* and seemed to be related to gene dosage, the pIJ361 derivative conferring much higher levels of resistance (S. G. Foster and J. A. Gil, personal communication). However, limited searches for Tn5-directed cointegrate formation or transposition within *S. lividans* have failed to reveal any evidence of transposon-like properties being expressed (S. G. Foster, personal communication).

IV. Applications of DNA Cloning in Streptomyces

A. Heterospecific Gene Expression

Heterospecific gene expression in *Streptomyces* was first demonstrated after construction of the shuttle vectors pSLP120 and pSLP125 (Schottel *et al.*, 1981). The *E. coli* antibiotic-resistance genes of pACYC177 and pACYC184 coding for Km phosphotransferase and CAT, respectively, were shown to be expressed in

S. lividans. These and later reports of the expression of the Tc-resistance gene of pBR322 in *S. albus* (Chater *et al.*, 1982b) and the Km phosphotranferase gene of Tn5 in *S. lividans* (Section III) indicate that streptomycetes possess the necessary charged tRNA molecules to translate probably all *E. coli* mRNA species, although in no case has the level of translation of the heterologous DNA been studied. The abundance of different tRNA species responsible for the recognition of different codons in *Streptomyces* has not been established. It is therefore difficult to predict any possible constraints on the efficiency of translation of foreign mRNA species in these organisms that may result from the limited availability of particular tRNA species.

In these early studies, the ability of *Streptomyces* RNA polymerase and ribosomes to recognize bona fide transcriptional and translational control signals of the gram-negative enterobacteria could not be accurately assessed. Transcriptional read-through and the production of fused proteins from *Streptomyces* control sequences present on the recombinant vectors remained a formal possibility. However, application of the promoter-probe plasmid vectors pSLP114 and pSLP124 has confirmed that the mycelium of vegetatively growing *S. lividans* contains at least one form of RNA polymerase that will recognize the transcription-initiation signals of the *E. coli lacUV5* and *recA* genes, of the *trp* operon of *Serratia marcescens,* and of the penicillinase gene *penP* of the gram-positive nonmycelial bacterium *Bacillus licheniformis* (Bibb and Cohen, 1982). Small DNA fragments containing these completely sequenced and characterized promoter regions were inserted in the unique *Bam*HI sites of both pSLP114 and pSLP124 without selecting for Cm resistance. In each case, many of the plasmids conferred Cm resistance on their *Streptomyces lividans* host. Restriction-enzyme analysis of the resulting plasmids indicated that each of these promoter sequences permits transcription of the CAT genes in *S. lividans* only when inserted in the correct orientation; expression of the CAT genes could not be detected when the promoter fragments were inserted in the opposite orientation. This demonstrated that the transcription initiated by a form of *Streptomyces* RNA polymerase at each of several heterologous promoters occurred only in the correct direction.

The predicted mRNA sequences immediately preceding the AUG translation-start codons of the CAT genes in each of the pSLP114 and pSLP124 derivatives that contained the inserted promoter fragments have been analyzed. The analysis indicates that the ribosomes of *Streptomyces lividans* recognize sequences for the initiation of translation (ribosome-binding sites) that show little complementarity to the 3′ end of their 16-S rRNA (5′ GAUCACCUCCUUUCU 3′; Bibb and Cohen, 1982). This differs from the observations made in *Bacillus* and *Staphylococcus* (other gram-positive bacteria) that led to the hypothesis that gram-positive organisms in general require an extensive degree of such complementarity (compared to *E. coli* and other gram-negatives) for translation to occur (McLaughlin *et al.*, 1981). These conclusions were proffered as the reason for the observed lack of expression of gram-negative genes in a gram-positive host, whereas the converse frequently occurred (Chang and Cohen, 1974; Ehrlich, 1978; Courvalin and Fiandt, 1980). Although the level of translation of the CAT mRNA required to give phenotypically detectable levels of Cm resistance in *Streptomyces lividans* is not known (and may be low), an extensive region of complementarity between the ribosome-binding site of mRNA and the 3′ end of 16-S rRNA is clearly not a prerequisite for translation to occur in *Streptomyces*.

Attempts to isolate sequences from *Streptomyces* DNA that would allow transcription and translation of the β-galactosidase gene of the promoter probe pMC874 (Casadaban *et al.*, 1980) in *E. coli* produced relatively few fragments compared to similarly treated *E. coli* and *B. subtilis* DNA preparations (Bibb and Cohen, 1982). Furthermore, the levels of expression of the β-galactosidase gene from these *Streptomyces* sequences were much lower than from the *Escherichia*- and *Bacillus*-derived fragments. A minority of these same *Streptomyces* fragments permit low levels of expression of the CAT gene in *S. lividans* when inserted into pSLP114 and pSLP124. More recent experiments indicate that fragments containing the promoter regions of the *aph* gene of *S. fradiae* and the *vph* gene of *S. vinaceus* do not promote transcription in *E. coli;* although these fragments permit expression of the CAT genes of pSLP114 and pSLP124 in *S. lividans,* they do not produce any detectable level of Cm resistance in *E. coli* after introduction of the pSLP114 and pSLP124 derivatives into *E. coli* as part of pBR322-derived recombinant plasmids (M. J. Bibb and S. N. Cohen, unpublished). Similar negative results have been obtained using *Streptomyces* promoter fragments of less well defined origin that were isolated from the plasmids SCP2* and pIJ2 and the phage φC31.

There are several examples of phenotypic expression of *Streptomyces* genes in *E. coli.* These include the *vph* gene of *S. vinaceus* (Kieser *et al.*, 1982), the endoglycosidase H gene of *S. plicatus* (Robbins *et al.*, 1981), the *aph* gene of *S. fradiae* (Rodgers *et al.*, 1982), and the *para*-aminobenzoic acid (PABA) synthetase gene of *S. griseus* (Section IV,G). Cloning of intact fragments that contain the *aph* and PABA-synthetase genes in *E. coli* does not lead to detectable levels of expression: phenotypic expression only results after deletion of *Streptomyces* sequences (for the *aph* gene, of sequences known to be adjacent to the 5′ end of the gene), presumably placing the coding sequences of the *Streptomyces* genes under the control of an *E. coli* promoter sequence. In the reported case of expression of Vio resistance in *E. coli,* transcription of the coding sequences almost certainly occurred from the promoter of the Tc-resistance gene of the *E. coli* vector pBR322; insertion of the same DNA fragment containing the *vph* gene at the same site in pBR322 but in the opposite orientation failed to produce any detectable increase in the levels of Vio resistance (J. M. Ward and M. J. Bibb, unpublished). The cloning and expression of the endoglycosidase H gene from *S. plicatus* in *E. coli* represents the first published example of the cloning of a gene for an enzyme that is secreted by *Streptomyces*. Although in most of the recombinant plasmids that have been constructed efficient transcription of the endoglycosidase H gene was obtained from *E. coli* promoters, there was some evidence from Charon 4 phage derivatives that transcription could have initiated at *Streptomyces* promoter sequences. The promoter region of the endoglycosidase H gene may therefore represent one of the apparently few *Streptomyces* promoter sequences that can be recognized by the RNA polymerase of *E. coli*, albeit at a low efficiency.

In summary, the vegetatively growing mycelium of *Streptomyces* appears to contain at least one form of RNA polymerase that will recognize promoter sequences similar to those of other prokaryotes (including *E. coli*, *Serratia marcescens*, and *Bacillus licheniformis*); hence, one would predict that at least some *Streptomyces* promoters would contain similar sequences. The relatively infrequent ability of randomly cloned *Streptomyces* fragments to function as promoters in *E. coli* tends to suggest that such promoter sequences are rare in *Streptomyces*. This is consistent with the observation that the few promoter-containing fragments isolated within *Streptomyces* all fail to function effectively in *E. coli*. Not surprisingly, determination of the sequence of these *Streptomyces* DNA fragments fails to reveal the presence of typical *E. coli*–like promoter sequences. These observations suggest that the majority of *Streptomyces* promoter sequences do not resemble those of *E. coli*, and they imply the existence of different classes of promoter sequences within *Streptomyces* recognized perhaps by either different or modified forms of RNA polymerase. The existence of multiple forms of *S. coelicolor* RNA polymerase, with different promoter specificity, has been demonstrated directly (J. Westpheling and R. Losick, personal communication).

B. Cloning and Sequence Analysis of an Aminoglycoside Phosphotransferase Gene from *Streptomyces fradiae*

Using SLP1.2 as a vector, Thompson *et al.* (1980) isolated DNA fragments from a Nm-producing *S. fradiae* strain that conferred resistance to Nm on *S. lividans*. Restriction-enzyme analysis of the cloned DNA fragments and biochemical characterization of cell-free extracts indicate the existence of two different Nm-resistance genes, coding for an aminoglycoside phosphotransferase (*aph*) and an aminoglycoside acetyltransferase (*aac*). The *aph* gene has been subjected to intense study. The position of the gene within the cloned fragment

Table I Distribution of Nucleotides in the *aph* and *vph* DNA Sequences

Gene	Sequences (length)	G + C (mol %) in codon position[a]			
		1	2	3	Mean
aph	Precoding (306 bp)	85	67	84	79
	Coding (804 bp)	78	43	97	73
	Postcoding (170 bp)	80	82	84	82
vph	Precoding (96 bp)	53	66	66	62
	Coding (861 bp)	79	51	91	74
	Postcoding (162 bp)	78	78	72	76

[a]The codon register is in frame with the presumptive ATG translation-start codons of the *aph* and *vph* genes.

was defined by subcloning and insertional inactivation (Kieser *et al.*, 1982; Thompson *et al.*, 1982c), and the corresponding region was sequenced (Thompson and Gray, 1983). SDS-Polyacrylamide gel electrophoresis of cell-free extracts had indicated that the phosphotransferase had a molecular weight of 32,000. An open translational reading frame was identified in the DNA sequence that could code for a protein of this size, which was found to have a predicted amino acid composition very similar to that determined experimentally from the purified enzyme. The N-terminal sequence of the protein was determined, and it confirmed the position of the *aph* gene within the cloned fragment, although it did not unambiguously reveal the position of the first translated codon of the *aph* mRNA.

Analysis of the coding sequence illustrates some interesting features. The high G + C composition (73%) of the genome of the organism is reflected mostly in the third codon position, which is to be expected from a consideration of the degeneracy of the genetic code (Table I). (Assuming a full complement of tRNA species for the recognition of all possible amino acid codons, even the complete absence of codons ending in A or T would pose no limitations on the amino acid composition of a protein.) The observed preference for G or C at the first codon position presumably partially reflects the amino acid content of the protein, influenced perhaps by a degree of degeneracy at this position for the arginine and leucine codons. Of 61 potential sense codons, 23 are absent from the *aph* structural gene, although tRNA molecules capable of translating all of these must be present in the cell (Section IV,A). However, this nonrandom choice of codon usage may reflect the occurrence of major tRNA populations. These may be essential for efficient translation of mRNA of highly expressed proteins such as the aminoglycoside phosphotransferase, which constitutes up to 10% of soluble cell protein. If this is so, then the large differences in codon usage between *Streptomyces* genes and those of other genera may pose limits on the efficiency of translation of foreign genes in these organisms.

Comparison of the amino acid sequence of the *aph* gene of *S. fradiae* with those of the *aph* genes of the *E. coli* transposons Tn5 and Tn903 indicates an overall 36–40% homology, with very similar regions at the C-terminus of each of the genes, presumably reflecting functionally important domains involving binding of enzyme substrates (Thompson and Gray, 1983).

Using the pSLP114 and pSLP124 vectors, a *Sau*3A-derived segment of 527 bp was isolated as a promoter-containing fragment from DNA segments containing the whole *aph* gene (M. J. Bibb and S. N. Cohen, unpublished). This fragment was purified and reinserted into each of the vectors in the absence of selection for Cm resistance and was shown to promote transcription of the CAT gene only in one orientation. Southern hybridization analysis, followed by determination of the entire sequence of this fragment, confirmed that it originates from the 5′ end of the *aph* coding sequence identified by Thompson and Gray (1983) and that the orientation in which it promotes transcription is the expected orientation. This functionally defined promoter region lacks sequences obviously similar to those responsible for initiation of transcription in other prokaryotes. However, the 75-bp region immediately preceding the likely ATG translation-start codon is considerably lower in G + C content (64 mol % G + C) than the coding sequences themselves (73 mol % G + C); this is a characteristic of several randomly cloned and sequenced *Streptomyces* fragments that contain promoter activity. Examination of the sequence also fails to reveal a potential ribosome-binding site that shows extensive complementarity to the 3′ end of the 16-S rRNA of *S. lividans*.

An inverted repeat of 18 bp (with a 1-base loop out; ΔG = 54.6 kcal/mol) positioned 19 nucleotides after the TAG translation-stop codon has been identified; in analogy with similar regions of dyad symmetry in other prokaryotes, it may represent a transcription-stop signal (Rosenberg and Court, 1979).

C. Cloning and Sequence Analysis of a Viomycin Phosphotransferase Gene from *Streptomyces vinaceus*

In cloning experiments similar to those just reported, DNA fragments were obtained from total DNA of *S. vinaceus* that, when inserted into SLP1.2, confered resistance to Vio in *S. lividans*. At least two and possibly three different Vio-resistance determinants have been identified by restriction-enzyme analysis of the cloned DNA fragments and biochemical characterisation of cell-free extracts obtained from representative Vio^R clones (Thompson *et al.*, 1982a,b). A *Bam*HI fragment of 1.9 kb coding for the production of Vio phosphotransferase was chosen for further study. This fragment has been used extensively in the construction of cloning vectors for *Streptomyces* (Thompson *et al.*, 1982c; Chater, 1982) and has been useful (together with fragments carrying the *aph* and *tsr* genes) in the analysis of structure–function relationships of the pIJ101 series of plasmids (Kieser *et al.*, 1982).

A promoter-containing *Sau*3A fragment of 121 bp was isolated from the *Bam*HI fragment by use of the pSLP114 and pSLP124 promoter-probe vectors (M. J. Bibb and S. N. Cohen, unpublished). Southern hybridization experiments identified this functionally defined promoter fragment as originating close to one end of the *Bam*HI fragment. The fragment was purified and reinserted into the promoter-probe plasmids in a nonselective manner and was found to promote transcription of the CAT gene only when inserted in one orientation. The fragment has been completely sequenced; it contains an ATG translation-start codon preceded by a potential ribosome-binding site that shows a reasonable degree of complementarity to the 3′ end of 16-S rRNA of *S. lividans*. Like all other presumptive *Streptomyces* promoter sequences, it has a lower G + C base composition (61 mol %) than one would expect of randomly cloned *Streptomyces* fragments (73 mol %) but sequences showing obvious similarity to promoter regions of other prokaryotes cannot be recognized.

This functionally defined promoter sequence would lead to transcription of the 1.9-kb *Bam*HI *vph* fragment in a left-to-right direction (as shown in Fig. 5). This is consistent with the direction of transcription of the *vph* gene predicted from the expression of Vio resistance in *E. coli* from the Tc promoter of pBR322 (Section IV,A). (The occurrence of three translation-stop codons in front of and in phase with the ATG start codon of the *vph* coding sequence suggests that the ribosome-binding site of the *vph* gene is used to initiate translation of the coding sequences in *E. coli* and, therefore, that the ribosomes of *E. coli* can recognize at least some *Streptomyces* ribosome-binding sites.) Further evidence that this region contains the *vph* promoter and not some other transcription-initiation signal also present in the 1.9-kb *Bam*HI fragment stems from DNA sequence analysis. This promoter region is followed by an ATG translation-start codon and an open reading frame that extends for 861 bp and that would code for a protein of molecular weight ~31,500 (M. J. Bibb and S. N. Cohen, unpublished). This presumed coding sequence of the *vph* gene falls entirely within the 1.1-kb *Sph*I fragment recognized by Kieser *et al.* (1982) as containing all of the nucleotide sequences necessary to code for a functional protein. Although there are no other long open reading frames extending in the same direction, one does occur on the opposite strand, but it lacks any recognizable ribosome-binding site. Moreover, both the potential N- and C-terminal coding sequences fall outside the *Sph*I fragment. Thus the data obtained from *E. coli* and the following consideration make it very unlikely that this potential translational unit represents the *vph* gene.

Analysis of the coding sequences of the presumptive *vph* gene shows, as for the *aph* gene, a remarkably nonrandom distribution of nucleotides at the third codon position; 17 of the possible 61 codons are not represented. The G + C compositions at each of the codon positions for precoding, coding, and postcoding sequences are shown in Table I. These observations and their potential consequences can be interpreted as was done for the *aph* gene. The distribution of G + C base pairs at each of the three positions within the codons of the *Streptomyces aph* and *vph* genes can be represented relative to each other as medium, low, and high G + C for codon positions 1, 2, and 3, respectively. This distribution is confined to coding sequences and is not found in the nucleotide sequences even immediately preceding and immediately following translation-start and -stop codons, respectively. It has also been observed, in a less dramatic form, in the potential coding sequence for the *tsr* gene of *S. azureus,* which has an overall average G + C content of 62 mol %, considerably lower than that of the *aph* and *vph* genes (M. J. Bibb and S. N. Cohen, unpublished). We may predict that this nonrandom distribution of nucleotides between the three codon positions will be a feature of most, if not all, *Streptomyces* genes and that it can be used in a diagnostic fashion to identify true protein-coding sequences from *Streptomyces.* Application of these criteria to the open reading frame running in the direction opposite to that of the proposed *vph* gene indicates that this sequence is unlikely to code for a protein.

The cloned *vph* promoter fragment occurs at the very end of the 1.9-kb *Bam*HI fragment. When inserted in front of the CAT gene of pSLP114 and pSLP124 in *S. lividans,* it confers relatively low levels of Cm resistance to the host. It is possible that sequences that normally participate in RNA polymerase recognition were lost during the initial cloning of the *vph* gene; attempts to study the sequences normally adjacent to the 1.9-kb *Bam*HI fragment at the 5′ end of the gene are under way.

An inverted repeat sequence of 18 bp ($\Delta G = -26.4$ kcal/mol) has been identified 28 bp past the translation-stop codon TAG of the *vph* gene; this may be involved in the termination of transcription.

D. DNA Involved in Glycerol Catabolism in *Streptomyces coelicolor*

Glycerol is catabolized in *S. coelicolor* by the action of glycerol kinase and glycerol-3-phosphate (G3P) dehydrogenase (Seno and Chater, 1983). The enzymes are coordinately induced at the transcription level by glycerol and are subject to coordinate carbon-catabolite repression. Mutants lacking only G3P dehydrogenase are sensitive to glycerol ($\mathrm{Gyl^S}$) when growing on certain other carbon sources (e.g., arabinose), probably because of the accumulation of toxic G3P. Other *gyl* mutants ($\mathrm{Gyl^-}$) are unable to utilize glycerol as sole carbon source, but they can grow on a mixture of arabinose and glycerol. These are nearly all defective in both glycerol kinase and G3P dehydrogenase. All mutations studied map close together, near *argA* on the *S. coelicolor* linkage map, with the exception of two giving an unusual phenotype Gyl^X; they have not been studied further. By shotgun cloning *S. coelicolor* DNA into the *Bgl*II and *Sst*I sites of pIJ702, plasmids able to restore the Gyl^+ phenotype to both a Gyl^- *and* a Gyl^S mutant were detected. All of the clones proved to have DNA in common. Surprisingly, there was no evidence of effects of high gene dosage on glycerol enzyme levels in cells carrying the hybrid plasmids, although the two enzymes were presumably being synthesized under the control of the *gyl* promoter because they were subject to normal induction by glycerol. This apparent paradox was resolved by assuming that the *gyl* DNA in the plasmids lacked the *gyl* promoter and was expressed only when recombined into the host chromosome. Moreover, the cloned *Bgl*II and *Sst*I fragments differed in the efficiency with which the Gyl^+ phenotype was restored; a much higher proportion of spores from cultures of Gyl^- or Gyl^S strains carrying the *Sst*I-derived cloned DNA were Gyl^+. This may mean that the *Sst*I fragment contained all the essential coding DNA for the G3P dehydrogenase gene, whereas the *Bgl*II fragment did not contain the DNA for the C-terminal sequences of the enzyme. As a result, the Gyl^+ phenotype could be generated by a single crossover between the *Sst*I-generated plasmid and the chromosome, but double crossovers would be needed between the *Bgl*II-generated plasmid and the chromosome. If this hypothesis is correct (and we show later that it is), an interesting consequence is that self-cloning into pIJ702 (and perhaps lower-copy-number plasmids) to allow the restoration of wild-type phenotype to mutant indicator strains does not require that the cloned sequence be a whole gene or that it be transcribed in the vector. This may be additionally advantageous in the case of gene products that would be deleterious to a host cell if produced in large amounts.

Verification of the hypothesis was obtained by use of the *vph*-containing ϕC31 vector KC400 (Section II,B,1,*c*). The subcloning of *gyl* DNA into this *att*-defective phage provides it with a region of homology through which it can recombine into the host chromosome, thereby stably transducing the host to Vio resistance. Consideration of the structure of such phage::chromosome cointegrates shows, as described earlier, that when the cloned DNA contains neither its promoter nor the coding sequences for the C-terminus of a product needed for a wild phenotype, a mutant phenotype will be generated. The Gyl phenotypes of transductants obtained with ϕC31 KC400 containing various segments of the *gyl* DNA were determined; Gyl^-, Gyl^S, and Gyl^+ were all obtained (Fig. 8). Careful interpretation of these results allowed the conclusions to be drawn that the essential coding sequence (*gylB*) for G3P dehydrogenase extends from somewhere right of the *Pst*I site 6 (Fig. 8) to a point between *Bgl*II site 9 and *Sst*I site 10 (E. T. Seno, K. F. Chater, and C. J. Bruton, unpublished). At least one other gene (probably the gene for glycerol kinase) must be read upstream of *gyl*B as part of the same polycistronic mRNA, from a promoter absent from the cloned DNA but situated (in the chromosomal *gyl* region) to the left of *Bgl*II site 1.

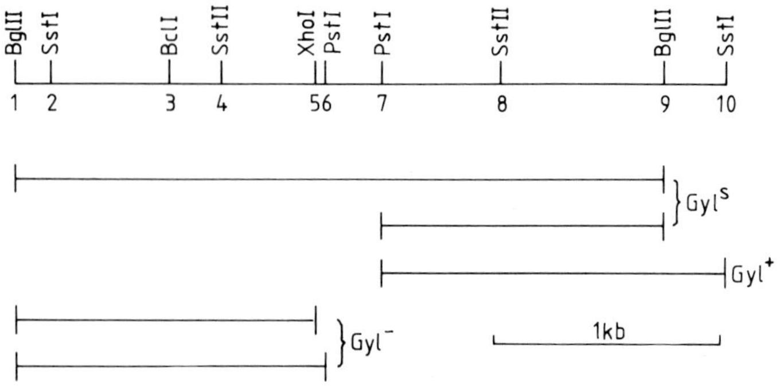

Fig. 8 Analysis of *gyl* DNA of *S. coelicolor* by gene disruption. The upper part of the diagram is a restriction map of a cloned part of the *gyl* operon. The segments indicated in the lower part were subcloned into φC31 KC400 (or a derivative with *Bgl*II and *Xho*I sites) and the resulting phages used to transduce a Gyl$^+$ (glycerol-utilizing) strain of *Streptomyces coelicolor* to viomycin resistance (see Fig. 6). Transduction required integration of the phage DNA into the region of the chromosome by homologous recombination. In all cases except one the transductant colonies were unable to grow on glycerol as sole carbon source, because of gene disruption (see Fig. 7). The nonutilizing colonies were of two phenotypes: Gyl^S, glycerol sensitive (unable to utilize glycerol or to grow on arabinose in the presence of glycerol); Gyl$^-$, unable to utilize glycerol (but not glycerol sensitive). From E. T. Seno, K. F. Chater, and C. J. Bruton (unpublished).

The cloned *gyl* DNA has been nick-translated and used to probe DNA of various streptomycetes for homologous sequences. Remarkably, there is extensive homology with *S. lividans* and *S. parvulus* DNA, the former DNA sharing all restriction sites tested with the *S. coelicolor gyl* region. This has allowed the use of the *S. coelicolor* φC31 KC400::!*gyl*! clones to transduce *S. lividans,* giving !*gyl*! insertional mutants. [The convention !*gyl*! has been proposed to denote mutants produced by the insertion of internal cloned fragments (Chater and Bruton, 1983).] An immediate benefit of this was that it allowed the use of pIJ41 (Thompson *et al.*, 1982c) as a vector for further *gyl* cloning (pIJ41 cannot be used as a vector in *S. coelicolor*). An *S. lividans* !*gyl*! mutant has been exploited in the isolation of a cloned *gyl* fragment of *S. griseus* DNA that has very low, but detectable, homology with the *S. coelicolor* clones.

E. Cloning of DNA Involved in Methylenomycin Production

Methylenomycin A represents the only well-established case of plasmid-specified antibiotic biosynthesis (Hopwood, 1979). However, exhaustive attempts to isolate the plasmid concerned (SCP1) from its host *S. coelicolor* A3(2) have met with little success (Westpheling, 1980; Aguilar and Hopwood, 1982). The availability of φC31 KC400 (Section II,B,1,*c*) provided a route to the identification of SCP1 fragments among all fragments cloned from the DNA of an SCP1$^+$ strain of *S. parvulus;* SCP1-fragment-containing phages could lysogenize *S. lividans* or *S. coelicolor* strains carrying SCP1, giving Vio-resistant growth by homologous recombination into the resident SCP1 DNA; on the other hand, phages carrying *S. parvulus* DNA do not have homology with *S. lividans* or *S. coelicolor* DNA sufficient to permit their efficient integration (but they could lysogenize an SCP1$^-$ strain of *S. parvulus*). By this means, 278 KC400 derivatives with inserted *Pst*I fragments of SCP1 DNA in the size range 2–6 kb were isolated. Transductants of an *S. coelicolor* NF (integrated SCP1) strain carrying these phages were tested for methylenomycin production, and 10 nonproducers were initially isolated. In 9 of these, nonproduction was shown to be owing to insertion of the phage. The simplest interpretation of these results was the presence in the phages of internal segments of DNA coding for enzymes of methylenomycin biosynthesis (as discussed in Section IV,D). Transductants carrying either of two such !*mmy*! fragments were able to cosynthesise methylenomycin when grown close to one of the *mmy* mutants isolated by Kirby and Hopwood (1977). A third fragment did not elicit this response in transductants. In all, more than 8 kb of *mmy* DNA has been isolated so far by this mutational cloning procedure.

F. A Tyrosinase Gene from *Streptomyces antibioticus*

Many streptomycetes produce melanin pigment on tyrosine-containing media by the action of tyrosinase. *Streptomyces antibioticus* IMRU 3720 is an example of such a Mel$^+$ strain, whereas *S. lividans* 66 is Mel$^-$. By cloning *Bcl*I fragments of *S. antibioticus* DNA into the SLP1.2-based vectors pIJ41 or pIJ37 and transforming into *S. lividans* it was possible to isolate Mel$^+$ clones carrying a 1.55-kb *Bcl*I fragment of *S. antibioticus* DNA (Katz *et al.*, 1983). Physiological studies were made on the original *S. lividans* clones carrying this fragment at low copy number and also after subcloning at high copy number in a pIJ101 derivative (to yield plasmids pIJ702, pIJ703, pIJ704, and pIJ705). In all of the clones, as in *S. antibioticus* itself, tyrosinase activity was very low unless induced by the addition of tryptone. The level of tyrosinase was enhanced in clones carrying the high-copy-number plasmids (compared to those of low copy number), but by a factor considerably lower than the copy-number ratio. An unexpected finding was that tyrosinase activity remained very largely intracellular in all the *S. lividans* clones, in contrast to the situation in *S. antibioticus,* where most of the tyrosinase was found in the medium after the first few hours of culture. Thus complete information for excretion of tyrosinase is not carried on the cloned fragment of *S. antibioticus* DNA, even though induction is apparently normal in the *S. lividans* clones.

G. A PABA-Synthetase Gene from *Streptomyces griseus*

Candicidin, made by *S. griseus* IMRU 3570, is a polyene macrolide antibiotic containing an aromatic moiety, a polyene chain, and a mycosamine sugar residue. The aromatic moiety, a 4-amino-acetophenone unit on which the polyene chain is assembled, is derived from para-aminobenzoic acid (PABA). It follows that PABA-synthetase, which converts chorismic acid into PABA, is a key enzyme in candicidin synthesis. Earlier studies established that the specific activity of this enzyme is low (undetectable) in young cultures of *S. griseus* but increases dramatically in the period immediately preceding the onset of candicidin synthesis. Moreover, there is a correlation between the activity of this enzyme and candicidin synthesis over a wide range (Gil *et al.*, 1980). A *Bam*HI restriction fragment of *S. griseus* DNA coding for PABA-synthetase was cloned into *S. lividans* by insertion at the *Bam*HI site of pIJ41 in two separate experiments (J. A. Gil and D. A. Hopwood, unpublished). In one, the donor was a sulfonamide-resistant mutant of *S. griseus,* which overproduced PABA, presumably by a mutation in the PABA synthetase gene or in a regulatory element, and the recipient was the wild-type *S. lividans* 66; selection was made for sulfonamide-resistant clones, and one was found among several thousand transformants. In the second experiment the donor was the wild-type *S. griseus* and the recipient a PABA-requiring auxotroph (*pab*) of *S. lividans*; selection for Pab^+ clones yielded one among about 5000 transformants. It was found that the two clones carried an apparently identical 4.5-kb segment of *S. griseus* DNA (the first clone also contained an additional 3.9-kb fragment that was shown to be irrelevant to the phenotypes under investigation). Moreover, the recombinant plasmid in the sulfonamide-resistant clone could also confer a Pab^+ phenotype on the *S. lividans pab* mutant; therefore, only this clone was analyzed further.

Cloning of the 4.5-kb *S. griseus* fragment into the *Bam*HI site of pBR322 in *E. coli* failed to confer a Pab^+ phenotype on *pabA* or *pabB* auxotrophic recipients. However, with both recipients, when the insert was in one orientation in pBR322, but not when it was in the other, rare Pab^+ colonies could readily be selected. The plasmids in such strains had suffered a deletion of about 1 kb at one end of the *S. griseus* DNA. The conclusion is that, whereas the 4.5-kb fragment did not carry a promoter sequence capable of expression in *E. coli,* the deletion allowed transcriptional read-through from the promoter of the Tc-resistance gene of pBR322. On recloning the deleted (~3.5 kb) *S. griseus* DNA segment in *S. lividans* in three different locations (the *Bam*HI site of pIJ41, the *Bcl*I site of pIJ41, and the *Bgl*II site of pIJ702), with selection for sulfonamide resistance, only one orientation of the inserted fragment was found in each case among 50 transformants. Although not definitive, this result suggests that the deletion that put expression of the *pab* gene under control of an *E. coli* promoter in *E. coli* removed the *Streptomyces* promoter, so that expression in *Streptomyces* could only occur by read-through from a promoter on the vector.

V. Concluding Remarks

Cloning in *Streptomyces* has now passed from its first stage—the development of cloning systems—into a second stage in which particular genes of interest have been isolated and are being subjected to detailed analysis. Early questions concerning the nature of vegetatively expressed *Streptomyces* promoters and ribosome-binding sites and their activity in other organisms are beginning to be answered. Thus *Streptomyces* can use promoters and ribosome-binding sites from *Escherichia coli* and other taxonomically distant bacteria; yet very few (if any) *Streptomyces* promoters can be efficiently expressed in *E. coli,* and among sequenced *Streptomyces* promoters, none has similarities to sequences at the -10 and -35 regions of typical *E. coli* promoters. We may speculate that promoters that *do* have such similarities will prove to be present in *Streptomyces* but that their expression in *E. coli* might be limited by other features of the whole promoter region, perhaps related to a higher G + C content. It should be remembered that many *E. coli* promoters with "typical" -10 and -35 regions require additional factors for efficient transcription (such as activated cAMP-binding protein). That typical promoters of other prokaryotes are expressed in *Streptomyces,* along with *Streptomyces* promoters (*vph, aph,* and *tsr*) that do not show any detectable sequence similarity to them, suggests that there should be several different forms of RNA polymerase present in the vegetative cells, as in *Bacillus subtilis* (Losick and Pero, 1981). Whatever the molecular reasons for the apparently rather versatile activity of the *Streptomyces* transcription– translation system, it may well prove of practical use for the expression of heterologous DNA, because *E. coli* and (even more so) *B. subtilis* are relatively fastidious in this respect. It remains to be seen whether different codon usage will present an obstacle to efficient heterologous DNA expression, because codons rich in G and C are clearly very abundant in *Streptomyces* DNA, and tRNA for some codons rich in A and T, though present, may conceivably be scarce.

If the promoters of the antibiotic-resistance determinants are indeed atypical of vegetative promoters, this may be related to their connection with antibiotic biosynthesis. (We may anticipate that more "typical" promoter sequences will be seen for the *gyl* operon, characterization of which will not only be of intrinsic interest but will also be of potential value in constructing vectors allowing regulated expression of cloned DNA.) As more antibiotic production and corresponding resistance genes are cloned, it will become possible to determine whether there is generally close physical linkage between them (which would, of course, aid cloning of production genes) and whether they share features of their promoter structure (or are even cotranscribed).

These kinds of studies will be greatly facilitated by the use of ϕC31 vectors in their mutational cloning mode, not only in the initial cloning of coding sequences and the analysis (by subcloning) of fragments isolated in other vectors, but also as a route to the insertion into the chromosome of DNA fragments manipulated *in vitro*, to give either mutations or gene fusions. In this connection, progress is now being made on the use of the *E. coli lacZ* gene in *Streptomyces*, although difficulties involving endogenous β-galactosidase activity and impermeability to the chromogenic β-galactosidase substrate 5-bromo-4-chloro-3-indoyl-β-D-galactoside (X-gal) have been encountered (Eckhardt and Fare, 1982; S. Grant and A. A. King, personal communications). These techniques will become particularly important in the analysis of genes for antibiotic production and differentiation, which represents perhaps the most interesting challenge to the *Streptomyces* geneticist.

References

Aguilar, A., and Hopwood, D. A. (1982). *J. Gen. Microbiol.* **128,** 1893–1901.

Baltz, R. H. (1978). *J. Gen. Microbiol.* **107,** 93–102.

Benigni, R., Antonov, R. P., and Carere, A. (1975). *Appl. Microbiol.* **30,** 324–326.

Bibb, M. J., and Cohen, S. N. (1982). *Mol. Gen. Genet.* **187,** 265–277.

Bibb, M. J., and Hopwood, D. A. (1981). *J. Gen. Microbiol.* **126,** 427–442.

Bibb, M. J., Freeman, R. F., and Hopwood, D. A. (1977). *Mol. Gen. Genet.* **154,** 155–166.

Bibb, M. J., Ward, J. M., and Hopwood, D. A. (1978). *Nature (London)* **274,** 398–400.

Bibb, M. J., Schottel, J. L., and Cohen, S. N. (1980). *Nature (London)* **284,** 526–531.

Bibb, M. J., Ward, J. M., Kieser, T., Cohen, S. N., and Hopwood, D. A. (1981). *Mol. Gen. Genet.* **184,** 230–240.

Casadaban, M. J., Chou, J., and Cohen, S. N. (1980). *J. Bacteriol.* **143,** 971–980.

Chang, A. C. Y., and Cohen, S. N. (1974). *Proc. Natl. Acad. Sci. USA* **71,** 1030–1034.

Chater, K. F. (1980). *Dev. Ind. Microbiol.* **21,** 65–74.

Chater, K. F. (1983). *Proc. Int. Symp. Genet. Ind. Microorg. 4th,* in press.

Chater, K. F., and Bruton, C. J. (1983). *Gene,* in press.

Chater, K. F., and Carter, A. T. (1979). *J. Gen. Microbiol.* **115,** 431–442.

Chater, K. F., Bruton, C. J., Springer, W., and Suarez, J. E. (1981). *Gene* **15,** 249–256.

Chater, K. F., Hopwood, D. A., Kieser, T., and Thompson, C. J. (1982a). *Curr. Top. Microbiol. Immunol.* **96,** 69–95.

Chater, K. F., Bruton, C. J., King, A. A., and Suarez, J. E. (1982b). *Gene* **19,** 21–32.

Courvalin, P., and Fiandt, M. (1980). *Gene* **9,** 247–269.

Eckhardt, T. G., and Fare, L. R. (1982). *Abstr. Int. Symp. Genet. Ind. Microorg. 4th,* p. 36.

Ehrlich, S. D. (1978). *Proc. Natl. Acad. Sci. USA* **75,** 1433–1436.

Enquist, L. W., and Bradley, S. G. (1971). *Dev. Ind. Microbiol.* **12,** 225–236.

Foor, F., and Morin, N. (1982). *Abstr. Int. Symp. Genet. Ind. Microorg. 4th,* p. 74.

Gil, J. A., Naharro, G., Villanueva, J. R., and Martin, J. F. (1980). *In* "Advances in Biotechnology" (C. Vezina and K. Singh, eds.), Vol. 3 (Fermentation Products), pp. 141–146. Pergamon, London.

Harris, J. E., Chater, K. F., Bruton, C. J., and Piret, J. M. (1983). *Gene* **22,** 167–174.

Hayakawa, T., Tanaka, T., Sakaguchi, K., Ōtake, N., and Yonehara, H. (1979). *J. Gen. Appl. Microbiol.* **25,** 255–260.

Hopwood, D. A. (1979). *J. Nat. Prod.* **42,** 596–602.

Hopwood, D. A. (1981). *Annu. Rev. Microbiol.* **35,** 237–272.

Hopwood, D. A., and Chater, K. F. (1982). *Genet. Eng.* **4,** 119–145.

Hopwood, D. A., Chater, K. F., Dowding, J. E., and Vivian, A. (1973). *Bacteriol. Rev.* **37,** 371–405.

Hopwood, D. A., Wright, H. M., Bibb, M. J., and Cohen, S. N. (1977). *Nature (London)* **268,** 171–174.

Hopwood, D. A., Thompson, C. J., Kieser, T., Ward, J. M., and Wright, H. M. (1981). *Microbiology (Washington, D.C.)* 376–379.

Hopwood, D. A., Wright, H. M., Kieser, T., and Bibb, M. J. (1983). *J. Gen. Microbiol.,* in press.

Isogai, T., Takahashi, H., and Saito, H. (1980). *Agric. Biol. Chem.* **44,** 2425–2428.

Katz, E., Thompson, C. J., and Hopwood, D. A. (1983). *J. Gen. Microbiol.,* in press.

Kieser, T., Hopwood, D. A., Wright, H. M., and Thompson, C. J. (1982). *Mol. Gen. Genet.* **185,** 223–238.

Kirby, R., and Hopwood, D. A. (1977). *J. Gen. Microbiol.* **98,** 239–252.

Kleckner, N. (1981). *Annu. Rev. Genet.* **15,** 341–404.

Krügel, H. J., Fiedler, G., and Noack, D. (1980). *Mol. Gen. Genet.* **177,** 297–300.

Lomovskaya, N. D., Chater, K. F., and Mkrtumian, N. M. (1980). *Microbiol. Rev.* **44,** 206–229.

Losick, R., and Pero, J. (1981). *Cell* **25,** 582–584.

Makins, J. F., and Holt, G. (1981). *Nature (London)* **293,** 671–673.

McLaughlin, J. R., Murray, C. L., and Rabinowitz, J. C. (1981). *J. Biol. Chem.* **256,** 11283–11291.

Nakano, M. M., Ishihara, H., and Ogawara, H. (1981). *J. Gen. Microbiol.* **122,** 289–293.

Okanishi, M., Manome, T., and Umezawa, H. (1980). *J. Antibiot.* **33,** 88–91.

Robbins, W. R., Wirth, D. F., and Hering, C. (1981). *J. Biol. Chem.* **256,** 10640–10644.

Rodgers, W. H., Springer, W., and Young, F. G. (1982). *Gene* **18,** 133–141.

Rosenberg, M., and Court, M. (1979). *Annu. Rev. Genet.* **13,** 319–353.

Schottel, J. L., Bibb, M. J., and Cohen, S. N. (1981). *J. Bacteriol.* **146,** 360–368.

Seno, E. T., and Chater, K. F. (1983). *J. Gen. Microbiol.* **129,** 1403–1413.

Sladkova, I. M., Vasilchenko, L. G., Lomovskaya, N. D., and Mkrtumian, N. M. (1980). *Mol. Biol. (USSR)* **14,** 910–915.

Stuttard, C. S. (1979). *J. Gen. Microbiol.* **110,** 479–483.

Suarez, J. E., and Chater, K. F. (1980a). *J. Bacteriol.* **142,** 8–14.

Suarez, J. E., and Chater, K. F. (1980b). *Nature (London)* **286,** 527–529.

Takahashi, H., Isogai, T., Morino, T., Kojima, H., and Saito, H. (1983). *Proc. Int. Symp. Genet. Ind. Microorg. 4th,* pp. 61–65.

Thompson, C. J., and Gray, G. S. (1983). *Proc. Natl. Acad. Sci. USA,* in press.

Thompson, C. J., Ward, J. M., and Hopwood, D. A. (1980). *Nature (London)* **286,** 525–527.

Thompson, C. J., Ward, J. M., and Hopwood, D. A. (1982a). *J. Bacteriol.* **151,** 668–677.

Thompson, C. J., Skinner, R. G., Thompson, J., Ward, J. M., Hopwood, D. A., and Cundliffe, E. (1982b). *J. Bacteriol.* **151,** 678–685.

Thompson, C. J., Kieser, T., Ward, J. M., and Hopwood, D. A. (1982c). *Gene* **20,** 51–62.

Walter, F., Hartmann, M., and Klaus, S. (1981). *Gene* **13,** 57–63.

Westpheling, J. (1980). Ph.D. Thesis, Univ. of East Anglia, Norwich, England.

CHAPTER 5

Vectors for High-Level, Inducible Expression of Cloned Genes in Yeast

JAMES R. BROACH
YU-YANG LI
LING-CHUAN CHEN WU
MAKKUNI JAYARAM

Department of Microbiology
State University of New York at Stony Brook
Stony Brook, New York

Experimental Manipulation
of Gene Expression

ISBN 0-12-372380-9

I. Introduction

In this chapter we describe the design and application of vectors that permit high-level, inducible synthesis of the product of a cloned gene in yeast. These vectors have been developed in recognition of the increasing need for obtaining reasonable amounts of the product of a cloned gene in research projects. For instance, given the variety of cloning strategies currently available, it is often easier to clone DNA encoding a particular gene than to purify the product of that gene. However, appreciation of the biological function of such a gene usually requires the subsequent identification and characterization of its product. This task is greatly facilitated by the vectors and techniques described here. Similarly, extensive physical and biochemical characterization of a particular biologically active protein can be accomplished only if significant quantities of the purified protein are available. The acquisition of such quantities of material is often technically impossible or prohibitively expensive if one is restricted to the native source of that protein. However, this difficulty is obviated by high-level expression of the protein from its cloned gene in an appropriate host–vector system.

As is evident from the other chapters in this volume, there are a number of host–vector systems currently available for the high-level expression of cloned genes. However, there are a number of reasons why yeast systems should be included in the arsenal of genetic engineers. First, the codon bias in yeast is significantly different from that in *Escherichia coli,* as is, undoubtedly, the spectrum of endogenous proteolytic enzymes (Bennetzen and Hall, 1982a; Jones, 1983). Thus for either of these reasons it is possible that a foreign gene that is expressed poorly in *E. coli* may in fact be expressed at a significantly higher level more readily in yeast. In addition, yeast is capable of promoting glycosylation of newly synthesized proteins (Schekman and Novick, 1982). Thus, to the extent that such modification is an essential or useful aspect of the biological activity of a protein, yeast would be the preferred host for its synthesis. A third consideration is the thorough genetic characterization of yeast. As a consequence, yeast can be manipulated genetically with great facility, either using classical methods or through recombinant DNA techniques in conjunction with transformation (Mortimer and Schild, 1981; Botstein and Davis, 1982). Thus one has the potential for extensive and facile manipulations of the host yeast strain to achieve optimum expression of a particular cloned gene. Finally, yeast has been well adapted to large-scale fermentation. As a consequence, one can grow standard laboratory strains of yeast to high density on a large scale with little effort and in media that cost only a few pennies per liter.

There are two classes of expression vectors available for use in yeast that we discuss in this chapter. The first are those designed to obtain high-level synthesis of the authentic product—or a relatively close variant of the product—of a cloned gene of interest. These vectors are used primarily to evaluate the biological activity of the product of a cloned gene or to produce reasonable quantities of the protein to facilitate its purification. However, the use of these vectors generally requires a reasonable knowledge of the sequence of at least the 5′ and 3′ ends of the gene. In addition, for genes from higher eukaryotes, successful application of these vectors most likely requires a full-length cDNA copy, rather than genomic clones, of the coding region. The second class of vectors are those designed to produce, through appropriate gene fusions, reasonable amounts of a hybrid protein consisting of a portion of the product of a cloned gene linked to, for example, β-galactosidase. The hybrid protein can be isolated and used to generate antisera against the gene product of interest. Such antisera constitute diagnostic tools for identification, characterization, and purification of the gene product. The use of these vectors requires only a limited knowledge of the structure of the gene, and they can be applied to genomic clones or partial cDNA copies of the gene of interest.

In Section III,A we discuss the individual component parts of yeast expression vectors and examine the considerations for optimizing each of these elements. In Section III,B we describe several expression vectors that are currently available and discuss procedures for using them. In Section III,C we present results that document the effective application of these vectors by describing their use in the identification and overproduction of the product of the *REP1* gene of yeast.

II. Materials and Methods

A. Strains and Media

DC04 [cir$^+$] and DC04 [cir^0] are isogenic *a ade1 leu2-04* Gal$^+$ yeast strains, respectively containing or lacking endogenous 2-μm circles. *Escherichia coli* strain C600 (*thr thi1 leuB6 supE44 relA1* r_K^- m_K^-) was used for construction, propagation, and amplification of hybrid plasmids. SD and SG media consist of 0.67% yeast nitrogen base supplemented with amino acids as described by Broach *et al.* (1979b) and either 2% glucose or 2% galactose, respectively.

B. Miscellaneous Methods

Transformation of *E. coli* was performed by the method of Mandel and Higa (1970). Yeast transformations were performed as described by Beggs (1978). Restriction enzymes were obtained from New England Biolabs or Bethesda Research Laboratory, and digestions were performed as recommended by the supplier. Yeast polyadenylated RNA was purified as described previously (Broach *et al.*, 1979a). RNA fractionation, transfer, and hybridization were accomplished as described by Thomas (1980).

C. Plasmid Constructions

A detailed description of the construction of plasmids described in this chapter is provided in the Appendix.

D. Preparation of *REP1* Antisera and Identification of the *REP1* Protein

A culture (1 liter) of strain DC04 [cir$^+$, SΔ6-5B] (SΔ6-5B is a plasmid containing a *GAL10–REP1–lacZ* fusion gene; see the Appendix) in SG minus leucine was grown to a density of 1.5×10^7 cells/ml. The cells were harvested, washed in 1.0 *M* sorbitol, resuspended in 10 ml of 1.0 *M* sorbitol, and converted to spheroplasts by incubation at 30°C with 1% glusulase for 30 min. The spheroplasts were harvested, washed twice with 1.0 *M* sorbitol, and then lysed in 10 ml of 10 m*M* Tris-Cl, pH 7.0, 0.1% Triton X-100. Cell debris was removed by centrifugation, and 0.2 ml of anti-β-galactosidase serum [obtained from a rabbit following one subcutaneous injection and three subsequent intravenous injections each of 0.1 mgm of *E. coli* β-galactosidase (Sigma) in Freund's adjuvant] was added to the supernatant. After 30 min of incubation at 0°C, 1 ml of formaldehyde-treated *Staphylococcus aureus* cells (Kessler, 1975) was added and incubation continued for an additional 30 min. Cells were harvested by centrifugation, and the supernatant was reextracted with an additional 0.2 ml of serum and 1.0 ml of *S. aureus* cells. The two cell pellets were combined and extracted with PAG sample buffer (Laemmli, 1970) and the extract fractionated by electrophoresis on a preparative 7.5% polyacrylamide gel. Following electrophoresis, the gel was stained with 0.1% Coomassie Blue in water. The portion of the gel containing the 150,000-dalton hybrid protein was excised and the protein electroeluted in 10 m*M* ammonium bicarbonate, 0.02% SDS. Approximately 30 μgm of hybrid protein was recovered per liter of culture. Hybrid protein (0.4 ml, 0.1 mgm/ml) prepared in this fashion was mixed with an equal volume of Freund's adjuvant and injected into both popliteal lymph nodes of a New Zealand rabbit. The rabbit was boosted every 2 weeks with an intragastocnemial injection of 0.04 mgm of hybrid protein in Freund's adjuvant. After 14 weeks the rabbit was sacrificed and its serum (anti-*REP1* serum) retained.

To identify *REP1* protein, appropriate yeast strains were grown in 100-ml cultures in the indicated media to $\sim 2 \times 10^7$ cells/ml. Cells were harvested, washed with 1.0 *M* sorbitol, resuspended in 5 ml of sorbitol, and converted to spheroplasts by incubation with 1.0% glusulase. Spheroplasts were harvested, washed three times with sorbitol, and then lysed in 0.5 ml of 10 m*M* Tris-Cl, pH 7.0, 0.1% Triton X-100, 1.0% SDS. Following removal of cell debris by centrifugation, a sample (0.04 ml) of the extract was fractionated by electrophoresis on a 12.5% analytical polyacrylamide gel. Following fractionation, the protein was electrotransfered to nitrocellulose and the filter incubated sequentially with anti-*REP1* serum and ^{125}I-labeled protein A (Amersham) as described by Towbin *et al.* (1979) and Renart *et al.* (1979).

III. Results and Discussion

A. Components of Yeast Expression Vectors

1. PROMOTERS

Available evidence indicates that, in yeast, RNA polymerase II promoters lie immediately 5′ to the sequences transcribed. That is, only a relatively small region immediately 5′ to the transcription-initiation site of a gene is required to determine the site and rate of transcriptional initiation of the gene. In addition, modulation of the level of transcription by regulatory components most often is also mediated through sequences within this vicinity. The significance of this observation in terms of expression vectors is that the 5′ sequence of one gene, when placed adjacent to the coding region of a second, will induce transcription of the second gene at approximately the same level and subject to the same regulatory influences as the gene from which the sequences were taken.

Results from several types of experiments suggest that RNA polymerase II promoters lie completely 5′ to the site of transcriptional initiation of a gene. First, a number of investigators have constructed in-frame fusions of the 3′ portion of the *lacZ* gene of *E. coli* to the 5′ coding region of various yeast genes. Such fusions yield hybrid proteins with functional β-galactosidase activity, the expression of which in yeast exhibits a level and regulation characteristic of the yeast gene to which the *lacZ* gene was fused. This has been true for all genes examined, regardless of the location in the coding region of the insertion of *lacZ* sequences (Rose *et al.*, 1981; Guarente and Ptashne, 1981; Casadaban *et al.*, 1983). Second, Faye *et al.* (1981) have demonstrated that deletion of the coding region of the yeast gene *CYC1* from position +8 to +251 (from the translation start) has no effect on the level or site of transcriptional initiation of the gene. Third, it is possible to fuse the 5′-flanking region of a yeast gene to the coding region of a second gene to yield transcription of the coding region at a level characteristic of the yeast gene. For example, Hitzeman *et al.* (1981) obtained high-level expression of human leukocyte interferon in yeast by fusion of its coding region to sequences lying 5′ to the coding region of *ADC1*, the gene coding for constitutive alcohol dehydrogenase, at a site 5 bp 3′ to the point of transcriptional initiation of *ADC1*. Similarly, as described later, we have found that sequences 5′ to *GAL10*, the gene encoding UDPgalactose 4-epimerase, can be fused at positions from +8 to +20 from the transcription-initiation site to yield transcription of adjacent sequences at the same level and subject to the same control as that of *GAL10* itself.

Determination of the 5′ limits of several yeast promotors as well as identification of essential sequences within these promoters necessary for *in vivo* transcriptional activity have been determined for a number of yeast genes. This has been accomplished primarily by constructing deletion derivatives *in vitro* of plasmids containing DNA spanning the gene of interest. The effects on expression of the removal of specific sequences could be assessed following reintroduction of the DNA into yeast by transformation. By this procedure, sequences lying 247–670, 139–242, and 75–100 bp upstream fom the translation start of *CYC1* have been shown to be necessary for full transcriptional activity of the gene (Faye *et al.*, 1981; Guarente and Ptashne, 1981). Similar experiments have shown that sequences lying between positions −33 and −170 with respect to the *ADC1* translation start site have full promoter activity (Hitzeman *et al.*, 1981). Additionally, Struhl (1981, 1982) has shown that there are only two regions—one from −112 to −155 and the other from −32 to −52 with respect to the translation start site—that are essential for *HIS3* expression. Even in light of these experiments, though, there is not as yet sufficient evidence to define the precise nature of yeast promoters, although elements common to a number of yeast promoters have been observed (Hall and Sentenac, 1982; Dobson *et al.*, 1982). However, for practical purposes these data warrant the assumption that a yeast polymerase II promoter resides in the region several hundred base pairs immediately 5′ to the site of transcriptional initiation.

On the basis of the foregoing discussion, it is evident that high-level expression of a cloned gene in yeast is best achieved by its fusion to the promoter of a yeast gene that is normally expressed at high levels. In *Saccharomyces*, highly expressed genes are almost exclusively those encoding glycolytic enzymes or enzymes otherwise involved in carbon metabolism. This is not surprising because *Saccharomyces* strains have been selected and optimized over the years for their ability to promote fermentation. Most glycolytic enzymes are present in yeast growing in conditions of fermentation at 0.5–5% of total cellular protein. In addition, the genes encoding them are transcribed to yield comparable relative levels of specific mRNAs in the cell (Holland and Holland, 1978; Fraenkel, 1982). The genes for a number of these enzymes have been cloned and sequenced. These include *ADC1*, encoding the alcohol dehydrogenase present in glucose-grown cells (Hitzeman *et al.*, 1981); *PGK1*, encoding phosphoglycerate kinase (Hitzeman *et al.*, 1980; Dobson *et al.*, 1982); *GAP1* and *GAP2*, encoding glycerlaldehyde-3-phosphate dehydrogenase (Holland and Holland, 1979, 1980); and *ENO1* and *ENO2*, encoding enolase (Holland *et al.*, 1981). The promoter region from *ADC1* has been appropriately tailored to allow its fusion to exogenous coding regions, and vectors constructed around this promoter have been used to induce reasonably high-level synthesis from a number of foreign genes in yeast (Hitzeman *et al.*, 1981; Ammerer, 1983). Similarly, Hitzeman *et al.* (1982) have described the construction and application of expression vectors tailored from the promoter region from *PGK1*.

Expression from promoters of genes encoding glycolytic enzymes is essentially constitutive. That is, although there is some fluctuation in the level of expression of various glycolytic genes depending on growth conditions, expression is always moderately high (Holland and Holland, 1978; Fraenkel, 1982). In constrast, the expression of genes encoding enzymes required for catabolism of fermentable carbon sources other than glucose (e.g., sucrose, galactose, and maltose) exhibits substantial alteration of expression in response to growth conditions. For example, synthesis of galactose metabolic enzymes is strictly inducible by galactose. That is, in the absence of galactose in the growth medium, expression of the gene encoding galactose metabolic enzymes is undetectable, whereas in the presence of galactose and the absence of glucose each of these genes is expressed at the level of approximately 1% of the total polyadenylated RNA in the cell (St. John and Davis, 1981). Expression of genes encoding maltose (*MAL1* through *MAL6*) and those encoding invertase (*SUC1* through *SUC7*) also show induction, by addition of maltose or sucrose to the culture medium, respectively, from an essentially undetectable basal level to a level constituting 1% of the cellular RNA (Mowshowitz, 1979; Fraenkel, 1982; Carlson *et al.*, 1982).

In our laboratory we have focused on the construction and application of expression vectors constructed around inducible promoters. We have done this for several reasons. First, an inducible promoter provides us with a transcriptional switch that allows us to turn on or off a gene of interest at will. This can be useful in correlating the presence or absence of a particular gene product with the presence or absence of a characteristic phenotype. A second advantage of inducible expression vectors is that the inducible expression of a gene can often enhance the absolute level of synthesis one can obtain from that gene. For example, if high-level expression of the gene of interest is deleterious to the cell, one can propagate that gene on an inducible expression vector under repressed conditions. High-level transcription of the gene can then be initiated after the host culture has grown to appropriately high density. In addition, because synthesis of the clone gene product can be activated after continued growth of the culture is

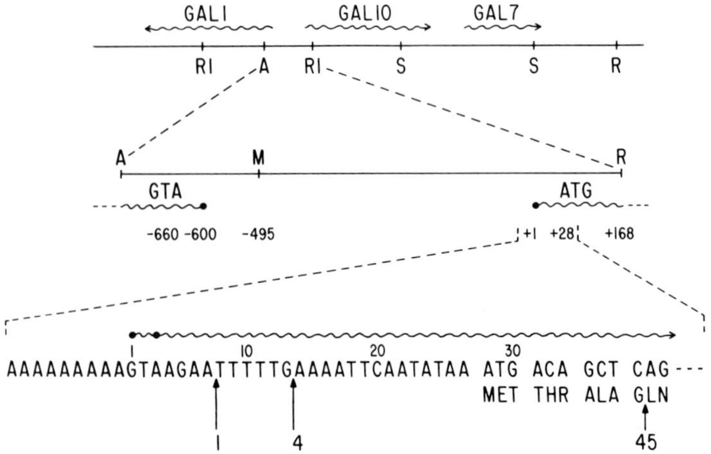

Fig. 1 The *GAL7, GAL10, GAL1* gene cluster of yeast. The upper line represents a 9-kb segment of DNA from chromosome II of *Saccharomyces* spanning three genes—*GAL1, GAL10,* and *GAL7*—required for metabolism of galactose. The relative positions of the genes and their direction of transcription are indicated by the wavy arrows above the line. The locations of a selected set of restriction sites are indicated [abbreviations: R, *Eco*R1; S, *Sal*I; A, *Ava*I; M, *Mbo*I (*Sau*3A)]. The 910-bp *Ava*I-to-*Eco*R1 restriction fragment spanning the *Gal1–GAL10* intercistronic region is represented on an expanded scale in the middle. The locations of the transcription-initiation sites and the initial ATG codons for the two genes are indicated. The exact nucleotide positions of these elements, as well as of several restriction sites, are indicated below the line, using the transcriptional initiation site of *GAL10* as position 1. The nucleotide sequence of DNA spanning the transcriptional and translational initiation sites of *GAL10* is shown at the bottom of the figure. The numbered arrows indicate the positions of the deletion endpoint in several different derivatives of plasmid pNN78. As described in the text, each derivative retains only those sequences to the left of the arrow, joined through a *Sal*I linker at the site of the arrow to the *Sal*I site of pBR322.

no longer required, one can in theory subvert 100% of the synthetic capacity of the cell to the synthesis of the gene product of interest. These considerations are certainly relevant to the production in yeast of DNA-binding proteins, such as those involved in regulation, because evidence has accumulated to suggest that overproduction of such proteins is often deleterious to a host strain.

To develop inducible cloning vectors, we have focused on the genes involved in galactose metabolism. Three genes (*GAL1, GAL10,* and *GAL7*) encoding the enzymes required for metabolism of galactose (galactose kinase, UDPgalactose 4-epimerase, and uridylyl transferase, respectively) are located in a cluster on chromosome II of yeast. St. John and Davis (1981) isolated DNA spanning these three genes and determined the genetic and transcriptional organization of the region. These results are summarized in Fig. 1. As indicated, *GAL10* and *GAL1* are transcribed divergently from within a 910-bp *Eco*R1–*Ava*I restriction fragment. The sequence of this fragment and the approximate location of the transcription-initiation sites of the two genes have been determined (R. Yocum and M. Johnston, personal communications) and are indicated in Fig. 1. For both *GAL10* and *GAL1* the first ATG codon downstream from the transcription-initiation site is followed by an open coding region that extends to the end of the fragment. In addition, the amino acid sequence predicted by the DNA sequence of *GAL1*, continuing from this ATG, agrees with the published amino acid sequence from the amino terminus of galactose kinase.

We have obtained a DNA fragment from the 5′ end of the *GAL10* gene that promotes galactose-inducible, high-level transcription and that can be readily fused to other coding regions. A pBR322 plasmid containing the 910-bp *Eco*R1–*Ava*I fragment (Fig. 1) was linearized at the *Eco*R1 site and the ends resected for variable distances with the double-stranded exonuclease *Bal*31. *Sal*I octonucleotide linkers were added to the ends, and the plasmid was then recircularized and recovered in *E. coli*. The position of the linker in a number of clones was determined by restriction analysis or by nucleotide-sequence analysis using the procedure of Maxam and Gilbert (1977). The positions of several of these are indicated in Fig. 1. As demonstrated in Section III,C, the fragment extending from the *Sau*3A site at −495 to the *Sal*I site in clones 1 or 4 promotes high-level, galactose-dependent transcription in yeast outward across the *Sal*I site. Thus this 500-bp fragment functions as a portable, inducible promoter.

The 5′ region of the *PHO5* gene of yeast, which encodes one form of acid phosphatase, can also be used as a portable inducible promoter. *PHO5* is not transcribed at a detectable level in strains growing in media containing high levels of inorganic phosphate. However, when inorganic phosphate is depleted from the medium, transcription of *PHO5* is induced to yield a level of *PHO5*-specific mRNA up to 5% of the total cellular polydenylated RNA (Oshima, 1982; R. Kramer, personal communication). The advantage of using this promoter in an expression system is that the medium can be designed such that inorganic phosphate is depleted when a strain harboring such an expression vector reaches an appropriately high density. Transcription from the *PHO5* promoter on the vector would then be induced to transcribe the gene to which it has been fused without any further manipulation of the culture medium. This is not a significant advantage in small-scale production used in a research situation. However, it is a significant consideration in very large-scale fermentation associated with commercial production of genetically engineered products. The *PHO5* gene has been cloned and suitably tailored for use in expression vectors (R. Kramer and A. Hinnen, personal communications).

Although there are significant advantages to using inducible promoters, such as those from *GAL10* or *PHO5*, for achieving high-level production of a cloned gene in yeast, there are certain constraints on their use that accrue from the nature of the regulation of these genes. The scheme of the regulatory circuit for genes encoding galactose metabolic enzymes is presented in Fig. 2. Transcription of

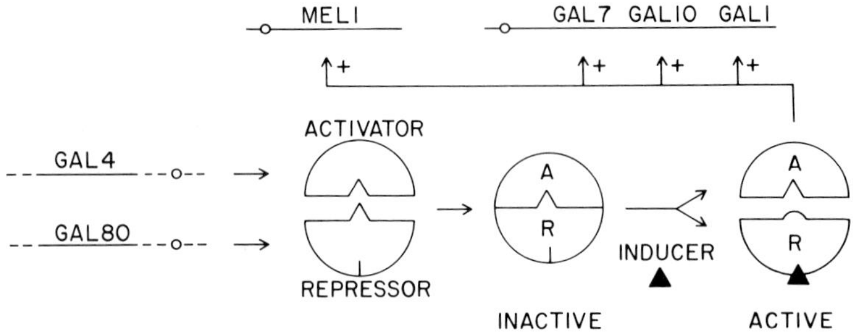

Fig. 2 Regulatory circuit for the control of galactose metabolism in yeast. The regulation of transcription of genes involved in galactose metabolism—including *GAL1*, *GAL7*, *GAL10*, and *MEL1*—is accomplished by the action of proteins encoded by two genes, *GAL4* and *GAL80*. *GAL4* synthesizes constitutively an activator protein the activity of which is absolutely and specifically required for transcription of the structural genes for galactose metabolic enzymes. *GAL80* constitutively synthesizes a repressor protein that modulates the activity of the *GAL4* gene product. In the absence of inducer, *GAL80* protein binds to the *GAL4* activator protein, preventing its functioning as a transcriptional activator. An unidentified metabolite of galactose serves as inducer by binding to the *GAL80* protein. This binding relieves *GAL80*-protein inactivation of *GAL4* protein, releasing it to activate transcription of galactose-controlled genes.

GAL1, *GAL7*, and *GAL10* as well as *GAL2* (which encodes the galactose-specific permease) and *MEL1* (which encodes an extracellular α-galactosidase) is absolutely dependent upon a transcriptional activator encoded by the gene *GAL4* (Oshima, 1982). In addition, the ability of the *GAL4* activator protein to promote transcription of these genes is reversibly inactivated by the product of the *GAL80* gene. This inactivation occurs through a direct protein–protein interaction of the two gene products. Galactose, or a metabolite thereof, can bind to the *GAL80* repressor protein. Binding of galactose to the repressor prevents its inactivation of *GAL4* protein and thus frees *GAL4* protein to promote transcription of the galactose-specific genes. Regulation of *PHO5* occurs through a similar mechanism, except that an additional two regulatory elements are involved that invert the regulatory consequences of the presence of the effector molecule, inorganic phosphate (Oshima, 1982).

Because transcription from galactose-controlled promoters is absolutely dependent upon *GAL4* protein, the maximal level of expression from a galactose-controlled promoter is limited by level of the *GAL4* protein in the cell. For instance, Klar and Halvorson (1976) have shown that the level of synthesis of galactokinase from *GAL1* is substantially reduced in cells containing one-fourth the normal gene dosage of *GAL4*. Thus it is likely that increasing the number of copies of an expression vector designed around galactose-regulated promoter will not necessarily increase production from the gene inserted in the vector, because at some point the limiting factor in synthesis becomes not the number of genes present in the cell but the level of the *GAL4* activator protein. It is possible to overcome this limitation by including the *GAL4* gene on the expression vector itself (the *GAL4* gene has been cloned and sequenced; Laughon and Gesteland, 1982). However, although the resultant overproduction of *GAL4* protein would undoubtedly enhance transcription from a *GAL10* promoter on the vector, it would also soon outstrip the level of the *GAL80* protein in the cell and thus short-circuit regulation by galactose. Therefore, in order to obtain maximal synthesis of a cloned gene using the *GAL10* promoter while maintaining its response to galactose regulation, it would be necessary either to design a vector in which all regulatory genes are included or to develop a strain in which the regulatory components are overexpressed. Similar considerations apply to the use of vectors designed around the *PHO5* promoter.

An additional constraint exists on the use of vectors derived from galactose-controlled promoters, namely, in the selection of strains and growth conditions. First, only those strains in which the galactose-regulatory apparatus is intact and which are not sensitive to galactose are suitable hosts. This eliminates *gal4*$^-$, *gal80*$^-$, *gal7*$^-$, or *gal10*$^-$ strains. In addition, even if the host strain is Gal$^+$, galactose is not a suitable carbon source for propagating the plasmid-bearing strain. That is, requiring the strain to grow on galactose would provoke a competition for the limited *GAL4* protein in the cell, between the galactose-controlled promoters adjacent to genes in the chromosome (the expression of which is required for metabolism of galactose) and the galactose-controlled promoter on the vector directing synthesis of the cloned gene. The result would be either reduced growth or reduced expression from the vector. The most suitable hosts for these vectors are *gal1*$^-$ strains. These strains exhibit normal induction by galactose (Broach, 1979), they are not sensitive to galactose, and they require only a minimal amount of galactose for induction, because the added galactose is not metabolized. Also, because the strain cannot use galactose for growth, cells with high copy number of the expression vector are not at a selective disadvantage. Thus *gal1* strains bearing the expression vector can be grown on a carbon source—such as raffinose, lactose, ethanol, or glycerol—that does not promote catabolite repression (Adams, 1972) and then induced at a proper cell density by addition of a small amount of galactose. This regime should yield the maximum expression possible from vectors designed around a galactose-controlled promoter.

2. TRANSCRIPTION-TERMINATION SITES

Sequences at the 3′ end of regions transcribed by RNA polymerase II in yeast direct polyadenylation of nascent RNA transcripts. It is not known whether the same site induces transcriptional termination as a concerted event in polyadenylation or whether, as in higher eukaryotes, these two processes are spatially

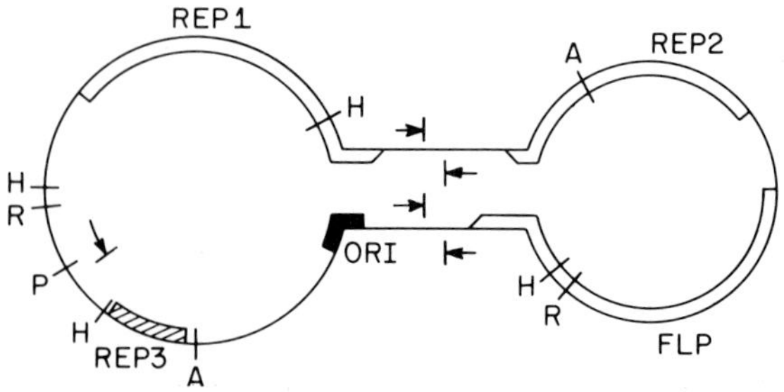

Fig. 3 Yeast plasmid 2-μm circle. The genome of one form (the B form) of the yeast plasmid 2-μm circle is represented as a dumbbell-shaped figure: the linear portions of the figure correspond to the two inverted repeat sequences, and the circular portions to unique regions. The three major open coding regions—corresponding to the genes *REP1, REP2,* and *FLP*—are indicated by open double segments, with a taper designating the 3′ end. The locations of the origin of replication (■), the cis-acting *REP3* locus (▨), a number of restriction sites (abbreviations: H, *Hin*dIII; R, *Eco*R1; A, *Ava*I; P, *Pst*I; Hp, *Hpa*I), and transcriptional polyadenylation sites (blocked arrows) are indicated.

distinct (Fraser *et al.*, 1979; Hofer and Darnell, 1981). The paucity of observable read-through transcription into sequences 3′ to the normal polyadenylation site of yeast genes suggests, but does not conclusively demonstrate, that the former interpretation is correct. In any event, evidence has accumulated to indicate that the absence of a polyadenylation site at the end of a gene significantly reduces the *in vivo* steady-state level of mRNA from that gene. Zaret and Sherman (1982) found that *CYC1*-specific RNA in a strain containing a 38-bp deletion in the 3′ noncoding region of *CYC1* is substantially longer than that from wild-type cells and, in addition, that the total level of *CYC1*-specific RNA in the strain is only 5–10% that in wild-type cells. Conversely, Hitzeman *et al.* (1981) found that the specific activity of human leukocyte interferon in a yeast strain programmed with a vector containing the interferon gene fused to the *ADC1* promoter could be increased significantly by placing the 3′ end of the *TRP1* gene of yeast at the end of the interferon-coding region. Thus it appears that inclusion of a yeast polyadenylation site at the end of a cloned gene is necessary to obtain optimal expression of that gene in yeast.

The specific sequences that constitute a yeast polyadenylation site have not been identified, even though comparative analyses of sequences spanning the site corresponding to the 3′ end of a number of yeast polymerase II transcripts have been performed (Zaret and Sherman, 1982; Bennetzen and Hall, 1982b). Nor have the limits of this region for any yeast gene been determined. However, it is unlikely that relevant sequences extend further than 100 bp to either side of the site corresponding to the 3′ end of a polymerase II transcript. Thus, for purposes of vector construction, a 200-bp fragment spanning the site corresponding to the 3′ end of any yeast polymerase II transcript sould prove suitable as a portable polyadenylation site. The approximate locations of the 3′ ends of a number of transcripts from the yeast plasmid 2-μm circle are shown in Fig. 3.

3. Translation-Initiation Sites

Substantial genetic analysis of the *CYC1* locus conducted by Sherman and Stewart (1982) has demonstrated that translation of a yeast transcript initiates at the most proximal AUG codon at the 5′ end of a mRNA and that reinitiation or primary initiation at an internal AUG does not occur. Confirmation that this observation is generally applicable to yeast mRNAs derives from the observation that for most yeast genes for which the transcription-initiation site has been determined, the first AUG codon present on the mRNA is the initial AUG of the reading frame encoding the gene product. Thus the rules determining the site of initiation of translation in yeast messenger RNAs conform, to a first approximation, to those for other eukaryotic cells (Kozak, 1981).

Factors that influence the rate of translational initiation of a particular mRNA in yeast have been afforded only minimal attention. Sherman and Stewart (1982) observed that certain alterations of sequences 5′ to the initiating AUG could apparently influence the translational efficiency of the mRNA for *CYC1*. However, altering the location of the initiating AUG over a 30-base region had little influence on the rate of initiation. Similarly, little difference was observed in expression of human leukocyte interferon in yeast from different fusions of the *ADC1* promoter to the interferon gene, even though the position of the fusion ranged from +8 to +38 of the DNA corresponding to the 5′ untranslated leader region of *ADC1* mRNA. Thus major changes in the nature of the untranslated leader do not necessarily result in changes in the translational efficiency of the resultant mRNAs. It is evident from sequence analysis that the sequences 5′ to the initial AUG in most yeast mRNAs are reasonably A + U rich. However, as indicated in the preceeding, it is not apparent to what extent this feature contributes to translational efficiency of yeast mRNAs.

4. Splicing

Saccharomyces contains several genes the coding regions of which are not contiguous and thus are not collinear with their respective mRNAs (Ng and Abelson, 1980; Gallwitz and Sures, 1980; Rosbash *et al.*, 1981). Therefore, yeast clearly has the capacity to remove intervening sequences from the primary transcripts of such genes to yield functional mRNAs. In addition, the sequences at the splice junctions of such intron-containing genes do not differ significantly from those found in transcripts from higher eukaryotes. Nonetheless, Beggs *et al.* (1980) did not observe synthesis of functional rabbit β-globin mRNA in yeast programmed with cloned DNA carrying the genomic, intron-containing copy of the β-globin gene. This was due in part to an apparent inability of the yeast cell to promote appropriate RNA splicing of the primary transcript. Therefore, it is unlikely that genomic clones of higher eukaryotic genes can be used directly in yeast to program efficient synthesis of the gene product. This is not a serious limitation to the use of yeast as a host system for expression. Because most yeast genes do not contain intervening sequences, excision of an intron is not an obligatory step in mRNA synthesis in yeast. Therefore, cDNA copies of higher eukaryotic genes can be fused directly to yeast promoters to yield reasonable expression of the gene product. This is in contrast to higher cells in which the presence of an intron is often necessary for efficient expression of a gene (Mulligan *et al.,* 1979; Hamer and Leder, 1979; Gruss and Khoury, 1981; Gething and Sambrook, 1981).

5. SIGNALS FOR SECRETION

There are several situations in which it might be desirable to direct secretion of the product of a cloned gene concomitant with its synthesis in yeast. This could be useful, for example, in facilitating its purification, in removing it from a potentially hostile intracellular environment, in inducing its glycosylation, or in injecting it into its normal domain of a lipid bilayer. However, little is known about the signals that determine the ultimate cellular location of a portein in yeast. As is true of transported proteins in prokaryotes and other eukaryotes (Kreil, 1981), most proteins in yeast that are transported across a membrane contain a stretch of 20 or so amino acids, known as the signal peptide, at the N-end of the protein. This is true of proteins transported into the lumen of the endoplasmic reticulum, as a preliminary step in secretion or vacuolar sequestration, or transported into the lumens of the mitochondria (Schekman and Novick, 1982). The signal peptide, which is endoproteolytically removed during cotranslational translocation across the membrane, possesses a characteristic structure: it is composed predominantly of hydrophobic amino acids, but often with one or more positively charged amino acids at or near the N-terminus. In addition, several short side-chain amino acids are often located at the site of cleavage. However, aside from the signal sequence, signals that target a protein to particular subcellular locations, such as the plasma membrane, the vacuole, or a secretory vesicle, have not been identified. In addition, because it is reasonable to assume that such addressing information is an integral element of the primary amino acid sequence of the protein itself, we may have only a limited ability to redirect the localization of a protein to a site other than that at which it resides in its normal host.

The above caveats notwithstanding, Hitzeman *et al.* (1982) have reported limited success in directing the secretion of a protein concomitant with its synthesis in yeast. These researchers fused the *ADC1* promoter to a cDNA copy of a human leukocyte interferon gene, either at a site corresponding to the amino end of the mature protein or at a site corresponding to the amino end of prointerferon, that is, the mature protein plus its signal peptide. The former construction yielded interferon limited exclusively to the cytpolasm of the host strain. However, a reasonable proportion, but by no means all, of the protein synthesized from the latter construction was located outside the cell membrane or in the cell-membrane fraction. In addition, a majority of the secreted protein was cleaved at the appropriate site. Thus the presence of a signal peptide, even from a nonyeast gene, can induce the secretion from yeast of a protein synthesized from a foreign gene. It should be noted, though, that the particular protein Hitzeman *et al.* examined is one that is normally secreted from the cell.

There are several cloned yeast genes, including *SUC2* and *PHO5*, that could be appropriately tailored to allow fusion of an exogenous coding region to sequences corresponding to the signal peptides present in these genes (Carlson *et al.*, 1983; R. Kramer, personal communication). However, no expression vectors so designed have been described.

6. Selectable Markers

In order to introduce an expression vector into yeast and assure its maintenance, the vector must carry a gene the presence of which in the cell can be selected by an appropriate choice of host strain and growth medium. Such selectable markers fall into one of two categories: a cloned yeast gene that complements a genetic defect in the host strain or a dominant drug-resistance marker. There are a number of cloned yeast genes for which strains carrying nonrevertable alleles in the corresponding loci are available. These include *LEU2, URA3, HIS3,* and *TRP1*. In addition, each of these genes has been sequenced, which facilitates vector construction. The second class of selectable markers includes the neomycin-resistance gene from the bacterial transposon, Tn601, which confers resistance to the drug G-418 when present in yeast (Jiminez and Davies, 1980), and the chloramphenicol acetyltransferase gene from an *E. coli* R factor, which confers to yeast resistance to chloramphenicol (Cohen *et al.*, 1980).

Both classes of selectable markers require specific media in order to be effective: synthetic media lacking a particular amino acid in the former and media containing the appropriate drug in the latter. This is of little consequence in a research context, but it could be significant in large-scale industrial fermentation. Therefore, it could be of value to develop selection regimes that do not require specific media. Such a scheme could consist of a cloned gene encoding an essential function in yeast, used in conjunction with a strain containing a lesion in that locus. For example, the gene for histone H2B in conjunction with a strain defective in both loci encoding the histone would constitute one such system (Rykowski *et al.*, 1981).

7. Sequences for High Copy Propagation in Yeast

As a first approximation, it is reasonable to assume a gene-dosage effect for the expression of cloned sequences in yeast. Thus maximizing expression of a cloned gene will most likely entail maximizing the cellular copy number of the expression vector on which it is carried. For reasons elaborated later, this consideration dictates that the expression vector be constructed around sequences derived from the yeast plasmid 2-μm circle.

The yeast plasmid 2-μm circle is a 6318-bp double-stranded DNA species present at 60–100 copies/cell in most *Saccharomyces cerevisiae* strains (Clark-Walker and Miklos, 1974; Hartley and Donelson, 1980). Although the replication of this plasmid during normal exponential growth is strictly under cell-cycle control, it can escape from this control in certain situations to increase its copy number from as low as a single copy per cell to its normal high level (Sigurdson *et al.*, 1981; Jayaram *et al.*, 1983). As a consequence of this amplification potential, many hybrid plasmids constructed from 2-μm-circle sequences can establish and maintain high copy number in yeast strains, following their introduction by transformation (Hicks *et al.*, 1978; Struhl *et al.*, 1979). In addition, these hybrid plasmids containing 2-μm-circle sequences are relatively stable during mitotic growth and meiosis. Although a number of specific yeast chromosomal sequences are also capable of promoting autonomous replication in yeast of plasmids in which they are incorporated (Stinchcomb *et al.*, 1979; Chan and Tye, 1980; Beach *et al.*, 1980)—a property that is presumably a reflection of their normal function in the cell as chromosomal origins of replication—the ability to establish and maintain high copy number and to propagate stably during mitotic growth is a property unique to plasmids derived from 2-μm-circle sequences.

A diagram of the structure of 2-μm circle is presented in Fig. 3, in which are indicated the salient features of the replication system. The plasmid contains two regions of 599 bp each that are precise inverted repeats of each other and that divide the molecule into approximately equal halves. Because recombination readily occurs between these repeated sequences, 2-μm circles isolated from yeast actually consist of a mixed population of two plasmids that differ only in the orientation of one unique region with respect to the other (Beggs, 1978). The origin of replication lies in a 100-bp region spanning the junction between one of the inverted repeats and the contiguous unique region (Broach and Hicks, 1980; Jayaram *et al.*, 1983). This site was initially identified as the sole sequence within 2-μm circle that would promote autonomous replication in yeast of hybrid plasmids into which it was inserted. That this region actually functions *in vivo* as the primary 2-μm-circle origin of replication has been confirmed (Newlon *et al.*, 1981; Kojo *et al.*, 1981).

Although autonomous replication is conferred by the 2-μm-circle origin of replication, propagation at high copy number requires, in addition, two proteins encoded by the plasmid itself and a cis-active site distinct from the origin of replication (Jayaram *et al.*, 1983). Hybrid plasmids containing 2-μm-circle sequences spanning only the origin of replication are present at low copy number in yeast strains lacking endogenous 2-μm circles ([cir^0] strains) and are lost rapidly during nonselective growth. However, in a [cir$^+$] strain (that is, a strain containing a normal complement of 2-μm circles), the same plasmids are present stably and at high copy number. Similarly, hybrid plasmids containing the entire 2-μm-circle genome are present stably and at high copy number even in [cir^0] strains. Thus, trans-acting functions encoded in sites away from the origin of replication are necessary for high-copy-number propagation of the plasmid in yeast. The genes encoding this function have been identified by genetic analysis of hybrid plasmids containing the entire 2-μm-circle genome. Mutations that interrupt either of two plasmid genes, designated *REP1* and *REP2* in Fig. 3, abolish efficient, high-copy-number maintenance of the plasmid. We have demonstrated that high copy number and stability also require sequences between the *Pst*I and *Ava*I sites in the large unique region (Jayaram *et al.*, 1983). This locus is active only in cis with respect to the 2-μm-circle origin and probably defines the site through which *REP* proteins act to promote amplification. We have designated this region *REP3*. Thus the requisite components for high-copy-number, stable propagation of a hybrid plasmid in yeast are the presence on the plasmid of a fragment from 2-μm circle spanning both the origin of replication and *REP3* and the presence in the cell of the *REP1* and *REP2* gene products. These proteins can be provided either by the endogeneous 2-μm circles in the cell or by the hybrid plasmid itself.

In practical terms, it is generally easier to construct a vector using only the origin of replication and *REP3* from 2-μm circle and then propagate the plasmid in a [cir$^+$] strain. This is generally not a limitation, because the vast majority of *Saccharomyces* strains are [cir$^+$]. In addition, one can readily recover any one of several different restriction fragments spanning the origin and *REP3* that could be inserted readily into a plasmid at a single restriction site. These include the 2242-bp *Eco*R1 fragment, the 2214-bp *Hin*dIII fragment, and the 1576-bp *Sau*3A fragment, all from the B form of the plasmid (for an extensive restriction map of the plasmid see Broach, 1981).

Hybrid plasmids constructed from 2-μm-circle sequences are present in yeast strains at approximately 30–50 copies/cell (Gerbaud and Guerineau, 1980; Jayaram *et al.*, 1983). It is possible, though, to develop hybrid plasmids that propagate in yeast at 100–200 copies/cell by using sequences from a hybrid 2-μm plasmid designated pJDB219 (Beggs, 1978, 1981; Broach, 1983). Plasmid pJDB219 consists of the entire 2-μm-circle genome cloned, at the *Eco*R1 site in the small unique region, into the *Eco*R1 site of the bacterial plasmid pMB9. It contains a fragment of yeast DNA spanning the *LEU2* gene—isolated from randomly sheared, total genomic DNA—inserted at the *Pst*I site in the 2-μm-circle moiety through complementary homopolymer extensions. This fragment of yeast DNA is 1400 bp or so in length, which is not much larger than the *LEU2* gene itself.

Plasmid pJDB219 (or any derivative of pJDB219 containing that portion of the plasmid spanning the origin, *REP3*, and the *LEU2* insert) displays substantially higher copy number in yeast, as well as a significantly greater mitotic stability, than other hybrid plasmids. The reason for this high copy number is unknown. At this point the high copy number is best ascribed to an as yet unexplained activation of the 2-μm-circle replication origin fortuitously induced by the singular nature of the *LEU2* insertion in the plasmid. However, the high copy number and stability of pJDB219 derivatives are dependent in part of 2-μm-circle *REP* functions. That is, plasmids derived from pJDB219 but lacking either *REP1* or *REP2* propagate at 200 copies/cell in a [cir^+] strain, but they are present at only 15 copies/cell in a [cir^0] strain (Broach, 1983).

B. Vectors for Inducible, High-Level Expression in Yeast

We have developed several vectors, based on the considerations raised in Section III,B, that promote inducible, high-level expression of cloned sequences in yeast. These vectors are divided into two classes: (1) those designed to obtain high-level, inducible expression of a cloned gene in yeast and (2) those designed to produce reasonable amounts of a hybrid protein from appropriate gene fusions in order to generate antisera against the gene product of interest.

1. EXPRESSION VECTORS

Plasmids YEp51 and YEp52 (Fig. 4) were designed to obtain high-level, inducible expression of a cloned gene in yeast. Both plasmids contain the *Col*E1 origin of replication and the gene for β-lactamase. Thus both plasmids can be selected and propagated in *E. coli*. In addition, both vectors contain the *LEU2* gene of yeast, the 2-μm-circle origin of replication, and the 2-μm-circle *REP3* locus. Consequently, these plasmids can transform yeast *leu2*$^-$ strains to Leu$^+$ at high frequency and propagate stably and at high copy number in the transformants (Broach and Hicks, 1980; Jayaram *et al.*, 1983). These plasmids also contain a 500-bp promoter fragment from the 5′ end of the yeast *GAL10* gene, extending from the *Sau*3A site 490 bp upstream from the site of transcriptional initiation to a *Sal*I site we inserted between the initiating ATG of the *GAL10* coding region and the site of transcriptional initiation (at position 4 in Fig. 1). Therefore, transcription initiates within this fragment and extends outward across the *Sal*I site. As indicated in Section III,C, transcription initiated within this fragment still exhibits complete control by galactose: transcription from the promoter fragment occurs at high level when a strain harboring YEp51 or YEp52 is grown on galactose, but no transcription occurs when the strain is grown in the

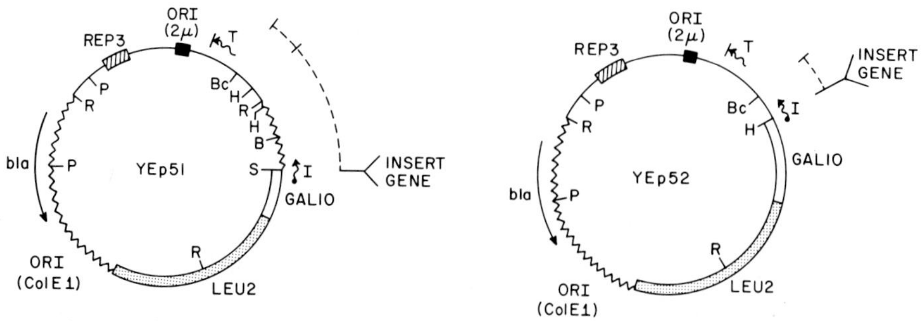

Fig. 4 Yeast expression vectors YEp51 (7.3 kb) and YEp52 (6.6 kb). The structures of two yeast expression vectors are diagrammed. Both are composed of sequences from the yeast plasmid 2-μm circle (smooth single line) spanning *REP3* (▨) and the origin of replication (■), from the bacterial plasmid pBR322 (jagged single line) spanning the *Col*El origin of replication and the gene conferring ampicillin resistance, from the yeast genome spanning the gene *LEU2* (▒), and from the region 5′ to the yeast *GAL10* gene (□), extending from the *Sau*3A site at −495 from the transcription-initiation site to the *Sal*I site present in plasmid pNN78-Δ4 at +13 (see Fig. 1). A cloned gene inserted in YEp51 in the *Sal*I, *Sal*I-to-*Bam*HI, *Sal*I-to-*Hin*dIII, or *Sal*I-to-*Bcl*I sites would be transcribed in a galactose-dependent fashion from the *GAL10* promoter sequences (at the point labelled I in the figure), terminating at a site in the 2-μm-circle sequences indicated by the blocked arrow (T). Similar transcription would be obtained with genes inserted in the *Hin*dIII or *Hin*dIII-to-*Bel*I sites of YEp52. Restriction enzymes: R, *Eco*R1; H, *Hin*dIII; B, *Bam*HI; S, *Sal*I; P, *Pst*I; Bc, *Bcl*I.

absence of galactose. In addition to the *GAL10* promoter, both plasmids contain a transcription-termination site from the 2-μm circle located distal to the promoter. Thus transcription initiated at the *GAL10* promoter will terminate at this site.

Plasmids YEp51 and YEp52 differ only by the restriction sites available for insertion of DNA fragments. This allows somewhat greater flexibility in cloning. However, to obtain optimum expression of a gene of interest using these plasmids, sequences 5′ to the gene must be removed so the first ATG codon at the 5′ end of the fragment is the initiating ATG for the coding region of the gene. Thus some knowledge of the sequence of the 5′ end of the gene of interest is generally a prerequisite for optimal use of these vectors.

2. Vectors for Synthesis of Trihybrid Proteins in Yeast

To complement the expression vectors just described we have developed several vectors that allow us to obtain antisera against the product of a cloned gene of interest in the absence of extensive knowledge about the sequence of the cloned gene. These vectors are diagrammed in Fig. 5. The plasmids have all the components necessary for selection and propagation in both *E. coli* and yeast. However, the salient feature of these plasmids is that they contain the 5′ moiety

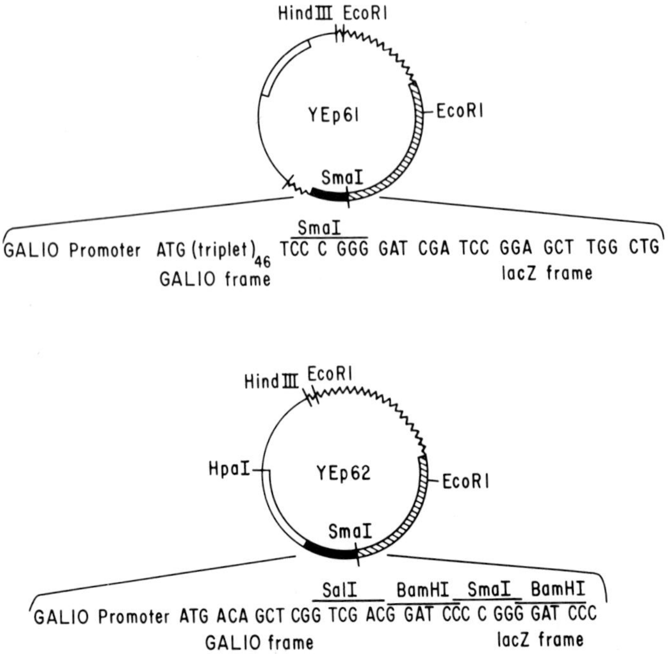

Fig. 5 Vectors for production of hybrid proteins in yeast. Plasmids YEp61 and YEp62 consist of sequences from pBR322 (jagged single line), the yeast plasmid 2-μm circle (smooth single line), and the *LEU2* gene of yeast ( ). The sequences permit their selection and propagation in *E. coli* and yeast. In addition, each contains the *lacZ* gene of *E. coli* ( )—missing the promoter and initial nine amino acids—fused to sequences from the 5′ portion of *GAL10* ( ), either at a position corresponding to the forty-seventh codon (for YEp61) or the fourth codon (for YEp62) of the *GAL10* coding region. The sequence across the *GAL10–lacZ* junction in each plasmid is shown. Because the fusion of the two coding regions is out of frame, no protein with β-galactosidase activity is produced in yeast from either plasmid. However, insertion of a DNA fragment with an open reading frame extending its entire length that upon insertion, is in frame both with the *GAL10* and with the *lacZ* regions yields a plasmid that programs galactose-dependent synthesis in yeast of a hybrid protein with the translational product of the cloned coding region fused to β-galactosidase. Discussions of the use and applications of these vectors are provided in the text.

of the *GAL10* gene, including a portion of the coding region, fused to the *lacZ* gene of *E. coli.* Because the fusion occurs at amino acid 9 in *lacZ,* any protein synthesized from this fusion will possess functional β-galactosidase activity. However, the fusion of *GAL10* to the *lacZ* gene is constructed in both vectors in such a fashion that the coding region of the *GAL10* gene is *not* in the frame of that of *lacZ.* Thus, although the entire *lacZ* sequence is transcribed from the *GAL10* promoter, it is not translated. Thus no β-galactosidase activity is produced in cells harboring either plasmid. At the site of the fusion of the two genes is a unique *Sma*I site. This serves as a cloning site for introducing DNA carrying an open reading frame from a gene of interest. If the fragment is sufficiently large and is from completely within the coding region of a gene, then the fragment will most likely have only one open reading frame extending its entire length: that corresponding to the coding region of the gene from which it was taken. In addition, if this fragment is inserted into the vector in such a fashion that the open reading frame is in frame both with the *GAL10* sequences preceeding it and the *lacZ* sequences following it, then a trihybrid protein—consisting of epimerase sequences at the N-end, a portion of the gene product of interest in the middle, and β-galactosidase at the C-end—will be produced in yeast cells harboring the plasmid. Such a trihybrid protein can be isolated from extracts of the strain, with anti-β-galactosidase antibodies, and then used to inject rabbits to raise antisera against all the epitopes on the trihybrid protein. Such polyclonal sera should contain antibodies directed at that portion of the product of the cloned gene present in the hybrid.

In practice the YEp60s vectors are used in the following manner. A fragment of a reasonable size from within the coding limits of a gene of interest is isolated. Following mild treatment of the fragment with double-stranded exonuclease *Bal*31, used to randomize the ends, the fragment is ligated into the *Sma*I site of YEp61 or YEp62. Because the ends have been randomized, a subset of the cloned fragments will be in frame with both the *GAL10* gene and the *lacZ* gene. These in-frame insertions can be readily identified following transformation of an appropriate yeast strain. The products of ligation are first recovered in *E. coli*. Plasmid DNA from the pooled bacterial clones, or protoplasts of the bacteria themselves (Broach *et al.*, 1979a), can then be used to transform a GAL^+ $leu2^-$ strain to Leu^+ on media lacking leucine and containing galactose as well as XG, an indicator of β-galactosidase activity (Miller, 1972; Casadaban *et al.*, 1983). Transformants obtained with the parental plasmid or with any out-of-frame insertions will yield white transformants on XG indicator plates, whereas the in-frame insertions, producing the trihybrid protein, will yield blue colonies on XG indicator plates.

There are several noteworthy features to the use of YEp60s vectors. First, knowledge of the precise location of the gene of interest within a cloned DNA segment is not required. That a DNA fragment, when cloned into YEp61 or YEp62, gives rise to XG-positive transformant in yeast indicates that the segment contains an open coding region extending its entire length. Thus, if the fragment was taken from that portion of the cloned DNA to which the gene of interest had been approximately localized, then the hybrid protein produced almost certainly carries a portion of the protein encoded by the gene of interest. Second, it should be apparent that in using these vectors, the complete coding region of a gene is not needed in order to generate a hybrid protein that will elicit antisera reactive against the product of the gene of interest. Thus partial cDNA clones or single exons of an interrupted gene are adequate starting materials for generating the fusions. Finally, by using different segments of a cloned gene to make the fusions it is possible to raise polyclonal antisera that recognize only a specific subsection of the protein. Thus one can develop a set of immunological reagents, each directed against a different subset of epitopes of the protein of interest without purifying the protein or going through the difficulty and expense of developing monoclonal antibodies against it.

Similar vectors for the production of trihybrid proteins in *E. coli* have been developed by T. Silhavy and G. Weinstock (personal communication). In fact, the multilinker junction between *lacZ* and *GAL10* sequences in plasmid YEp62 is derived from one of their vectors. The relative merits of production of trihybrid proteins in yeast versus their production in *E. coli* have not been thoroughly evaluated. However, as we indicate in Section III,C, we have had success using our YEp60s vectors in yeast to generate hybrid proteins that induce antibodies against the products of genes in which we are interested.

C. Use of Yeast Expression Vectors for the Identification of the Yeast *REP1* Gene

We have been interested in the mechanism by which the yeast plasmid 2-μm circle can propagate stably and at high copy number in yeast. By genetic means we determined that this property requires two 2-μm-circle genes, which we have designated *REP1* and *REP2*. To proceed in our analysis of the replication properties of 2-μ circle, we have obtained antibodies against *REP1* protein, using a YEp60 vector essentially as described previously. In addition, we have placed the expression of the *REP1* gene under control of the *GAL10* promoter, which, in conjunction with the antiserum, has allowed us to identify and overproduce its gene product.

A portion of the *REP1* coding region was fused between the *GAL10* promoter region and the *lacZ* coding region of a YEp60 vector as outlined in Section II and described in detail elsewhere (M. Jayaram, S. Sumida, L. Wu, and J. Broach, unpublished). The resultant plasmid promotes the synthesis in yeast of a hybrid protein that can be precipitated by anti-β-galactosidase antisera and that migrates on SDS polyacrylamide gels at a position expected for a protein of 150,000 daltons, the size predicted for the *REP1–lacZ* fusion protein (data not shown). In addition, the protein is present in galactose- but not glucose-grown cells. We calculate, on the basis of the level of β-galactosidase activity in extracts of a galactose-grown strain harboring the plasmid that the hybrid protein constitutes approximately 0.3% of total soluble protein. This calculation assumes that the hybrid protein has approximately the same specific activity as authentic β-galactosidase. The hybrid protein was purified from such extracts by immunoprecipitation followed by preparative acrylamide-gel electrophoresis and then was injected into rabbits to elicit antisera.

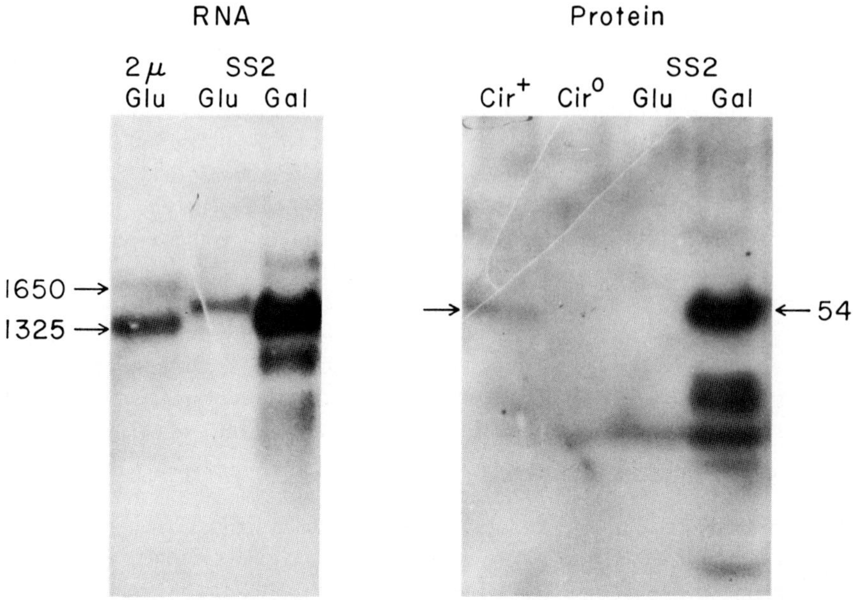

Fig. 6 Galactose-dependent expression of the *REP1* gene of 2-μm circle. Strain DC04, harboring either no plasmids [Cir^0], the wild-type 2-μm-circle plasmid (2μm or [Cir$^+$]), or plasmid pSS2, in which the *REP1* locus is fused to the *GAL10* promoter, was grown in either glucose-containing (Glu) or galactose-containing (Gal) media to ~2×10^7 cells/ml. Cells were harvested and total polyadenylated RNA or total protein was extracted as described in Section II. Polyadenylated RNA was fractionated on a formaldehyde–agarose gel, transferred to nitrocellulose, and hybridized with the ^{32}P-labeled 1315-bp *Hin*dIII fragment from the *REP1* gene of 2-μm circle (see Fig. 3). The autoradiogram of the filter is shown on the left, on which the 1325- and 1650-base transcripts from wild-type 2-μm circle are identified. Samples of total protein from the strains were fractionated on a 12.5% SDS-polyacrylamide gel, transferred to nitrocellulose, and incubated with anti-*REP1* antiserum and ^{131}I-labeled protein A as described in Section II. The autoradiogram of the filter is shown on the right. The size of the protein indicated by the arrows (kilodaltons) was determined by comparison of its position of migration with that of proteins of known molecular weight fractionated on the same gel.

Concurrent with the construction of the *REP1–lacZ* fusion, we inserted the entire coding region of *REP1* into an expression vector essentially identical to YEp51 (M. Jayaram, S. Sumida, L. Wu, and J. Broach, unpublished). This construction required resection of DNA corresponding to the 5′-flanking region of the coding region so that the first ATG codon in the fragment fused to the *GAL10* promoter was the initiating codon for the *REP1* coding sequence. The resulting plasmid, when introduced into yeast, yields high-level, galactose-inducible synthesis of the *REP1* gene product. These results are documented in Fig. 6. Extracts were obtained from a yeast strain harboring either no copies of the *REP1* gene (a [cir^0] strain, i.e., one lacking any 2-μm-circle plasmids), a normal complement of 2-μm circles ([cir$^+$] strain), or the plasmid with *REP1* coding sequences fused to the *GAL10* promoter (pSS2). The extracts were fractionated by electrophoresis on SDS-polyacrylamide gels and then electrotransferred to nitrocellulose, using a procedure that maintains the fractionation pattern of the protein extract (Towbin *et al.*, 1979; Renart *et al.*, 1979). We then incubated the filter with the antiserum obtained against the *REP1–lacZ* hybrid protein. Proteins recognized by antibodies in the antiserum were identified by subsequent incubation with ^{125}I-labeled protein A from *Staphylococcus aureus,* followed by autoradiography. As is evident in Fig. 6, the antibodies in the serum bind to a single 54-kdalton protein in an extract from a strain containing 2-μ circles (i.e., a strain containing the normal *REP1* gene) and binds to no proteins in an extract from a strain lacking 2-μm circles (i.e., a strain lacking the *REP1* gene). In addition, there is a protein of the same size in cells containing the *GAL10–REP1* fusion that is recognized by antibodies in this serum. This protein is produced in very large amounts when those cells are grown in the presence of galactose, but it is not produced when the cells are grown in the absence of galactose. The minor, smaller proteins that are recognized by the antiserum raised against the hybrid protein are proteolytic breakdown products of the 54-kdalton protein; they are not observed if suitable steps are taken to avoid proteolysis (data not shown).

We have also examined *REP1*-specific mRNA isolated from these strains, the results of which are also shown in Fig. 6. Total polyadenylated RNA was isolated from a [cir$^+$] strain or a [cir^0] strain containing pSS2 after growth either on glucose or galactose. Samples were fractionated on a denaturing agarose gel and transferred to nitrocellulose and the filter hybridized with a labeled DNA probe corresponding to the *REP1* coding region. The 1325-base *REP1* mRNA is evident in RNA isolated from the [cir$^+$] strain. In addition, a 1650- base transcript derived from the opposite strand (Broach *et al.*, 1979a) is also evident. In RNA isolated from the pSS2-containing strain grown on glucose, no *REP1* mRNA is seen. The only transcript from this region seen is the prematurely terminated 1650-base transcript derived from the noncoding strand (Fig. 6 and other data not shown). However, the 1325-base transcript is present at high levels in RNA from the galactose-grown, pSS2-containing strain. We calculate that the level of the 1325-base transcript in this strain is at least 20 times that found in the [cir$^+$] strain.

The results presented in Fig. 6 thus demonstrate that (1) the *REP1–lacZ* hybrid protein elicited antibodies against the *REP1* moiety of the hybrid, (2) the *REP1* mRNA and *REP1* protein are substantially overproduced when placed under control of the *GAL10* promoter, and (3) *REP1* expression from the *REP1–GAL10* fusion exhibits tight control by galactose. Thus the *GAL* promoter fragment we have isolated can be used as an effective transcriptional switch as well as a means for overproduction of specific gene products. In addition, the protocol for production of antisera from hybrid proteins synthesized in yeast is efficient and specific.

IV. Summary

In this chapter we have described and evaluated approaches to expression of cloned genes in yeast. In Section III,A of the chapter we examined those elements that influence expression of genes in yeast, primarily in the context of designing optimally efficient expression vectors. These elements include promoter sequences, polyadenylation and/or transcription-termination sequences, sequences influencing sites and rates of translation initiation, intervening sequences, sequences promoting secretion of proteins from yeast, sequences for autonomous replication of yeast plasmids, and sequences for selection in yeast. In Section III,B we described two classes of yeast expression vectors, both designed around an inducible promoter derived from *GAL10,* a gene required for the metabolism of galactose in yeast. The first class of vectors, represented by plasmids YEp51 and YEp52, are designed to yield high-level, galactose-inducible synthesis of the authentic product of a cloned gene in yeast. The second class, represented by plasmids YEp61 and YEp62, are designed to produce reasonable quantities in yeast of a trihybrid protein, following insertion of a portion of the coding region of the gene of interest into the plasmid, between sequences from the proximal portion of *GAL10* and sequences from the distal portion of the *lacZ* gene of *Escherichia coli.* The trihybrid protein synthesized in yeast from such a fusion can purified with anti-β-galactosidase antiserum and then used to raise antibodies against that portion of the product of the gene of interest present in the hybrid. Thus these vectors provide a rapid and convenient method for developing an immunological assay for the product of a cloned gene, which provides for its identification and purification. In Section III,C we documented the efficacy of these vectors by describing their use in the identification and hyperexpression of the product of the *REP1* gene of yeast.

Appendix: Plasmid Constructions

A. Isolation of *GAL10* Promoter Fragments

A sample (20 μgm) of plasmid pNN78 DNA, which consists of a 2-kb *Eco*R1 fragment spanning the *GAL1–GAL10* intercistronic region inserted in the *Eco*R1 site of pBR322 and oriented with the *GAL10* sequences nearer Tet^R (see Fig. 1), was digested with *Hin*dIII and then incubated with a titrated amount of *Bal*31 sufficient to remove an average of 150 bp from each end. *Sal*I 8-nucleotide linkers were ligated to the resected ends. The plasmids were then recircularized at low DNA concentration, following digestion with *Sal*I, and used to transform *E. coli* strain C600. The approximate locations of the *Sal*I linker in plasmids from a number of transformants were determined by restriction analysis of plasmid DNA recovered from 1.5-ml overnight cultures of individual colonies (Birnboim and Doly, 1979). For several of the plasmids the precise location of the *Sal*I site was determined by sequence analysis using plasmid DNA labeled at the *Sal*I site (Maxam and Gilbert, 1977).

B. Construction of Plasmids YEp51 and YEp52

The construction of these plasmids is diagrammed in Fig. 7. Plasmid YEp51 was obtained from the simultaneous ligation of three purified DNA fragments: one spanning the *LEU2* gene, one containing the *GAL10* promoter, and one consisting of pBR322 and 2-μm-circle sequences. The yeast *LEU2* gene is located on a *Sal*I-to-*Xho*I genomic restriction fragment, which has been sequenced (Sutcliff, personal communication). In a plasmid carrying this fragment, we converted the *Sal*I site to a *Bam*Hl site, using *Bam*Hl 8-nucleotide linkers, to generate plasmid pC4B (Broach, 1983). From pC4B we could recover a 1900-bp *Bam*Hl-to-*Hpa*I restriction fragment completely encompassing the *LEU2* gene. The *GAL10* promoter could be recovered as a 500-bp *Sau*3A-to-*Sal*I fragment from plasmid pNN78-Δ4 (see Fig. 1). Finally, from plasmid pCV4, which consists of pBR322 containing the small *Eco*R1 fragment of the B form of 2-μm circle (Broach *et al.*, 1979b; see Fig. 3), we could recover a 4450-bp *Pvu*II-to-*Sal*I fragment that spans the *Col*E1 origin, the β-lactamase gene, and the entire 2-μm-circle *Eco*R1 fragment. Therefore, we disgested 5 μgm each of pNN78-Δ4, pC4B, and pCV4, respectively, with *Sal*I plus *Sau*3A, *Bam*H1 plus *Hpa*I, and *Pvu*II plus *Sal*I and fractionated the samples by electrophoresis on a Sigma type-VII low-melting-point agarose gel. Following electrophoresis, an agarose slice containing the desired fragment from each sample was suspended in 2 volumes of TE buffer plus 0.2 *M* NaCl and incubated at 65°C until melted. The three solutions were mixed and extracted with 65°C buffer-saturated phenol. The aqueous phase was then extracted twice more with phenol at room temperature. Two volumes of ethanol were added to the aqueous phase, and, after incubation at −70°C for 0.5 hr or more, the DNA was recovered by centrifugation at 50 k*g* for 30 min. The DNA fragments were dissolved in 0.02 ml of ligation buffer and incubated with T4 DNA ligase for 12 hr at 12°C. The ligation mixture was used to transform *E. coli* strain C600 to ampicillin resistance and leucine prototrophy, and the structure of plasmids in individual transformants was determined by restriction analysis. One plasmid with the structure diagrammed in Figs. 4 and 7 was retained and designated plasmid YEp51.

Plasmid YEp52 was obtained by digestion of 5 μgm of YEp51 DNA with *Sal*I

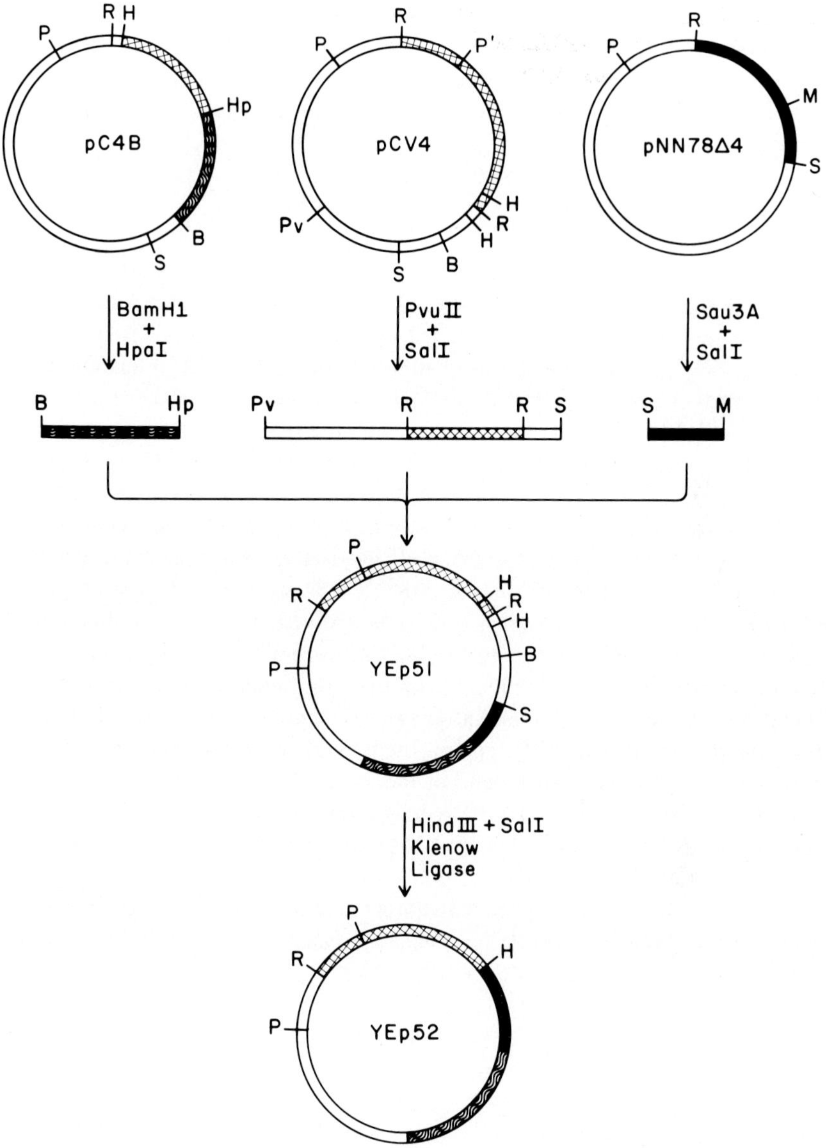

Fig. 7 Protocol for the construction of plasmids YEp51 and YEp52. A detailed description of these constructions is provided in the text. Restriction sites: R, *Eco*R1; H, *Hind*III; B, *Bam*Hl; S, *Sal*I; M, *Mbo*I (*Sau*3A), Hp, *Hpa*I. Sequence sources: pBR322, ; 2-μm circle, ; *LEU2*, ; *GAL10*, .

plus *Hin*dIII followed by incubation with the Klenow fragment of *E. coli* DNA polymerase I (Boehringer–Mannheim) and all four deoxynucleotide triphosphates. Following ligation at low DNA concentration, plasmids were recovered in *E. coli,* and one having the structure indicated in Figs. 4 and 7 was retained.

C. Construction of YEp61

The construction of this plasmid is diagrammed in Fig. 8. Plasmid pSI5, which consists of pBR322 carrying a 3.9-kb *Hin*dIII fragment from pJDB219 spanning the origin, *REP3,* and *LEU2* (Broach, 1983) was digested with *Pst*I plus *Bam*Hl, mixed with similarly digested pLG200 DNA (Guarante *et al.*, 1980), and incubated with T4 DNA ligase. Plasmid pLG200 carries the 3′ moiety of *lacZ,* extending from a *Bam*Hl site—inserted at codon 9—through the end of the gene. From this ligation a plasmid conferring ampicillin resistance and leucine prototrophy and carrying *lacZ* sequences was recovered and designated pJRB10. Concurrent with this construction, the 2-kb *Eco*R1 fragment spanning the *GAL1–GAL10* region was inserted into the *Eco*R1 site in the polylinker region of phage M13-mp8 (Messing, 1981). One such recombinant phage, with *GAL10* in the same orientation as *lacZ,* was retained. As a result of this construction, the *Eco*R1 site at codon 46 of *GAL10* lies immediately adjacent to a *Sma*I site in the M13 vector, which itself is immediately adjacent to a *Bam*H1 site. We therefore digested the recombinant phage with *Sau*3A to recover a 660-bp fragment containing the *GAL10* promoter, the first 46 codons of *GAL10,* and *Eco*R1, *Sma*I, and *Bam*H1 sites at the end of the fragment. This fragment was inserted into *Bam*H1-digested pJRB10 DNA to generate plasmid pJRB17. Because the sequences of *GAL10,* pLG400, and M13-mp8 predicted that this construction would yield an in-frame fusion of *GAL10* to *lacZ,* we were able to confirm the structure of the plasmid by demonstrating that Leu$^+$ yeast transformants obtained with it were blue on XG + galactose plates and white on XG + glucose plates. Plasmid YEp62 was then obtained by digesting pJRB17 with *Bam*Hl, filling in the 4-bp 5′ extensions with the Klenow fragment of *E. coli* DNA polymerase, and resealing the plasmid with T4 DNA ligase.

D. Construction of YEp62

This construction is outlined in Fig. 9. Plasmid pC4B consists of pBR322 in which the *Eco*R1 to *Bam*H1 sequences are replaced by *LEU2* sequences and sequences from 2-μm circle spanning *REP3* and the origin of replication. We replaced the *Bam*H1-to-*Sal*I sequences in pC4B with the 530-bp *Sau*3A-to-*Sal*I fragment from pNN78-Δ45 (see Fig. 1), which spans the *GAL10* promoter and

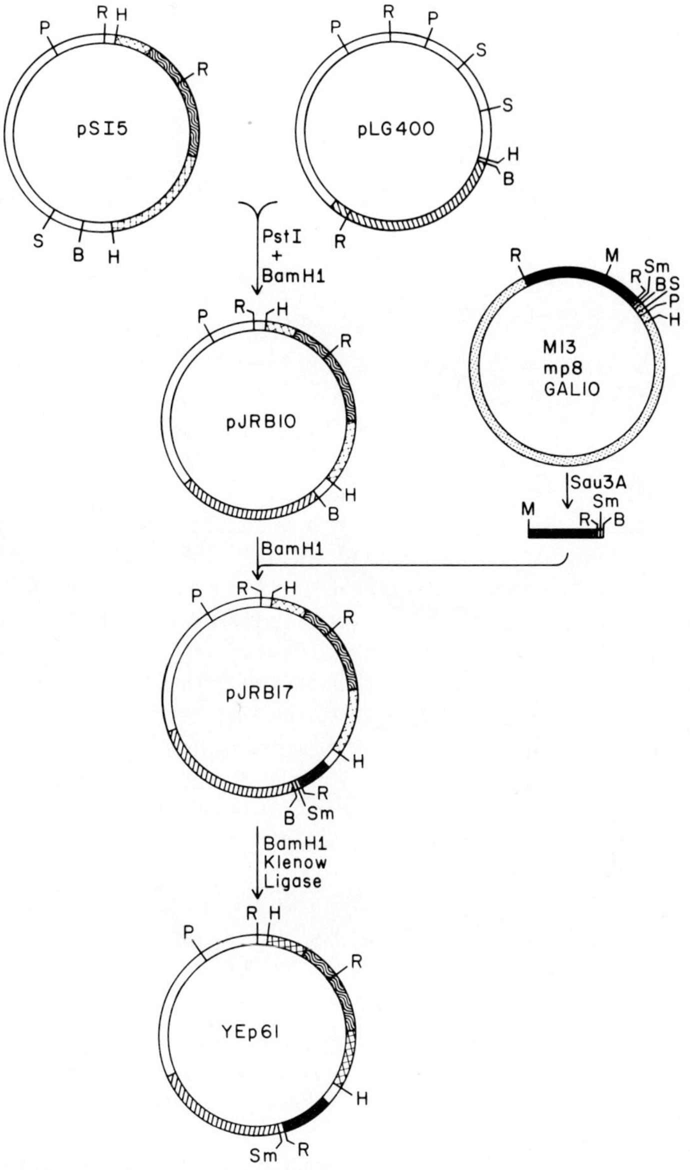

Fig. 8 Protocol for construction of plasmid YEp61. A detailed description of this construction is provided in the text. Restriction sites: R, *Eco*RI; H, *Hin*dIII; B, *Bam*HI; S, *Sal*I; P, *Pst*I; M, *Mbo*I (*Sau*3A); Sm, *Sma*I. Sequence sources: pBR322, ; 2-μm circle, ; *LEU2* ; *lacZ*, ; *GAL10*, ; M13, .

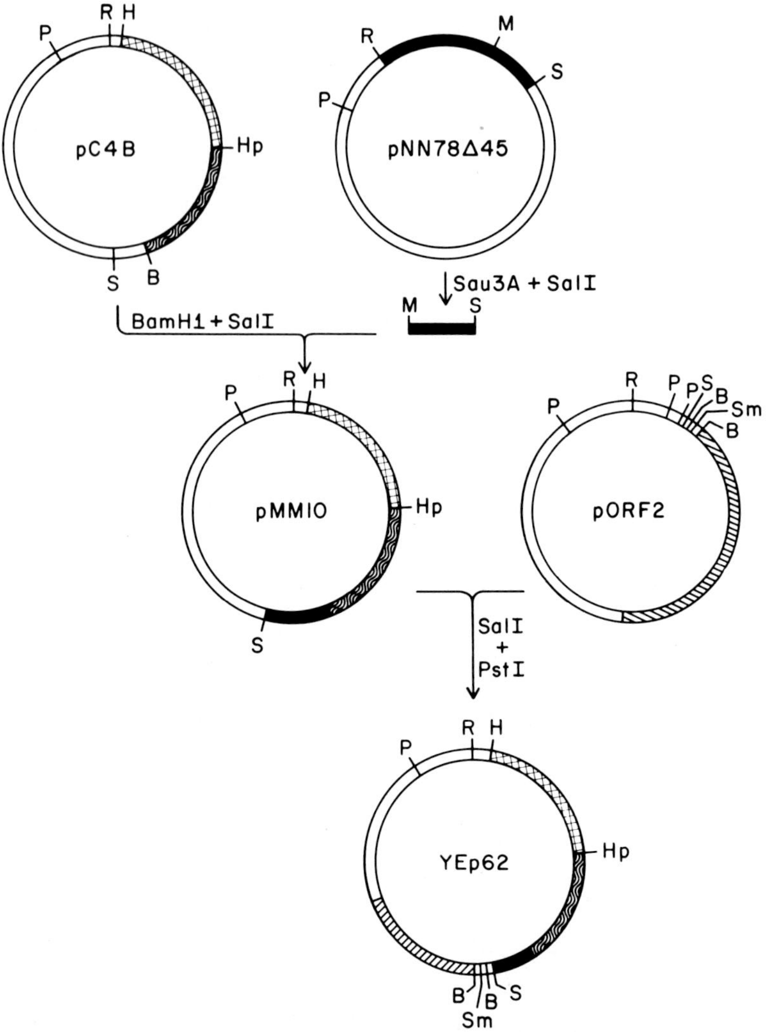

Fig. 9 Protocol for construction of plasmid YEp62. A detailed description of this construction is provided in the text. Restriction sites: R, *Eco*R1; H, *Hin*dIII; B, *Bam*HI; S, *Sal*I; P, *Pst*I; M, *Mbo*I (*Sau*3A); Sm, *Sma*I, Hp, *Hpa*I. Sequence sources: pBR322, ; 2-μm circle, ; *LEU2*, ; *lacZ*, ; *GAL10*, .

the first two codons of the *GAL10* coding region. The resulting plasmid was designated pMM10. We recovered plasmid YEp62 by ligating *Pst*I-plus-*Sal*I-digested pMM10 DNA with similarly digested pORF-2 DNA obtained from G. Weinstock, followed by *E. coli* transformation and restriction analysis of plasmids from individual clones.

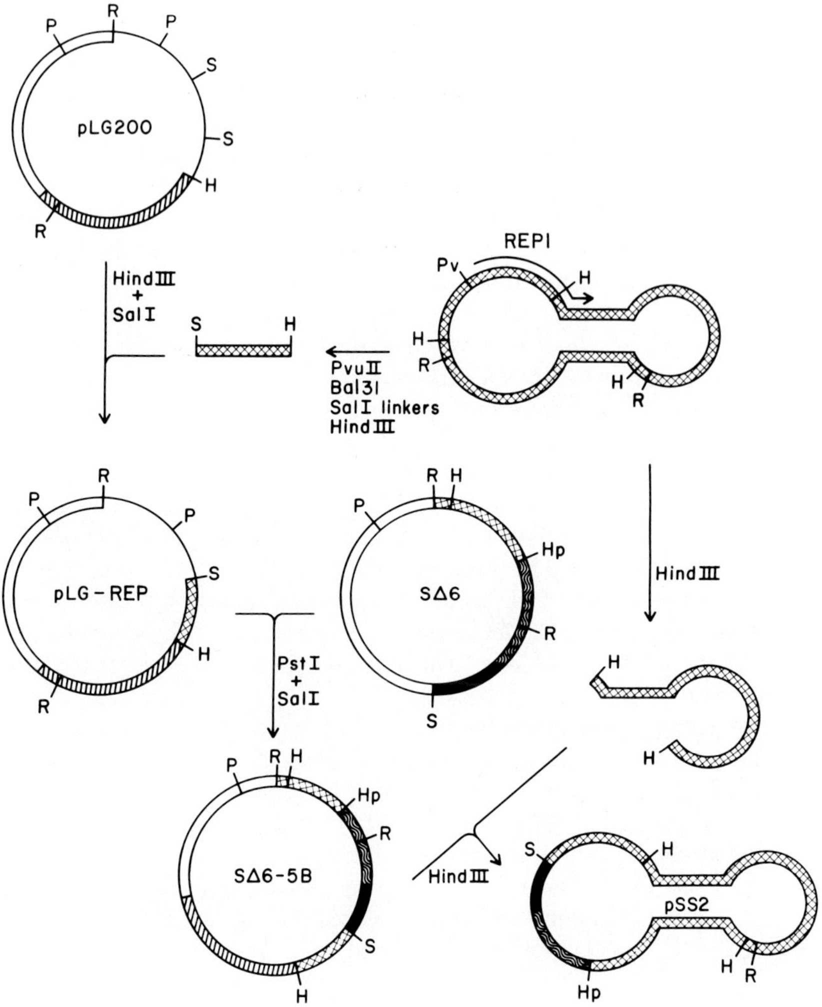

Fig. 10 Protocol for construction of plasmid pSS2. A detailed description of this construction is provided in the text. Restriction sites: R, *Eco*RI; H, *Hin*dIII; S, *Sal*I; P, *Pst*I; Hp, *Hpa*I; Pv, *Pvu*II. Sequence sources: pBR322, ▭; 2-μm circle, ▨; *LEU2*, ▨; *lacZ*, ▨; *GAL10*, ■.

E. Construction of *GAL10–REP1–lacZ* and *GAL10–REP1* Fusions

These constructions are outlined in Fig. 10. The first stage involved insertion of a *Sal*I site immediately upstream from the *REP1* coding region and fusion of the *REP1* coding region to *lacZ*. This was accomplished by cloning the 800-bp

*Pvu*II-to-*Hin*dIII fragment of 2-μm circle, containing the proximal half of *REP1*, into the *Pvu*II/*Hin*dIII sites of pBR322. The *Pvu*II site lies only 40 bp upstream from the first ATG of the *REP1* coding region. However, a second ATG codon lies between the *Pvu*II site and the initiating ATG, translation from which would terminate after four codons. Therefore, we digested the plasmid with *Pvu*II, resected with BAL31, and resealed in the presence of *Sal*I linkers. From plasmids recovered by this procedure we identified one in which the *Sal*I site has been inserted between the two ATG sequences. We then inserted this fragment into *Sal*I-plus-*Hin*dIII-digested pLG200 DNA (Guarente *et al.*, 1980). From the sequence both of *REP1* (Hartley and Donelson, 1980) and of the region of pLG200 spanning the *Hin*dIII site we could predict that this ligation would yield an in-frame fusion of *REP1* to *lacZ*.

To provide a vehicle from propagation and expression of the *REP1–lacZ* fusion, we constructed plasmid SΔ6. Plasmid SΔ6 is identical to plasmid pMM10 except that *GAL10* sequences were obtained from pNN78-Δ6, in which the *Sal*I site lies upstream from the *GAL10* ATG. Thus, although transcription can be initiated from the promoter fragment in this plasmid, translation cannot. Plasmids pLG-*REP* and SΔ6 were digested with *Pst*I plus *Sal*I, mixed, and ligated. From this ligation we recovered plasmid SΔ6-5B, with the structure indicated in Fig. 10. In yeast this plasmid yields galactose-dependent transcription of *REP1–lacZ*, translation of which begins at the first ATG of the *REP1* coding region.

From SΔ6-5B we obtained a plasmid with a complete *REP1* gene fused to the *GAL10* promoter. The *Hin*dIII fragment from plasmid SΔ6-5B spanning *REP1*, *GAL10*, *LEU2*, and the remaining 2-μm-circle sequences was recovered and ligated to the purified 2700-bp *Hin*dIII fragment from 2-μm circle. The products of this ligation were recovered by transformation of the yeast strain DC04 [cir^0] to leucine prototrophy. One such transformant was shown to contain a plasmid, designated pSS2, with the structure indicated in Fig. 10 (M. Jayaram, S. Sumida, L. Wu, and J. Broach, unpublished).

Acknowledgments

We thank Drs. George Weinstock and Tom St. John for their generous gifts of various plasmid and phage samples. We would also like to thank them and Drs. Roger Yokum, Mark Johnston, Malcolm Casadaban, and Allan Laughon for communicating unpublished results. Vicki Guarascio, Sajida Ismail, Maureen McCloud, and Mark Marshall provided excellent technical assistance with various projects described in this communication. This work was supported in part by an NIH grant to JRB, who is an Established Investigator of the American Heart Association.

References

Adams, B. G. (1972). *J. Bacteriol.* **111,** 308–315.

Ammerer, G. (1982). *Methods Enzymol.* **101,** 192–201.

Beach, D., Piper, M., and Shall, S. (1980). *Nature (London)* **284,** 185–187.

Beggs, J. D. (1978). *Nature (London)* **275,** 104–109.

Beggs, J. D. (1981). *Alfred Benzon Symp.* **16,** 383–390.

Beggs, J. D., van den Berg, J., Van Ooyen, A., and Weissman, C. (1980). *Nature (London)* **283,** 835–840.

Bennetzen, J. M., and Hall, B. D. (1982a). *J. Biol. Chem.* **257,** 3026–3031.

Bennetzen, J. M., and Hall, B. D. (1982b). *J. Biol. Chem.* **257,** 3082–3091.

Birnboim, H. A., and Doly, J. (1979). *Nucleic Acids Res.* **7,** 1513–1523.

Botstein, D., and Davis, R. W. (1982). *In* "The Molecular Biology of the Yeast *Saccharomyces:* Metabolism and Gene Expression" (J. N. Strathern, E. W. Jones, and J. R. Broach, eds.), pp. 607–638. Cold Spring Harbor Lab., Cold Spring Harbor, New York.

Broach, J. R. (1979). *J. Mol. Biol.* **131,** 41–53.

Broach, J. R. (1981). *In* "The Molecular Biology of the Yeast *Saccharomyces:* Life Cycle and Inheritance" (J. N. Strathern, E. W. Jones, and J. R. Broach, eds.), pp. 445–470. Cold Spring Harbor Lab., Cold Spring Harbor, New York.

Broach, J. R. (1983). *Methods Enzymol.* **101,** 307–325.

Broach, J. R., and Hicks, J. B. (1980). *Cell* **21,** 501–508.

Broach, J. R., Atkins, J. A., McGill, C., and Chow, L. (1979a). *Cell* **16,** 827–839.

Broach, J. R., Strathern, J. N., and Hicks, J. B. (1979b). *Gene* **8,** 121–133.

Carlson, M., Taussig, R., Kustu, S., and Botstein, D. (1983). *Mol. Cell. Biol.* **3,** 439–447.

Casadaban, M. J., Martinez-Arias, A., Shapira, S. K., and Chou, J. (1982). *Methods Enzymol.*

Chan, C. S. M., and Tye, B.-K. (1980). *Proc. Natl. Acad. Sci. USA* **77,** 6329–6333.

Clark-Walker, G. D., and Miklos, G. L. G. (1974). *Eur. J. Biochem.* **41,** 359–372.

Cohen, J. D., Eccleshall, T. R., Needleman, R. B., Federoff, H., Buchferer, B. A., and Marmur, J. (1980). *Proc. Natl. Acad. Sci. USA* **77,** 1078–1082.

Dobson, M. J., Tuite, M. F., Roberts, N. A., Kingsman, A. J., Kingsman, S. M., Perkins, R. E., Conroy, S. C., Dunbar, B., and Fothergill, L. A. (1982). *Nucleic Acids Res.* **10,** 2625–2637.

Faye, G., Leung, D. W., Tatchell, K., Hall, B. D., and Smith, M. (1981). *Proc. Natl. Acad. Sci. USA* **78,** 2258–2262.

Fraenkel, D. G. (1982). *In* "The Molecular Biology of the Yeast *Saccharomyces:* Metabolism and Gene Expression" (J. N. Strathern, E. W. Jones, and J. R. Broach, eds.), pp. 1–38. Cold Spring Harbor Lab., Cold Spring Harbor, New York.

Fraser, N. W., Nevins, J. R., Ziff, E., and Darnell, J. E. (1979). *J. Mol. Biol.* **129,** 643–656.

Gallwitz, D., and Sures, I. (1980). *Proc. Natl. Acad. Sci. USA* **77,** 2546–2550.

Gerbaud, C., and Guerineau, M. (1980). *Curr. Genet.* **1,** 219.

Gething, M. J., and Sambrook, J. (1981). *Nature (London)* **293,** 620–625.

Gruss, P., and Khoury, G. (1981). *Proc. Natl. Acad. Sci. USA* **78,** 133–137.

Guarente, L., and Ptashne, M. (1981). *Proc. Natl. Acad. Sci. USA* **78,** 2199–2203.

Guarente, L., Lauer, G., Roberts, J. M., and Ptashne, M. (1980). *Cell* **20,** 543–553.

Hall, B. D., and Sentenac, A. (1982). *In* "The Molecular Biology of the Yeast *Saccharomyces:* Metabolism and Gene Expression" (J. N. Strathern, E. W. Jones, and J. R. Broach, eds.), pp. 561–606. Cold Spring Harbor Lab., Cold Spring Harbor, New York.

Hamer, D. H., and Leder, P. (1979). *Cell* **18,** 1299–1302.

Hartley, J. L., and Donelson, J. E. (1980). *Nature (London)* **286,** 860–865.

Hicks, J. B., Hinnen, A., and Fink, G. R. (1978). *Cold Spring Harbor Symp. Quant. Biol.* **43,** 1305–1312.

Hitzeman, R. A., Clarke, L., and Carbon, J. (1980). *J. Biol. Chem.* **255,** 12073–12080.

Hitzeman, R. A., Hagie, F. E., Levine, H. L., Goeddel, D. V., Ammerer, G., and Hall, B. D. (1981). *Nature (London)* **293,** 717–722.

Hitzeman, R. A., Leung, D. W., Perry, L. J., Kohr, W. J., Hagie, F. E., Chen, C. Y., Lugovoy, J. M., Singh, A., Levine, H. L., Wetzel, R., and Goeddel, D. V. (1982). *In* "Recent Advances in Yeast Molecular Biology: Recombinant DNA" pp. 173–190. Univ. of California Press, Berkeley.

Hofer, E., and Darnell, J. E. (1981). *Cell* **23,** 585–593.

Holland, J. P., and Holland, M. J. (1979). *J. Biol. Chem.* **254,** 9839–9843.

Holland, J. P., and Holland, M. J. (1980). *J. Biol. Chem.* **255,** 2596–2605.

Holland, M. J., and Holland, J. P. (1978). *Biochemistry* **17,** 4900–4907.

Holland, M. J., Holland, J. P., Thill, G. P., and Jackson, K. A. (1981). *J. Biol. Chem.* **256,** 1385–1395.

Jayaram, M., Li, Y.-Y., and Broach, J. R. (1983). *Cell,* in press.

Jiminez, A., and Davies, J. (1980). *Nature (London)* **287,** 869–871.

Jones, E. W. (1983). *In* "Yeast Genetics: Fundamental and Applied Aspects" (J. Spenser, D. Spenser, and A. Smith, eds.), pp. 167–203. Springer-Verlag, New York.

Kessler, J. (1975). *J. Immunol.* **115,** 1617–1628.

Klar, A. J. S., and Halvorson, H. O. (1976). *J. Bacteriol.* **125,** 379–381.

Kojo, H., Greenberg, B. D., and Sugino, A. (1981). *Proc. Natl. Acad. Sci. USA* **78,** 7261–7265.

Kozak, M. (1981). *Curr. Top. Microbiol. Immunol.* **93,** 81.

Kreil, G. (1981). *Annu. Rev. Biochem.* **50,** 317–348.

Laemmli, U. (1970). *Nature (London)* **227,** 680–685.

Laughon, A., and Gesteland, R. F. (1982). *Proc. Natl. Acad. Sci. USA* **79,** 6827–6831.

Mandel, M., and Higa, A. (1970). *J. Mol. Biol.* **53,** 159–170.

Maxam, A. M., and Gilbert, W. (1977). *Proc. Natl. Acad. Sci. USA* **74,** 560–564.

Messing, J. (1981). *Proc. Cleveland Symp. Macromol. 3rd,* pp. 143–153.

Miller, J. (1972). "Experiments in Molecular Biology." Cold Spring Harbor Lab., Cold Spring Harbor, New York.

Mortimer, R. K., and Schild, D. (1981). *In* "The Molecular Biology of the Yeast *Saccharomyces:* Life Cycle and Inheritance" (J. N. Strathern, E. W. Jones, and J. R. Broach, eds.), pp. 11–26. Cold Spring Harbor Lab., Cold Spring Harbor, New York.

Mowshowitz, D. B. (1979). *J. Bacteriol.* **137,** 1200–1207.

Mulligan, R. C., Howard, H., and Berg, P. (1979). *Nature (London)* **277,** 108–114.

Newlon, R. G., Devenish, R. J., Suci, P. A., and Roffis, C. J. (1981). *ICN-UCLA Symp. Mol. Cell. Biol.* **22,** 501–512.

Ng, R., and Abelson, J. (1980). *Proc. Natl. Acad. Sci. USA* **77,** 3912–3916.

Oshima, Y. (1982). *In* "The Molecular Biology of the Yeast *Saccharomyces:* Metabolism and Gene Expression" (J. N. Strathern, E. W. Jones, and J. R. Broach, eds.), pp. 159–180. Cold Spring Harbor Lab., Cold Spring Harbor, New York.

Renart, J., Reisen, J., and Stark, G. R. (1979). *Proc. Natl. Acad. Sci. USA* **76,** 3116–3120.

Rosbash, M., Harris, R. K. W., Woolford, J. L., and Teem, J. L. (1981). *Cell* **24,** 679–686.

Rose, M., Casadaban, M. J., and Botstein, D. (1981). *Proc. Natl. Acad. Sci. USA* **78,** 2460–2464.

Rykowski, M. C., Wallis, J. W., Choe, J., and Grunstein, M. (1981). *Cell* **25,** 477–487.

St. John, T. P., and Davis, R. W. (1981). *J. Mol. Biol.* **152,** 285–316.

Schekman, R., and Novick, P. (1982). *In* "The Molecular Biology of the Yeast *Saccharomyces:* Metabolism and Gene Expression" (J. N. Strathern, E. W. Jones, and J. R. Broach, eds.), pp. 361–398. Cold Spring Harbor Lab., Cold Spring Harbor, New York.

Sherman, F., and Stewart, J. W. (1982). *In* "The Molecular Biology of the Yeast *Saccharomyces:* Metabolism and Gene Expression" (J. N. Strathern, E. W. Jones, and J. R. Broach, eds.), pp. 301–334. Cold Spring Harbor Lab., Cold Spring Harbor, New York.

Sigurdson, D. C., Gaarder, M. E., and Livingston, D. M. (1981). *Mol. Gen. Genet.* **183,** 59–65.

Stinchcomb, D. T., Struhl, K., and Davis, R. W. (1979). *Nature (London)* **282,** 39–45.

Struhl, K. (1981). *Proc. Natl. Acad. Sci. USA* **78,** 4461–4465.

Struhl, K. (1982). *Proc. Natl. Acad. Sci. USA* **79,** 7385–7389.

Struhl, K., Stinchcomb, D. T., Scherer, S., and Davis, R. W. (1979). *Proc. Natl. Acad. Sci. USA* **76,** 1035–1039.

Thomas, P. (1980). *Proc. Natl. Acad. Sci. USA* **77,** 5201–5202.

Towbin, H., Staehelin, T., and Gordon, J. (1979). *Proc. Natl. Acad. Sci. USA* **76,** 4350–4354.

Zaret, K. S., and Sherman, F. (1982). *Cell* **28,** 563–573.

CHAPTER 6

Genetic Engineering of Plants by Novel Approaches

JOHN D. KEMP

Agrigenetics Advanced Research Laboratory
Madison, Wisconsin

I. Introduction

One of the most important events in all of human history occurred perhaps 10,000 yr ago when man stopped merely gathering food and became a farmer. That was probably the beginning of our genetic engineering. Through breeding programs, man domesticated countless plants and animals that have been of service in any number of ways through the centuries. A classical breeding program can be divided into two parts: (1) the creation of genetic diversity by merely crossing diverse individuals and (2) a conscious effort on the part of the breeder to select those offspring that show some useful trait. However, a classical breeding program has one major limitation. It depends on sexual compatibility. Therefore, successful crosses usually occur only within a species. This limitation narrows the gene pool available to the breeder to that of the species.

Experimental Manipulation
of Gene Expression

ISBN 0-12-372380-9

Table I Amino Acids Used by Humans for Protein Synthesis

Essential	Nonessential
Phenylalanine	Serine
Leucine	Proline
Isoleucine	Alanine
Valine	Tyrosine
Methionine	Histidine
Cysteine	Glutamate
Threonine	Aspartate
Tryptophan	Arginine
Lysine	Glycine

There are obvious disadvantages to limiting oneself to the individual species. An example of a major disadvantage concerns human nutrition. We are aware that as human beings we require a number of amino acids in our diet. There are 11 essential amino acids that must be provided by the proteins in our diet (Table I). The remaining amino acids needed for protein synthesis are synthesized from metabolites. Not only must the 11 essential amino acids be provided in the human diet, they must be provided in just the right proportions. If the 11 amino acids are provided in a protein that is of high quality (e.g., egg or beef protein), an adult human requires approximately 30 gm of protein per day. The results are very different if dietary protein is provided by plant material. If the 30 gm of protein are provided by wheat, that will not fulfill our daily protein requirement. This is because wheat protein is very low in the essential amino acid lysine. As a consequence, it will limit the use of all the other essential amino acids; 30 gm of wheat protein is the equivalent of about 12–13 gm of high-quality protein. All of the classical breeding programs together will not improve this situation because within wheat species, wheat storage proteins are all low in lysine.

Similar amino acid deficiencies occur in all plant species. However, it is not always the same amino acid that is deficient. As a further example, bean protein is sufficient in lysine but is very low in the sulfur amino acids methionine and cysteine. Again, 30 gm of bean protein have an effective utilization yield of only 14 gm. There is not a single plant storage protein that provides a high-quality source of dietary protein.

Man has circumvented this problem by mixing protein from various sources. A good combination is wheat and bean protein because each will complement the deficiency of the other. In fact, a mixture of 15 gm of wheat protein and 15 gm of bean protein (total 30 gm) is a high-quality source of protein for nutrition. The result is the efficient use of plant proteins. It does, however, require a conscious effort to mix the proper proteins from the proper sources. A more detailed treatment of protein nutrition has been presented by Bressani and Elias (1974), Hegsted (1976), Scrimshaw (1976), and Scrimshaw and Young (1976).

II. Novel Approaches to Creating Genetic Diversity

The preceding discussion illustrates how a classical breeding program cannot solve a serious deficiency. There are many other examples of how difficult or impossible it is to improve certain qualities of plants through a classical breeding program. Thus it becomes necessary to think in terms of engineering plants by novel approaches.

A. Cell Culture

Some of the novel approaches that are being considered are discussed in the following sections. Most of the approaches take advantage of a unique property of plants. That property is totipotency, the ability of a single plant cell growing in tissue culture to generate a whole plant.

1. Cell Culturing

Many plant cells can be grown in tissue culture on a defined medium containing phytohormones (Helgeson, 1968). Such cells usually grow as an undifferentiated mass or callus. If the callus is treated with the proper ratio of hormones, it will organize into shoots. By changing the hormones the shoots will root, thereby producing a normal plant (Helgeson, 1968). Occasionally, plants will result that have altered chromosome numbers. These plants are usually abnormal, not showing any useful properties.

2. Regeneration of Protoplasts

Protoplasts are plant cells that have had their cell walls removed by treatment with cellulytic enzymes (Nagata and Takebe, 1970). They are analogous to yeast spheroplasts. Merely making plant protoplasts and then regenerating those protoplasts will occasionally give recombinations that may be useful. An example is the work of Shepard *et al.* (1980). They made protoplasts from potato cells, then regenerated those protoplasts back to whole potato plants. Some of the regenerated plants had increased resistance to *Phytophthora infestans* that normally could not have been acquired by a classical breeding program. The mechanism by which the potato plants acquired the resistance is uncertain, as is the applicability to other plants. Cell-culturing techniques can release phenotypes that may not normally be released by classical breeding. These techniques, however, will not introduce new genes.

B. Protoplast Fusions

There was considerable excitement when protoplasts from two different species were first fused (Melchers, 1977). It was thought that this technique might be used to create new genetic diversity. However, stable fusion products occur only between closely related species. Fusions of distantly related species results in elimination of one of the chromosome sets; the end result is one of the original species. Thus incompatibilities appear to exist even at the chromosome level. Some success at crossing the species barrier has occurred starting with a protoplast in which the chromosomes have been fragmented (Jinks *et al.,* 1981). Another possibility is to fuse protoplasts with organelles. These techniques have not been fully explored, but they do hold some promise.

C. Recombinant DNA Vehicles

Ultimately, the isolation and transfer of individual genes will circumvent the incompatibility barrier. This approach requires both the identification and isolation of useful genes, as well as a vehicle for their transfer to another plant species. There are a number of potentially useful vehicles for the transfer of DNA to plant cells.

1. Circular Mitochondrial DNA

Researchers have found small, circular DNA molecules replicating in the mitochondria of corn (*Zea mays;* Kim *et al.,* 1982). These autonomously replicating molecules could serve as a vehicle for the transfer of characterized genes. However, there are many hurdles that must be overcome. These DNA molecules must be isolated and characterized. A location within them must be identified where foreign DNA can be introduced. The location must allow the foreign gene to be expressed. Finally, a procedure must be available for reintroduction of the vehicle back into the plant cell. The advantages are that these vehicles are small, easily manipulatable pieces of DNA, and they replicate naturally in the plant cell without producing adverse effects, as do some of the following vehicles.

2. Chloroplast DNA

The chloroplast genome is not as small and potentially as manipulatable as mitochondrial DNA, but it has been well characterized (Bedbrook and Kolodner,

1979). It may be possible to identify areas within the chloroplast DNA that can be replaced with genes of our choice. Again, the problems are where to introduce our gene and how to reintroduce the chloroplast DNA back into the plant.

3. Transformation

As pointed out previously, reintroduction of DNA vehicles back into the plant cell has yet to be accomplished. Most of the effort has centered on reintroduction through transformation of plant protoplasts in a manner similar to transformation of yeast spheroplasts. The advantage that yeast transformation systems have is possession of a selectable marker. In other words, a rare transformant can be identified by selecting for a marker carried on the transforming DNA. A selectable marker in higher plants has not been developed. Auxotrophy is often considered, but stable auxotrophy in higher plants is very difficult to identify.

Another approach is to use antibiotic resistance, because these genes have been cloned. They, however, are of bacterial origin and are not expressed in higher plant cells. We are modifying the neomycin phosphotransferase II (NPTII) genes that instill resistance to the antibiotic kanamycin (Km). The modification includes attaching a eukaryotic promoter to the resistance gene, creating a vehicle that can be used in protoplast transformations.

4. Plant Viruses

The problems of reintroduction can be circumvented by using a vehicle that naturally introduces itself into the plant cell. Such vehicles are the DNA plant viruses (e.g., caulimoviruses and gemini viruses). These molecules have the disadvantage of causing disease.

The gemini viruses are small, single-stranded, circular DNA viruses (Goodman, 1981). However, we do not know enough about them to say whether or not they can be used as a vehicle for transferring DNA.

The caulimoviruses are the only known double-stranded circular DNA viruses in plants. The best-known virus in this group is cauliflower mosaic virus (CaMV). CaMV has been extensively studied; its DNA has been sequenced, and many of the functions in the virus have been identified (Shepherd, 1979). The DNA has been cloned, and it is small enough to make a good vehicle for engineering. Unfortunately, CaMV appears to be a highly conserved virus in that its DNA can tolerate little increase in size without losing its virulence (Shepherd, 1979). Researchers have introduced about 200–400 bp of added DNA to the CaMV genome without losing virulence (Gronenborn *et al.*, 1981). This is less than one-third the size of the average gene. Even that small amount of added DNA is not stable in the CaMV genome.

5. Ti Plasmid of *Agrobacterium tumefaciens*

The remainder of this chapter focuses on the Ti plasmid, the most advanced of all model systems for transfer of genetic information. As with plant viruses, the Ti plasmid is a natural transformation system (Smith and Townsend, 1907). It differs from the viruses in that the transformation results in integration of DNA into plant chromosomes (Chilton *et al.*, 1980). When dicotyledonous plants are wound inoculated with a virulent strain of *A. tumefaciens,* a tumorous growth called a crown gall results after 4–6 weeks. The crown-gall tissue can be removed from the plant, freed of the bacteria, and grown indefinitely in tissue culture.

Among the unique characteristics of crown-gall cells is the ability to grow independently of plant hormones. In fact, hormone autonomy is often used as a selection mechanism to identify transformed cells in a population of non-transformed cells. Another characteristic of crown-gall cells is the ability to synthesize a unique group of amino acid derivatives called opines (Kemp, 1982). These compounds are synthesized in crown-gall cells by enzymes that are coded for by genes that are transferred to the plant by the Ti plasmid (Kemp, 1982). The portion of the Ti plasmid that is transferred to the plant has been mapped (Fig. 1) as a contiguous piece of DNA (T-DNA) between 12 and 14 kb in size.

The T-DNA is integrated into plant nuclear DNA during the transformation process, and genes carried on the T-DNA are expressed in the crown-gall cells. There are functions other than opine synthesis carried on the T-DNA. On the left side of pTi-15955 (Fig. 1), there are three or four areas that are involved in oncogenicity. If any of these locations is mutated by either deletion or by transposon insertion, virulence is modified.

The center of the T-DNA carries the gene for opine synthesis. In the case of pTi-15955, the opine gene codes for the enzyme octopine synthase (Hack and Kemp, 1980). This enzyme catalyzes the synthesis of the opine octopine. Chemically, octopine is the reduced condensation product of arginine and pyrurate. Octopine is not found in any plant cells but crown-gall cells. Further, it is accumulated in crown-gall cells as a source of food for *A. tumefaciens.*

The T-DNA of a second type of Ti plasmid (pTi-C58) is also illustrated in Fig. 1. The opine synthesized by this plasmid is nopaline, not octopine. Nopaline is the reduced condensation product of arginine and α-ketoglutarate. Because Ti plasmids are incompatible (Hooykaas *et al.*, 1980), octopine and nopaline are never found in the same crown-gall cell.

The right-hand side of the T-DNA of pTi-15955 (Fig. 1) is replaceable. That is, all of the T-DNA from the octopine synthase gene to the right border of the T-DNA can be deleted without interfering with tumor formation (Thomashow *et al.*, 1980). This is the area of the T-DNA toward which we directed our engineering.

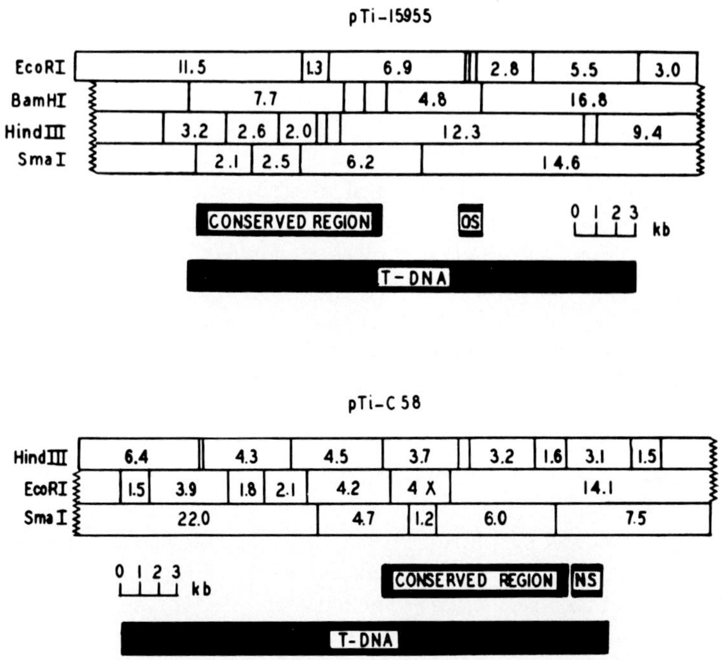

Fig. 1 Restriction-endonuclease map of the T-DNA regions of pTi-15955 and pTi-C58. Each fragment is designated by its size in kilobase pairs. OS, Octopine synthase gene; NS, nopaline synthase gene.

There are a number of general approaches to engineering the Ti plasmid. The most direct approach is to shrink the plasmid from its normal size of 200 kb to a manageable size of perhaps 10–20 kb. Such attempts have not been successful because regions involved in tumorigenesis are found in many locations around the plasmid (DeGreve *et al.*, 1981). Another approach has included site-specific linearization of the plasmid through R- or D-loop formation. A linear single-stranded piece of DNA can pair with its complement in superhelical DNA to form a triple-stranded structure called a D loop (Shibata *et al.*, 1980). The unpaired strand of the triplet can then be digested, opening the molecule at a specific site. D-Loop formation is strictly dependent on the plasmid molecule's remaining supercoiled. This is very difficult to do with a plasmid as large as the Ti plasmid. Therefore, D and R looping have not proven fruitful.

The most practical approach to engineering the Ti plasmid involves cloning a fragment of the Ti plasmid, engineering that fragment, and relying on *in vivo* recombination to move the engineered fragment back to the Ti plasmid. Before describing in detail the methods for engineering, useful genes for engineering must be described.

D. Genes and Their Transfer to Sunflower Cells

1. Isolation and Characterization of the Phaseolin Gene

As discussed earlier, the combination of wheat and bean storage proteins results in a high-quality source of nutritional protein. At present, it is not possible to engineer the grains. However, sunflower (*Helianthus annuus*) storage proteins are similar to those of wheat (*Triticum* spp.) in that they are low in lysine. Therefore, the transfer of the gene coding for bean storage protein to sunflower may result in a new combination that is useful.

Phaseolin is the major storage protein of beans (*Phaseolus* spp.), accounting for 50% of the protein in the dry seed (Ersland *et al.*, 1982). A gene coding for phaseolin has been isolated as a λ clone from a phaseolin DNA library (Sun *et al.*, 1981). The clone was isolated by probing the library with radioactive cDNA synthesized *in vitro* from purified phaseolin mRNA. The purification of phaseolin mRNA was simplified when it was discovered that maturing bean seeds are rich in this mRNA. The phaseolin gene was subcloned from the λ clone into pBR322 and then sequenced (Slightom *et al.*, 1983). Consensus promoter sequences are present upstream from the start of the gene. There is a cap site and a short untranslated region before the first ATG. The gene is composed of six exons and five short introns totaling 1777 bp. There is an untranslated region at the 3′ end of the gene as well as a poly(A) addition site.

There are three useful restriction-endonuclease sites in and around the phaseolin gene. There is a *Bgl*II site 800 bp upstream from the first ATG, an *Eco*RI site 32 bp downstream from the ATG, and a *Bam*HI site 1000 bp downstream from the poly(A) addition site. These sites were used to make two constructions: (1) one from the *Bgl*II-to-*Bam*HI site, including the entire phaseolin gene, and (2) another from the *Eco*RI-to-*Bam*HI site, including the coding sequence minus the first 32 bp and the promoter sequences.

2. Insertion and Expression of the Phaseolin Gene in Sunflower Cells

The first construction that introduced the phaseolin gene into sunflower cells used the *Eco*RI/*Bam*HI fragment of the phaseolin gene inserted into the 5.5-kb *Eco*RI fragment found on the right side of the T-DNA of pTi-15955; the engineering strategy is shown in Fig. 2. The T-DNA fragment was first cloned into the *Eco*RI site of the broad-host-range plasmid pRK290, giving pKSIII. The T-DNA was then engineered by cloning the phaseolin DNA and the NPTII gene (Rothstein *et al.*, 1980) into the *Hin*dIII sites of the T-DNA, giving pKSIII-KB.

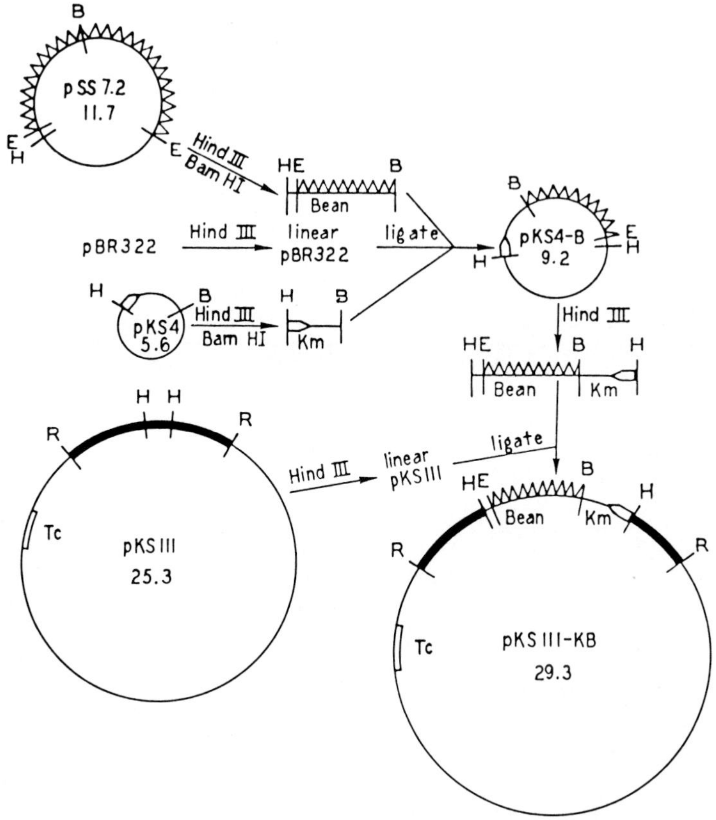

Fig. 2 Construction of pKSIII-KB. A 3.1-kb *Hind*III/*Bam*HI DNA fragment containing the *Eco* RI/*Bam*HI fragment of the phaseolin gene was isolated from a pBR322 clone containing the large 7.2-kb *Eco*RI fragment of bean DNA. The 3.1-kb fragment contains 32 bp of pBR322 DNA from the *Hin*dIII-to-*Eco*RI site of pBR322. This 3.1-kb fragment and a 2.1-kb *Hin*dIII/*Bam*HI fragment containing the neomycin phosphotransferase II gene (from pRZ102) were cloned into the *Hin*dIII site of pBR322. The resulting plasmid (pKS4-B) served as a source of the kanamycin/bean DNA. The 4.9-kb *Hin*dIII fragment from pKSIV-B was ligated into the *Hin*dIII site of pKSIII, constructing the "shuttle" vector pKSIII-KB.

The NPTII gene was cloned from the bacterial transposon Tn5 as the *Bam*HI/*Hin*dIII fragment pKS4. The *Bam*HI sites of the phaseolin and NPTII genes were ligated and the pair cloned into the *Hin*dIII sites of the T-DNA. It was necessary to clone the pair first in pBR322 (pKS4-B) as an intermediate source of the genes because of the difficulties of cloning directly into pRK290. Note that there are 32 bp of pBR322 DNA from the *Eco*RI site of the phaseolin gene to the *Hin*dIII cloning site. These 32 bp replace the 32 bp of phaseolin DNA that were excluded by using the *Eco*RI site. Further, the 32 bp of pBR322 DNA are in the same

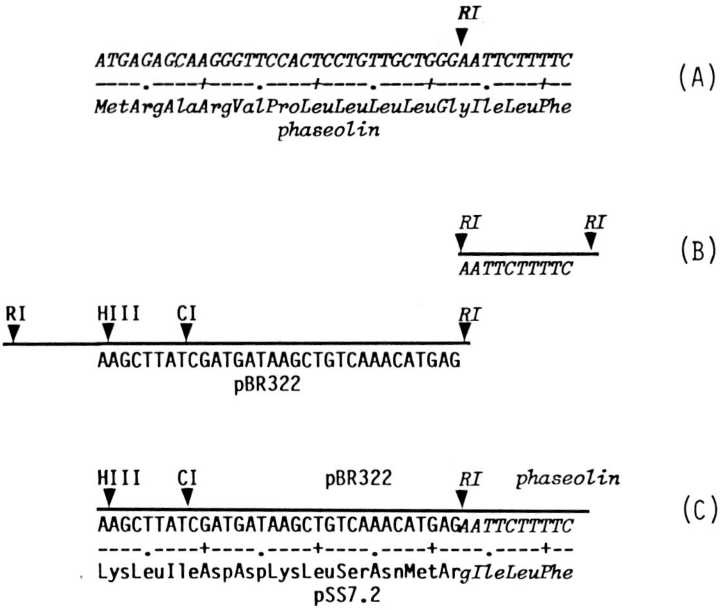

Fig. 3 Construction of pSS7.2. The partial base sequence and amino acid sequence of the phaseolin gene is shown in italics, starting at the 5′ end of the coding sequence (A), the *Eco*RI site (B), or ligated into the *Eco*RI site of pBR322 (roman type) (C).

amino acid reading frame as the phaseolin gene (Fig. 3). This assures that translation will not be prematurely terminated before the phaseolin gene is read. However, it does not assure that the correct reading frame will be used if translation starts within the T-DNA.

The "shuttle" plasmid pKSIII-KB, carrying the engineered T-DNA was introduced into *Agrobacterium tumefaciens* strain 15955 by transformation. Once in the same cell with pTi-15955, the shuttle vector DNA recombines with its homologous counterpart of the Ti plasmid (Kemp *et al.*, 1983). Because the phaseolin and NPTII sequences are contained within the homologous region, they too will be carried onto the Ti plasmid. The double homologous-recombination event is rare (1 of 10^6 cells). Therefore, the isolation of a cell in which recombination occurred was effected by first conjugation with *E. coli* carrying pPH1J1. This plasmid is of the same incompatibility group (P1) as pRK290. Thus simultaneous selection for gentamycin resistance (Gm^R; conferred by pPH1J1) and Km^R results in cells that do not have pRK290-derived plasmids but that have preserved Km^R by transferring it to pTi-15955.

Agrobacterium tumefaciens harboring the engineered plasmid pTi-15955-KB was used to infect sunflower plants, and the resulting crown galls were established in tissue culture (PSCG-15955-KB). The infection process and the cultured crown-gall cells do not appear different from those incited by non-engineered Ti plasmids. Therefore, insertions into the right side (Fig. 1) of the T-DNA up to 3.5 kp in size do not affect the infection or transformation process. PSCG-15955-KB DNA was analyzed for phaseolin DNA. The *Eco*RI/*Bam*HI fragment of the phaseolin gene is stably integrated, and no reorganization of the phaseolin DNA was noted during the year following its insertion.

PSCG-15955-KB tissue appears to synthesize an RNA that has sequence homology to the *Eco*RI/*Bam*HI fragment of the phaseolin gene, but antibodies to phaseolin do not detect this protein in the tissue (Hall *et al.*, 1983). A possible reason that the phaseolin gene is only partly expressed may rest in the way the gene was tailored and the position at which it was inserted into T-DNA. The gene was tailored to exclude its promoter region and the first few amino acids of the coding sequence (Fig. 3). It does, however, contain its introns, stop signal, and the poly(A) addition site. Furthermore, it was inserted into an active T-DNA gene. Without its own promoter, the phaseolin gene may have been transcribed as part of the active T-DNA gene but was not translated because the reading frame of the phaseolin DNA did not match that of the T-DNA gene. A final explanation will have to wait until the T-DNA gene has been sequenced and the reading frames matched.

The second phaseolin gene construction was the engineering of the entire gene, including its own promoter (*Bgl*II/*Bam*HI fragment), into the Ti plasmid. For this experiment we used a different Ti plasmid, pTi-C58, and a well-characterized T-DNA gene, that for nopaline synthase (NS). The crown galls from this experiment are being established in tissue culture, but they have not been tested for phaseolin protein. These and similar experiments should elucidate the mechanism of foreign gene expression in the crown-gall system.

3. Isolation and Characterization of the Nopaline Synthase Gene

The efficacy of our method of engineering was demonstrated by isolating, characterizing, and transferring the gene coding for NS (Kemp *et al.*, 1983). It has been presumed that the NS gene is carried on the right side of the T-DNA of pTi-C58 (Fig. 1). The evidence merely allows one to conclude that the T-DNA gene *controls* the synthesis of NS in crown-gall cells. One cannot conclude that it actually codes for the enzyme until its DNA is sequenced and the derived amino acid sequence compared to that of NS.

We have determined the base sequence for the region of T-DNA controlling NS. This region of DNA contains an open reading frame for translation of 1239 bp. This length could code for a protein of 413 amino acids (MW 45,000). The region appears to be a functional gene of eukaryotic character because it contains consensus promoter sequences upstream from the 5′ end of the coding sequence

Table II Comparison of Amino Acid Composition of Nopaline Synthase[a]

Residue	Nopaline synthase: Hydrolysates	Nopaline synthase: Sequence
Lysine	20	19
Histidine	16	17
Arginine	19	22
Aspartic acid }	44	23
Asparagine }		21
Glutamic acid }	43	26
Glutamine }		7
Threonine	19	22
Serine	29	30
Glycine	41	21
Proline	22	23
Alanine	39	42
Cystine	10	11
Valine	22	26
Methionine	5	5
Isoleucine	25	32
Leucine	31	32
Tyrosine	9	8
Phenylalanine	19	21
Tryptophan	—	3
	413	413
Estimated molecular weight	45,000	45,600

[a]Calculated from sequence data and measured from protein hydrolysates.

and a poly(A) addition sequence at the 3′ end of the coding sequence. Further, crown-gall cells synthesize a polyadenylated RNA that is homologous to the DNA in this region.

Direct evidence that this sequence codes for NS is presented in Table II. A putative amino acid composition for the T-DNA gene was calculated from its nucleotide sequence and then compared to the measured amino acid composition of purified NS. The two compositions compare quite favorably. There are two discrepancies, however, that should be pointed out. The glycine levels in the protein appear higher than predicted by the DNA sequence. This may be the result of amino acids breaking down to glycine during the acid hydrolysis of the protein. No attempt was made to correct for breakdown. The other discrepancy is in the levels of glutamic acid. We have noted variability in glutamic acid content of hydrolyzed plant protein other than NS. This may reflect a technical problem, not a difference between the two compositions. If, indeed, the amino acid composition of NS matches the composition predicted from its gene, then the four subunits that compose NS must be identical.

4. Insertion and Expression of the Nopaline Synthase Gene in Sunflower Cells

The NS gene normally functions in the plant in a T-DNA environment. However, the NS gene has no sequence homology nor is it ever found associated with opine genes residing on octopine-type Ti plasmids (the type of plasmid used in our engineering experiments). For these reasons, NS seemed a logical choice to demonstrate whether an engineered gene could be fully expressed when transferred by pTi.

The cloning strategy was to transfer the 7.5-kb *Sma*I fragment from the T-DNA of pTi-C58 to the *Hin*dIII sites in the 5.5-kb *Eco*RI fragment of the T-DNA of pTi-15955 (Fig. 1). Sunflower plants were inoculated with the engineered *A. tumefaciens* cells, and the resulting crown galls were placed in tissue culture. The cultures were then assayed for octopine and nopaline as well as the enzyme NS. Normal levels of nopaline and NS suggest that such an engineering strategy can be fully successful.

The cloning details for the 7.5-kb *Sma*I fragment have been published elsewhere (Kemp *et al.*, 1983). The engineered *A. tumefaciens* is virulent, and the resulting crown-gall tissue culture from sunflower crown galls has been designated PSCG-15955-N. PSCG-15955-N tissue cultures synthesize both octopine and nopaline, whereas the control tissues PSCG-15955 synthesize only octopine, and PSCG-C58 only nopaline. The amounts of these products in PSCG-15955-N was approximately the same as in the control tissue. Both octopine and nopaline synthases were also detected in PSCG-15955-N. This evidence indicates that not only is the engineered NS gene functional in the plant cell, it is functional at levels similar to that found in its native environment.

A complication introduced by the cloning of the 7.5-kb *Sma*I fragment is that this fragment contains the right-hand border of the pTi-C58 T-DNA (Fig. 4). The question that arises is whether the engineered pTi uses its natural right-hand border for integration of T-DNA into the plant genome or whether it uses the engineered border found on the *Sma*I fragment (Fig. 4). Genomic Southern blot experiments have demonstrated that both right-hand borders were used. Is NS synthesized only when *its* border (pTi-C58) is used for integration? This question was answered by engineering with only a part of the *Sma*I fragment—the part that contains the NS gene but not the border. The 7.5-kb *Sma*I fragment was subcloned by cutting it at a *Bcl*I site that lies between the start of the NS gene and the T-DNA border. Again, the resulting crown galls were established in tissue culture, and normal levels of nopaline were detected. Thus it appears that the NS gene will function normally in the environment of foreign T-DNAs and that position in the T-DNA is not critical. From more recent experiments, we also know that the NS gene will function normally if engineered into T-DNA in the reverse orientation. Taken together, these experiments clearly demonstrate that T-DNA can be engineered and that engineered genes can be fully functional.

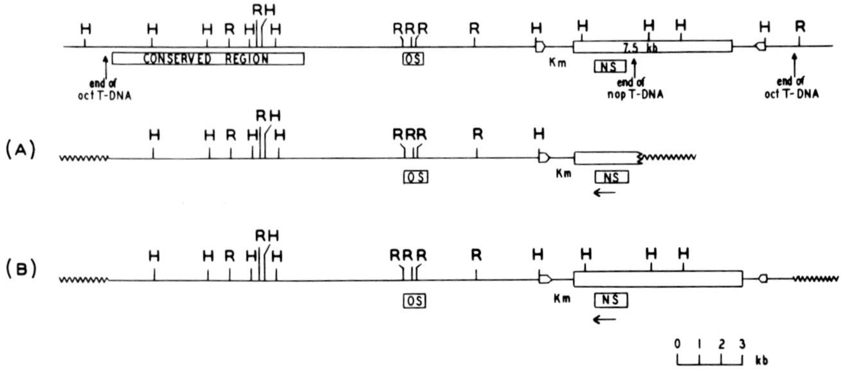

Fig. 4 Possible integration sites for pTi-15955-N. The restriction-endonuclease map of the T-DNA portion of pTi-15955-N is shown at the top. The left- and right-hand ends of the T-DNA are labeled as octopine T-DNA (oct T-DNA). The introduced nopaline T-DNA is indicated by the box labeled 7.5 kb. Within the nopaline T-DNA (nop T-DNA) is the gene for nopaline synthase (NS) and the right-hand end of the nopaline T-DNA. (A, B) Two possible integrations. The jagged line represents plant DNA. Integration at the right-hand end of T-DNA can occur at either the nopaline T-DNA (A) or the octopine T-DNA ends (B). OS, Octopine synthase; Km, neomycin phosphotransferase II gene. Arrow indicates the direction of transcription.

E. Regeneration of Transformed Plant Cells

1. Transformation Using the Ti Plasmid

The regeneration of transformed cells is the key to using recombinant DNA engineering as a useful tool for creating new recombinants. There are few plant species that have been successfully regenerated from tissue culture. This list is even shorter when the tissues cultured are crown galls. More recent work on the Ti plasmid has identified a number of T-DNA genes that are involved in tumorigenesis (Garfinkel *et al.*, 1981). These genes are located on the left side of pTi-15955 (Fig. 1), in the conserved region, and they can be inactivated one at a time by transposon insertion (Garfinkel *et al.*, 1981). When the gene located on the 1.3-kb *Eco*RI fragment is inactivated, the crown galls have a tendency to form shoots. It has been proposed that this gene may be involved in auxin synthesis. These shoots can occasionally be regenerated into whole plants. The inactivation of a second gene, located slightly left of center of the 6.9-kb *Eco*RI fragment, appears to produce crown galls with a tendency to root. This gene may be involved in cytokinin synthesis. These root tissues have not been tested for their ability to regenerate into plants. A very interesting experiment might be to inactivate both genes. The result might be a Ti plasmid that still transfers T-DNA, but the resulting transformed plant cells would then show none of the tumorigenic properties of crown-gall cells. Further, these cells may be as easily regenerated as normal plant cells. A third gene involved in tumorigenesis is located in the center of the 4.8-kb *Bam*HI fragment. Tumors comprising cells in which this gene is inactivated are larger than normal. Inactivating the large-tumor gene appears to be of no advantage for our engineering experiments. Nevertheless, it may be of interest to inactivate the large-tumor gene as well as the shooting and/or rooting genes.

Once the tumorigenic phenotype has been removed, the problem of recognizing transformed cells becomes paramount. A solution is to incorporate into the T-DNA a selectable plant marker. The one we have been studying is resistant to the antibiotic G418, which is known to inhibit plant cells. A gene for G418 inactivation is carried on a number of bacterial transposons. One of these transposons (Tn601) is known to be active in yeast. Therefore, we propose that it may be active in higher eukaryotes. We have engineered this transposon into the T-DNA of pTi-15955, and the crown-gall tissues are to be tested for resistance to G418. If successful, this marker will be used to conjunction with inactivation of the shooting and rooting genes to produce a useful plant transformation system.

2. Transformation Using the Ri Plasmid

The hairy-root disease incited by *A. rhizogenes* (Riker *et al.*, 1930) is very similar to the crown-gall disease. It involves the transfer and apparent integration of T-DNA and the expression of opines in the transformed tissues (Tepfer and Tempe, 1981; Chilton *et al.*, 1982). It differs from crown gall in that the result of transformation is root proliferation. These roots appear to be autonomically normal. The only abnormal behavior is that they tend to branch more than normal. A major advantage of this transformation is that the roots are easily regenerated into plants. Further, the T-DNA remains intact and is expressed in the regenerated plants. More recent results suggest that the T-DNA will pass through meiosis and be inherited as a single dominant factor in a Mendelian manner (Tepfer, 1983).

Engineering of the T-DNA of the hairy-root bacteria has not been attempted because our understanding of the system is not as advanced as that for the crown-gall system. As the fundamental studies are completed, we shall begin engineering the hairy root system to determine its utility.

III. Concluding Remarks

It would appear that we are very close to having transferred a foreign gene to a sunflower cell and to having that gene fully expressed. Once the problems of regeneration have been overcome, we can begin studying the stability of the foreign gene through successive generations as well as its regulation in its new environment. T-DNA genes appear to be expressed at all times in all tissues. This may be a disadvantage for transferred storage-protein genes because they are normally expressed only during seed development. As we begin to understand gene regulation, we may be able to include those DNA sequences that turn genes on and off at different stages of development. Such flexibility will make this novel technique of engineering a truly useful tool.

References

Bedbrook, J. R., and Kolodner, R. (1979). *Annu. Rev. Plant Physiol.* **30,** 593–620.

Bressani, R., and Elias, L. G. (1974). *In* "New Protein Foods" (A. M. Altschul, ed.), Vol. 1A, pp. 230–297. Academic Press, New York.

Chilton, M.-D., Saiki, R. K., Yadav, N., Gordon, M. P., and Quetier, F. (1980). *Proc. Natl. Acad. Sci. USA* **77,** 4060–4064.

Chilton, M.-D., Tepfer, D. A., Petit, A., David, C., Casse-Delbart, F., and Tempe, J. (1982). *Nature (London)* **295,** 432–434.

DeGreve, H., Decraemer, H., Seurinck, J., Van Montagu, M., and Schell, J. (1981). *Plasmid* **6,** 235–248.

Ersland, D. R., Brown, J. W. S., Casey, R., and Hall, T. C. (1982). *In* "The Genetics and Biochemistry of Seed Proteins" (W. Gottschalk, and H. Muller, eds.). Nijhoff, The Hague, in press.

Garfinkel, D. J., Simpson, R. B., Ream, L. W., White, F. F., Gordon, M. P., and Nester, E. W. (1981). *Cell* **27,** 143–153.

Goodman, R. M. (1981). *In* "Handbook of Plant Virus Infections: Comparative Diagnosis" (E. Kurstak, ed.), pp. 879–910. Elsevier/North-Holland Biomedical Press, New York.

Gronenborn, B., Gardner, C., Schaefer, S., and Shepherd, R. J. (1981). *Nature (London)* **294,** 773–776.

Hack, E., and Kemp, J. D. (1980). *Plant Physiol.* **65,** 949–955.

Hall, T. C., Slightom, J. L., Ersland, D. R., Murray, M. G., Hoffman, L. M., Adang, M. J., Brown, J. W. S., Ma, Y., Matthews, J. A., Cramer, J. H., Barker, R. F., Sutton, D. W., and Kemp, J. D. (1983). *In* "Structure and Function of Plant Genomes" (L. Duhr, and O. Ciferri, eds.). Plenum, New York, in press.

Hegsted, D. M. (1976). *J. Am. Diet. Assoc.* **66,** 13–21.

Helgeson, J. P. (1968). *Science (Washington, D.C.)* **161,** 974–981.

Hooykaas, P. J. J., Dulk-Ras, H. D., Ooms, G., and Schilperoort, R. A. (1980). *J. Bacteriol.* **143,** 1295–1306.

Jinks, J. L., Caligari, P. D. S., and Ingram, N. R. (1981). *Nature (London)* **291,** 586–588.

Kemp, J. D. (1982). *In* "Phytopathogenic Prokaryotes" (M. S. Mount, and G. Lacy, eds.), Vol. 1, p. 443. Academic Press, New York.

Kemp, J. D., Sutton, D. W., Fink, C., Barker, R. F., and Hall, T. C. (1983). *Beltsville Symp. Agric. Res.* **7,** in press.

Kim, B. D., Mans, R. J., Cande, M. F., Pring, D. R., and Levings, C. S. (1982). *Plasmid* **7,** 1–14.

Melchers, G. (1977). *In* "Recombinant Molecules: Impact on Science and Society" (R. F. Beers, Jr., and E. G. Bassett, eds.), pp. 209–227. Raven, New York.

Nagata, T., and Takebe, I. (1970). *Planta* **92,** 301–308.

Riker, A. J., Banfield, W. M., Wright, W. H., Keitt, G. W., and Sagen, H. E. (1930). *J. Agric. Res.* **41,** 507–540.

Rothstein, S. J., Jorgensen, R. A., Postle, K., and Reznikoff, W. S. (1980). *Cell* **19,** 795–805.

Scrimshaw, N. S. (1976). *New Engl. J. Med.* **294,** 136–142, 198–203.

Scrimshaw, N. S., and Young, V. R. (1976). *Sci. Am.* **235,** 50–64.

Shepard, J. P., Bidney, D., and Shahin, E. (1980). *Science (Washington, D.C.)* **208,** 17–24.

Shepherd, R. J. (1979). *Annu. Rev. Plant Physiol.* **30,** 405–423.

Shibata, T., Das Gupta, C., Cunningham, R. P., and Radding, C. M. (1980). *Proc. Natl. Acad. Sci. USA* **77,** 2606–2610.

Slighton, J. L., Sun, S. S., and Hall, T. C. (1983). *Proc. Natl. Acad. Sci. USA* **80,** 1897–1901.

Smith, E. F., and Townsend, C. O. (1907). *Science (Washington, D.C.)* **25,** 671–673.

Sun, S. M., Slightom, J. L., and Hall, T. C. (1981). *Nature (London)* **289,** 37–41.

Tepfer, D. (1983) *In* "Genetic Engineering in Eukaryotes—The Genetic Engineering of Higher Plants: Nature Got There First" (P. F. Lurquin, and A. Kleinhofs, eds.). Plenum, New York, in press.

Tepfer, D., and Tempe, J. (1980). *C.R. Hebd. Seances Acad. Sci. Ser. III* **292,** 153–156.

Thomashow, M. F., Nutter, R., Montoya, A. L., Gordon, M. P., and Nester, E. W. (1980). *Cell* **19,** 729–739.

CHAPTER 7

λSV2, a Plasmid Cloning Vector that Can Be Stably Integrated in *Escherichia coli*

BRUCE H. HOWARD
MAX E. GOTTESMAN

Laboratory of Molecular Biology
Division of Cancer Biology and Diagnosis
National Cancer Institute
National Institutes of Health
Bethesda, Maryland

ISBN 0-12-372380-9

I. Introduction

Central to progress in recombinant DNA technology is the development of new cloning vehicles designed to maximize the efficiency, versatility, and convenience with which foreign DNA segments may be cloned and characterized. Cloning vehicles derived from plasmids (e.g., pBR322) or from phages M13 or λ replicate autonomously in bacterial cells. Autonomous replication permits the isolation, in high yields and in relatively purified form, of foreign DNA segments. However, it also imposes certain risks in maintaining these segments without rearrangements or deletions. These problems are especially acute when the cloned DNA is of large size or carries repeated sequences. Loss of cloned DNA often imparts upon the vector a growth advantage, and the deleted derivative will rapidly overgrow its parent. Recombination between repeated sequences to generate deletions is at least partially independent of the host's generalized recombination system; growth in *recA* hosts, therefore, does not obviate the problem (Chia *et al.*, 1982).

The pressures exerted on autonomous replicons appear to be absent or reduced when DNA is carried as an inserted fragment in the bacterial chromosome, propagated by the host replication machinery. Thus two λ prophages, integrated in tandem in the *E. coli* chromosome, are stably maintained, even though the bacterium is charged with an extra 94 kb of DNA. This 2% increase in cell DNA content is, evidently, well tolerated by the cell. Even in $recA^+$ hosts, loss of a prophage by recombination between the two 47-kb DNA repeats occurs infrequently (M. E. Gottesman, unpublished).

In this chapter we describe the development of a new plasmid vector system in which recombinant genomes may be propagated either extrachromosomally or as a single copy integrated in the *E. coli* chromosome at the λ attachment site. The integrated state is maintained as long as the host is grown at 32°C; when the culture is shifted to 42°C, excision and amplification of the recombinant occurs. Potential uses of this vector system are discussed.

II. Materials and Methods

A. Bacterial Strains and Plasmids

The λ strain Y976:λ*cI857bio936int*$^+$*red3b515b519* was generously supplied by R. Weisberg (NIH) and converted to the λi^{21} derivative by standard genetic crosses. Strain λi^{21}*ctsSam7*Δ*S-X* was a gift of S. Rogers (Johns Hopkins Univer-

Table I Bacteriophage λ and *Escherichia coli* Strains[a]

Strain	Structure[b]	Parent and/or source	Selection and/or screening[c]
N38	C600 $r_k^- m_k^-$	NIH collection	—
N4956	C600 $r_k^- m_k^+$	NIH collection	—
N4903	F^- *his ilv* gal^+ su^- *Δ8* str^R *relA1*, "SIG"[d]	Gottesman *et al.* (1980)	—
N4830	SIG (λΔBAMΔHI)	Gottesman *et al.* (1980)	—
N4741	SIG (λΔH1)	Gottesman *et al.* (1980)	—
N5285	SIG blu^+ *galE*::Tn10 (λΔBAMΔH1)	N4830 + P1·DB1447-319	Tet^R, Blu^+
N5704	SIG (λΔ*H1*; λi^{21}*cts*Δ*S-X*)[e]	N4741 + λi^{21}*cts*Δ*S-X*	*21* Immunity
N5705	SIG (λΔ*S-X*Δ*H1*)	N5704	t^R
N5812	SIG (λΔ*S-X*; λi^{21}*bio936*)[f]	N5705 + λi^{21}*bio936*	*21* Immunity
N5817	SIG (λ*bio936*Δ*S-X*Δ*H1*)	N5812 + λi^4*b*Δ*2*	*21* Sensitive
N6102	N38 *galE*::Tn10 (λΔ*BAM*Δ*H1*)	N38 + P1·N5285	Tet^R, Bio^-
N6106	N38 (λ*bio936*Δ*S-X*Δ*H1*)	N6102 + P1·N5817	Gal^+, DTB^+
N6361	N4956 *galE*::Tn10 (λΔ*BAM*Δ*HI*)	N4956 + P1·N5285	Tet^R, Bio^-
N6377	N4956 (λ*bio936*Δ*S-X*Δ*H1*)	N6361 + P1·N5817	Gal^+, DTB^+

[a]For detailed methods of construction, media, and strain descriptions, see Gottesman *et al.* (1980).
[b]Phages designated λ are *cI857*.
[c]DTB^+, Ability to grow on desthiobiotin; t^R, temperature resistant (grows at 41°C); Tet^R, resistant to 15 μgm/ml tetracycline.
[d]"SIG," F^- su^- his^- *ilvA1* str^R.
[e]λi^{21}*cts*Δ*S-X* was a gift of S. Rogers; the *S-X* deletion extends from the *Sal*I site in *redB* to the *Xho*I site in *cIII*.
[f]λ*bio936* was obtained from R. Weisberg and converted to the *imm21* derivative by standard crosses; *bio936* is a substitution of *bio* for λ DNA between *att* and *int*.

sity); an S^+ derivative was constructed for this work. The *S-X* deletion was constructed by *in vitro* exposure of λ DNA to restriction endonucleases *Sal*I and *Xho*I and subsequent removal of the λ DNA between 67.4 and 69.0% λ. The *S-X* deletion removes the phage *kil* cistron but leaves *int* and *xis*. Plasmid pWR1 DNA (Hsu *et al.*, 1980) was obtained from U. Schmeissner (NIH). Strain DB1447-319 *galE*::Tn10 was a gift of D. Berg (Washington University). The recombinant plasmid pSV2*gpt* was provided by R. Mulligan (Stanford). Bacteriophage λ*cI857Sam7* DNA was purchased from Bethesda Research Laboratories. The construction of strains used in this study is outlined in Table I.

B. Enzymes

Restriction endonucleases were purchased from New England Biolabs, Bethesda Research Laboratories, or Boehringer–Mannheim. *Escherichia coli* DNA polymerase I (Kornberg polymerase) was from Boehringer–Mannheim; AMV reverse transcriptase was from Life Sciences, Inc., St. Petersburg, Florida; and T4 DNA ligase was from New England Biolabs.

C. Construction of Recombinant Plasmids

DNA restriction fragments were separated by agarose-gel electrophoresis in 0.04 *M* Tris acetate and 0.002 *M* EDTA (pH 8.1) containing 10 μgm/ml ethidium bromide, visualized by longwave UV illumination (Ultraviolet Products, San Gabriel, California), and eluted from excised gel slices as described by Chen and Thomas (1980). The 5′ recessed restriction-fragment ends were converted to "blunt" ends by incubation with DNA polymerase I (Seeburg *et al.*, 1977); 3′ recessed restriction-fragment ends were converted to blunt ends by incubation with AMV reverse transcriptase (1 unit/nmol fragment ends) for 30 min at 37°C. Purified restriction fragments were joined by incubation with T4 DNA ligase overnight at 14°C. Bacterial transformations of *E. coli* strain HB101 were carried out by the $CaCl_2$ method (Mandel and Higa, 1970). For transformation of *E. coli* strain N6106, bacteria were grown in LB broth supplemented with 0.003% biotin and shifted between 32 and 42°C as described in Section III. Transformants were selected for growth in agar containing 50 μgm/ml ampicillin or 15 μgm/ml chloramphenicol.

D. Characterization of Recombinants

To screen transformants of HB101, plasmid DNA was prepared by the lysozyme–Triton X-100 lysis method (Katz *et al.*, 1973); 2-ml overnight cultures yielded sufficient DNA for about 10 restriction digests. To screen transformants of N6106, 2-ml overnight cultures were diluted 1:100 into 5 ml of superbroth supplemented with 0.003% biotin, then induced by sequential incubation at 32°C for 2 hr, 42°C for 20 min, and 38–39°C for 2 hr.

Extrachromosomal DNA was prepared by the lysozyme–Triton X-100 method or by alkaline-SDS lysis (Birnboim and Doly, 1979). The 5-ml cultures induced by this protocol yielded sufficient DNA for about five restriction digests.

The DNA segments from which λSV1 and λSV2 have been constructed are for the most part well characterized. The complete DNA sequences of pBR322 (Sutcliffe, 1979), SV40 (Fiers *et al.*, 1978; Reddy *et al.*, 1978), and chloramphenicol acetyltransferase (Alton and Vapnek, 1979) have been published. In addition, Blattner and collaborators have collected published and unpublished sequence data on bacteriophage λ, organized this information, and made it available to other workers. We have used extensively the computer program of Queen and Korn (1979) to locate restriction-endonuclease sites in these DNA sequences and to assist in restriction analysis of both intermediate constructs and the final λSV recombinants.

E. Preparation of DNA and Southern Hybridization Analysis

Plasmid DNA was prepared by lysozyme–Triton X-100 lysis and cesium chloride–ethidium bromide equilibrium centrifugation (Radloff *et al.*, 1967). Restriction fragments generated by digestion of total *E. coli* DNA were separated by agarose-gel electrophoresis and transferred to GeneScreen hybridization membrane (New England Nuclear) according to Southern (1975). ^{32}P-Labeled hybridization probe was synthesized by nick-translation (Rigby *et al.*, 1977); hybridization was carried out as described previously.

III. Results

The λSV system consists of two components, the λSV vector and a specialized λ lysogen. The λSV vector is a double-stranded circular DNA with unique restriction sites in nonessential regions, used for insertion of foreign DNA sequences. Two versions of the vector have been constructed: λSV1 and λSV2, 7.7 and 8.6 kb in length, respectively. λSV is derived from elements of (1) pBR322, including *amp*; (2) the bacteriophage λ replication region and attachment site; (3) the *cat* cistron from Tn9, present in λSV2 but not in λSV1; (4) SV40; and (5) the *E. coli* cistron *guaX*, encoding xanthine-guanine phosphoribosyltransferase, in-

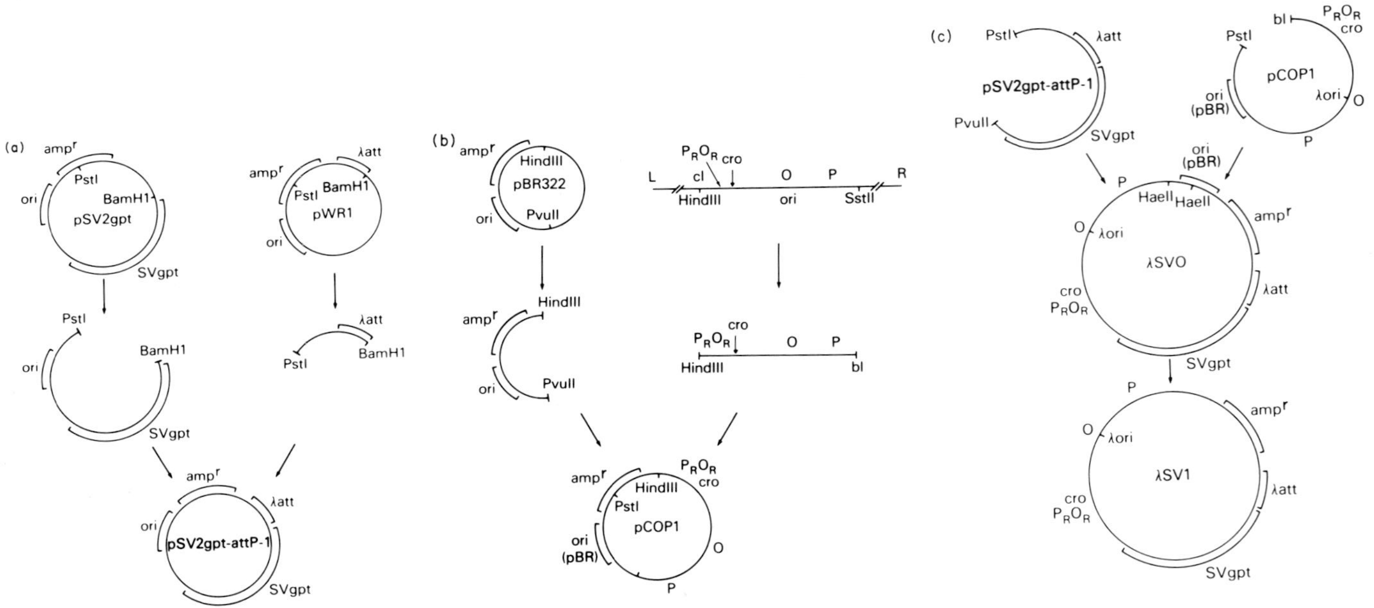

Fig. 1 Construction of λSV1. (a) pSV2*gpt* carries the ampicillin-resistance cistron (*amp*^{r}) and origin of replication (*ori*) from pBR322. The SV *gpt* segment, which contains the *E. coli* xanthine-guanine ribosyltransferase gene embedded in a modified SV40 early region, is used as a dominant selectable marker in mammalian cells (Mulligan and Berg, 1981). pWR1 carries the λ-phage-attachment site (λ*att*), inserted between the *Hin*dIII and *Bam*HI sites of pBR322. (b) λ*cI857Sam7* DNA was digested with *Sst*II, incubated with polymerase I to create a blunt (bl) end, heated to inactivate the polymerase I, and digested with *Hin*dIII. The resulting 2800-bp fragment carries the minimal essential functions required for replication as a λdv plasmid. The *Pvu*II end on the pBR322 fragment is blunt and thus matches the end produced by polymerase I. (c) pCOP1 DNA was digested with *Hin*dIII, incubated with AMV reverse transcriptase to create a blunt (b1) end, heated to inactivate the reverse transcriptase, and digested with *Pst*I. λSVO was digested to completion with *Hae*II; the resulting large vector fragment was isolated by agarose-gel electrophoresis and recircularized to generate the vector λSV1.

serted within SV40 sequences to provide a dominant selectable marker in mammalian cells (Mulligan and Berg, 1981).

A. The λSV1 Vector

The starting recombinant genome for the construction of λSV1 was pSV2 *gpt* (Mulligan and Berg, 1980), a derivative of pSV2 β*G* (B. H. Howard and P. Berg, unpublished). pSV2*gpt* was first converted to pSV2*gpt-attP*-1 (Fig. 1a) by introduction of a fragment containing the λ *attP* site from the plasmid pWR1 (Hsu *et al.*, 1980). pSV2*gpt-attP*-1 was then modified by replacing the pSV2 origin of replication with the λ replication region. The replacement was performed in three steps:

1. A *Hin*dIII–*Sst*II fragment from λ*cI857Sam7,* blunted at the *Sst*II end with DNA polymerase, was inserted between the *Hin*dIII and *Pvu*II sites of pBR322 to yield pCOP1 (Fig. 1b) This λ fragment extends from within *cI* to a *Sst*II site ~106 bp to the right of gene P (Kröger and Hobom, 1982); it thus carries λ genes *oR-pR, cro, cII, ori,* O, and P. The expression of λ replication genes O and P as well as the activation of the λ origin of replication requires transcription from the *pR* promoter (Dove *et al.*, 1971). The fragment will, therefore, support autonomous replication only in the absence of *cI* repressor.
2. pCOP1 was treated sequentially with *Hin*dIII, AMV reverse transcriptase, and *Pst*I to generate a large *Hin*dIII–*Pst*I fragment blunted at the *Hin*dIII end. The fragment was then substituted between the *Pvu*II and *Pst*I sites of pSV2*gpt-attP*-1 to generate λSV0 (Fig. 1c).
3. A 370-bp *Hae*II fragment overlapping the pSV2 origin of replication was deleted from λSV0 to create λSV1 (Fig. 1c). λSV1 was first isolated as a plasmid conferring ampicillin resistance to *E. coli* strain HB101.

B. The λSV2 Vector

The vector λSV2 differs from λSV1 in having both ampicillin- and chloramphenicol-resistance markers. First, the chloramphenicol acetyltransferase element (*cat*) was excised from the recombinant pSV2 *cat*^{r} (C. Gorman *et al.*, 1982) by complete digestion with *Hin*dIII and partial digestion with *Taq*I. Under these conditions *Hin*dIII and *Taq*I cut 5′ and 3′ to *cat,* respectively. The resulting 960-bp fragment was inserted between the *Hin*dIII and *Cla*I sites in pSV2*gpt-attP*-1, and recombinants with the structure of pSV2*gpt-attP*-2 were selected by screening for chloramphenicol resistance (Fig. 2a). Second, the smaller *Pst*I–*Bam*H1 fragment from pSV2*gpt-attP*-2 was recombined with the

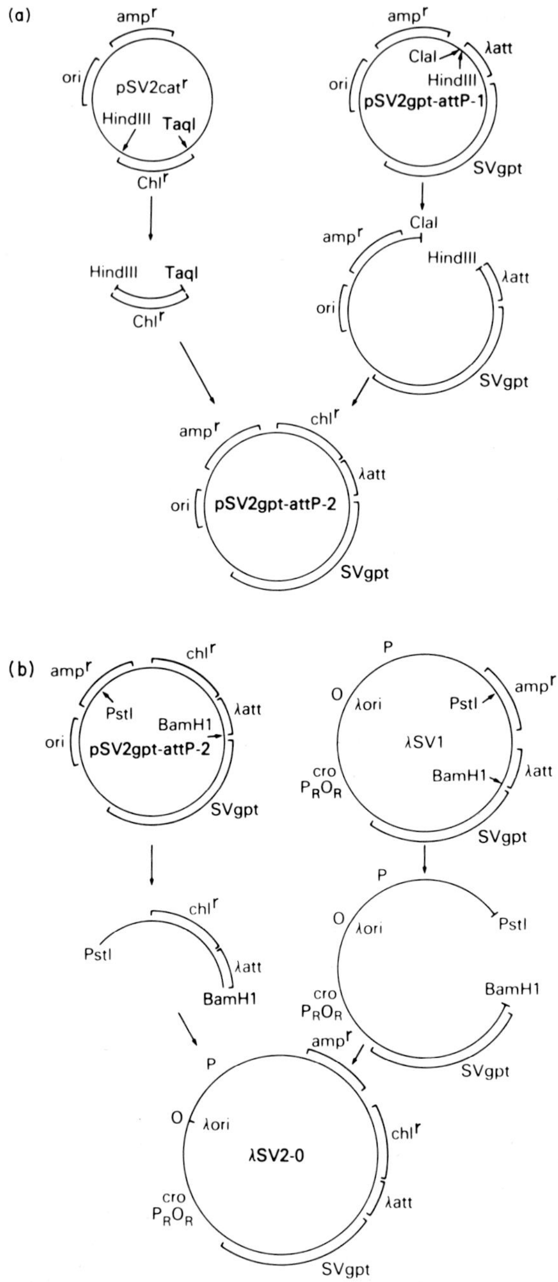

Fig. 2 Construction of λSV2. (a) A 960-bp fragment containing the chloramphenicol-resistance cistron (*chl*^{r}) was isolated from pSV2*cat*^{r}. Because *Taq*I and *Cla*I produce identical cohesive ends, the 960-bp fragment was readily inserted into the *Hind*III–*Cla*I fragment from pSV2*gpt-attP*-1. (b) Ligation of *Pst*I–*Bam*HI fragments from λSV1 and pSV2*gpt-attP*-2 yielded λSV2-0. The steps required for insertion of a *Sal*I site between the *amp*^{r} and *chl*^{r} cistrons, converting λSV2-0 to λSV2, are not shown.

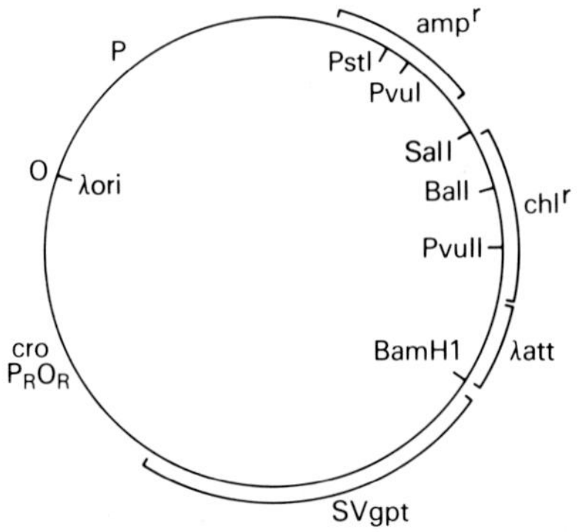

Fig. 3 Unique restriction sites and functional regions in λSV2. From Howard and Gottesman (1982). Copyright 1982 Cold Spring Harbor Laboratory.

larger *Pst*I–*Bam*H1 fragment from λSV1 to generate λSV2-0 (Fig. 2b). Finally, λSV2-0 was subjected to partial cleavage with *Eco*R1, incubated with AMV reverse transcriptase to create blunt ends, and ligated with *Sal*I synthetic oligonucleotide linkers. Recombinants were screened by restriction analysis with *Sal*I and *Eco*R1 to detect derivatives (λSV2) in which a unique *Sal*I site replaces the *Eco*R1 site between the *amp* and *cat* markers in λSV2-0. A summary schematic of the structure of λSV2 (Fig. 3) shows the unique restriction sites available for insertion of foreign DNA segments: *Pst*I and *Pvu*I sites in *amp,* a *Sal*I site between *amp* and *cat, Bal*I and *Pvu*II in *cat,* and a *Bam*H1 site between *attP* and the SVgpt marker.

C. The λ Lysogens N6106 and N6377

By a series of crosses between lysogens and superinfecting phage, (Table I), we constructed lysogen N6106, which carries the prophage genes required for repression and for phage integration and excision; most other prophage genes have been deleted. The relevant λ markers remaining in N6106 are the bacterial λ-insertion site, *attB;* the *int* and *xis* genes; a *Sal*I–*Xho*I deletion that eliminates *red*β, *gam, kil* and cIII; the *N* antitermination and *cI857* repressor genes; and the *H1* deletion, which eliminates *cro* and all prophage genes to the right, including replication and packaging functions. This lysogen was originally constructed in a W3102 background and subsequently transferred to the C600 $r_k^- m_k^-$ derivative, N38. A comparable construction yielded the C600 $r_k^- m_k^+$ derivative, N6377. A comparison of the structures of the prophage in N6106 and a wild-type lysogen is shown in Fig. 4.

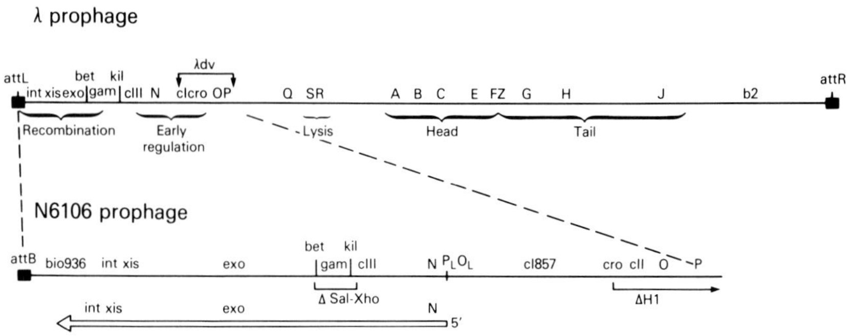

Fig. 4 Comparison of λ and N6106 prophage structures. From Howard and Gottesman (1982). Copyright 1982 Cold Spring Harbor Laboratory.

D. Transformation by λSV1 or λSV2

λSV1 or λSV2 may be introduced into the chromosome of N6106 or N6377 by the following protocol (summarized in Fig. 5). First, an overnight culture is diluted 1:100 and grown at 32°C in LB broth to mid-log phase (O.D. 0.650 = 0.3). Second, the culture is shifted to 41–42°C for 15 min. This step inactivates the temperature-sensitive *cI857* repressor, allowing leftward transcription from the *pL* promoter and the synthesis of Int and Xis proteins. Third, the culture is cooled to 32°C for 15 min. At 32°C, the repressor renatures, and leftward transcription and the expression of *int* and *xis* stops. Xis is a labile protein and

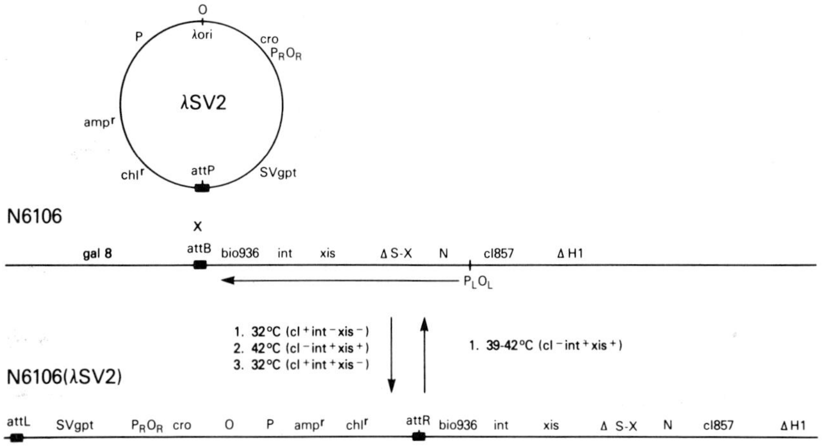

Fig. 5 Summary of steps involved in integration and excision of λSV2.

Table II Efficiency of Transformation by λSV2

		Colonies/ngm plasmid DNA	
Host[a]	Plasmid	32°C	32→42→32°C[b]
N4956	pBR322	8800	2900
	λSV2	9100	3200
N6377	pBR322	4700	3700
	λSV2	14	1870

[a]N4956 is C600 $r_k^- r_m^+$; N6377 is N4956(λ*cI857bio936ΔS-XΔH1*).
[b]log-Phase cultures growing in LB medium (Miller, 1972) were transferred from 32 to 42°C for 15 min and then returned for an additional 15-min incubation at 32°C prior to transformation.

rapidly decays; Int, in contrast, is stable, and considerable Int activity can be detected after the 15-min cooling step. The net effect of these manipulations is to establish, at the time of transformation, a cI^+, Int^+ condition. Vector λSV DNA is introduced by the standard $CaCl_2$ method (Mandel and Higa, 1970). The transformed bacteria are incubated at 32°C for 1 hr, then plated and selected for growth at 32°C in agar containing 50 μgm/ml ampicillin or 15 μgm/ml chloramphenicol.

The results of such an experiment are shown in Table II. The ability of λSV2 to transform the parental, nonlysogenic strain N4956 is roughly equivalent to that of pBR322; one obtains 3200 colonies/ngm λSV2 DNA. The frequency of transformation is increased about three-fold when the heating step is eliminated. This result indicates that λSV2 can establish itself as an autonomous replicon in a nonlysogen with an efficiency approximating that of pBR322. Transformation of N6377 by the standard protocol yields 1870 colonies/ngm λSV2 DNA and 3700 colonies/ngm pBR322 DNA. Because under repressed conditions only integrated λSV2 DNA can be stably inherited, this result indicates that the plasmid DNA integrates into the host chromosome with an efficiency of about 50%.

If the thermal-induction step is omitted, transformation with λSV2 DNA falls to 14 colonies/ngm, consistent with the postulated role for Int in the reaction. In contrast, the efficiency of transformation with pBR322 is 4700 colonies/ngm. Similar results were obtained with N6106 transformed with λSV1.

E. Structure of an N6106(λSV1) Transformant

Confirmation that N6106 transformed with λSV1 carries the plasmid at *attB* is presented in Fig. 6. DNA from λSV1, N6106, and a N6106(λSV1) transformant was isolated, digested with *Hinc*II and *Bam*H1, separated on a 1% agarose gel, and transferred to a GeneScreen membrane. ^{32}P-Labeled pWR1 DNA was pre-

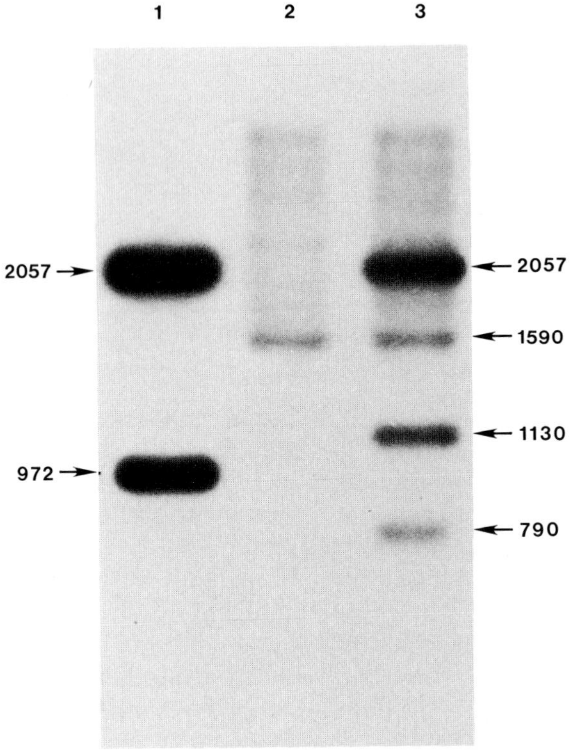

Fig. 6 Analysis of λSV1 DNA in an N6106(λSV1) transformant. DNAs from λSV1 (lane 1), N6106 (lane 2), and N6106(λSV1) (lane 3) were digested with *Hinc*II and *Bam*H1, separated by agarose-gel electrophoresis, transferred to a GeneScreen membrane, and hybridized to nick-translated ^{32}P-labeled pWR1 DNA.

pared by nick-translation and hybridized to the blotted DNA. The pWR1 probe shares homologies with λSV1 at *amp* and at *attP*; it also is homologous with a short region of the *int* cistron of the N6106 prophage.

Hybridization of the pWR1 probe to *Hinc*II- and *Bam*H1-digested circular λSV1 DNA (lane 1), detects a 2057-bp fragment containing a portion of *amp* and a 972-bp fragment carrying *attP*. With total N6106 DNA, the probe detects a 1590-bp fragment, which presumably contains *int,* as well as a fainter, heterogeneous background (lane 2). In lane 3, total N6106(λSV1) DNA is hybridized. We observe the 2057-bp fragment bearing *amp,* but not the 972-bp *attP* fragment. Instead, two fragments (1130 and 790 bp) are detected. These sizes are consistent with fragments carrying the *attL* and *attR* sites flanking an integrated copy of λSV1.

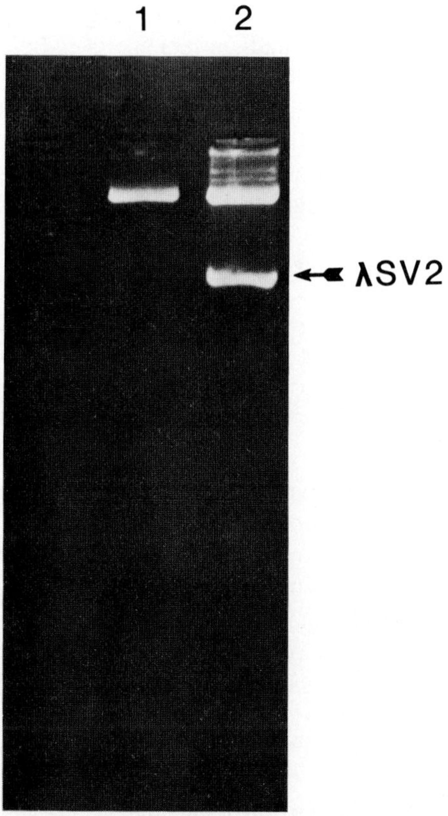

Fig. 7 Excision and amplification of λSV2 DNA. Supernatants cleared with lysozyme and Triton X-100 were prepared from cultures of N6106(λSV2) grown to mid-log phase at either 32°C (lane 1) or 42°C (lane 2). Aliquots were fractionated on a 1% agarose gel and stained with 0.5 μgm/ml ethidium bromide. The band in lane 1 and the more slowly migrating band in lane 2 represent heterogeneous higher molecular weight *E. coli* DNA.

F. Excision of λSV2 from an N6106(λSV2) Lysogen

To facilitate the purification and structural analysis of λSV recombinants, it may be preferable to convert the vector from a single-copy integrated form to the multicopy, extrachromosomal mode. This is readily achieved by thermally inducing an N6106(λSV) lysogen. Denaturation of the *cI857* repressor results in expression of the prophage *int* and *xis* genes and the consequent excision of the λSV genome. Figure 7 shows such an experiment. A log-phase culture of N6106(λSV2) was shifted from 32 to 42°C for 20 min and further incubated at

37–39°C for 2 hr. A supernatant cleared with lysozyme and Triton X-100 was then prepared and analyzed by ethidium bromide–agarose-gel electrophoresis. Heterogeneous higher molecular weight DNA migrates slowly in these gels; free plasmid migrates considerably faster. When extracts from uninduced and induced cultures are compared (Fig. 7, lanes 1 and 2, respectively), extrachromosomal, monomeric form-I λSV2 plasmid DNA is detected only in the latter. We estimate that the induction amplifies the vector λSV2 to 5–10 copies/cell (data not shown).

IV. Discussion

The bacteriophage λ Int protein, in concert with the *E. coli* IHF protein, mediates the site-specific integration of λ viral DNA into the host chromosome (Nash and Robertson, 1981). More than 90% of cells surviving phage infection can be stable lysogens. The experiments described here demonstrate that this integration mechanism can be modified and incorporated into a recombinant DNA vector system. Our system includes a λSV1 or a λSV2 plasmid, 7.7 or 8.6 kb, respectively, both of which carry a repressible λ replication region and the λ *attP* attachment site. The λSV plasmids are capable of autonomous replication in sensitive *E. coli*. We have also constructed specialized lysogenic host strains (N6106 and N6377), which provide Int and Xis after thermal induction. These strains may also be manipulated to provide Int alone by applying a period of thermal induction followed by growth at the noninducing temperature. The λSV vectors integrate into N6106 or N6377 after transformation; the overall efficiency of the reaction is 500–1800 colonies/ng closed circular vector DNA.

As is the case with wild-type λ prophage, the λSV plasmids will be stably maintained in the integrated form by λ *cI* repressor; on inactivation of the repressor, excision and amplification occur. The prophage in N6106 and N6377 carries a temperature-sensitive repressor. Thus the λSV-transformed cells maintain a single-copy, integrated vector at 32°C that can be converted to an autonomous, multicopy vector by shifting the culture to 42°C. The *red* and *gam* cistrons, together with most of the other prophage functions, are deleted in the lysogens. This means that the autonomous λSV vectors replicate as closed monomeric circles rather than as the concatamers characteristic of late λ replication (Enquist and Skalka, 1973). We estimate that after 2 hr at 41°C, about 5–10 molecules of closed circular λSV vector accumulate.

There are several potential advantages to the λSV system. The regulation of *E. coli* genes may be abnormal when these genes are present on multicopy plasmids. Our system permits cloned genes to be carried either in the multicopy mode, or as a single copy, integrated at *attB*. In the former state, structural studies are facilitated, whereas in the latter, gene control can be studied under physiological conditions.

Integration via homologous recombination can, in principle, be accomplished when λSV carrying cloned bacterial DNA is used to transform int^- lysogens at 32°C. Such integrants are useful for transferring mutations to the bacterial genome from cloned DNA fragments. At 32°C, the transformants are merodiploid; shift of the cells to 42°C selects cells that have lost the integrated plasmid and may have acquired the mutations.

The possibility of stably maintaining very large cloned DNA fragments, in the range of 30 to perhaps 100 kb, was a major motivation for the development of our system. Some mammalian genes or gene clusters are of this size. It might, for example, be possible to isolate the 65-kb human β-like globin locus (Fritsch *et al.*, 1980) on a single recombinant λSV clone. In addition, the capacity to clone large DNA pieces would both reduce the number of recombinants required to form complete genomic DNA libraries and simplify the process of mapping extended genomic regions.

The potential of the λSV system for isolating large DNA fragments is based on the following considerations. In a multicopy plasmid, a mutation that facilitates replication is rapidly amplified. Similarly, a mutant λ vector with an increased efficiency of packaging will overgrow other phages. In the case of small mutlicopy plasmids carrying large foreign DNA segements, these mutations are often deletions in the cloned fragment. DNA substitutions, deletions, and other rearrangements also occur, especially when the fragment carries symmetrical sequences. Propagation of a λSV recombinant as a single copy integrated in the *E. coli* chromosome eliminates competition between recombinant molecules and reduces the number of mutation targets. Furthermore, deletions that would confer selective growth advantage to an autonomous replicon would seem less likely to do the same for a bacterial cell; the contribution of an integrated λSV2 recombinant carrying a 50-kb insert to the total *E. coli* chromosome, which is about 5000 kb in size, is only about 1%.

The cloning of large DNA fragments will require a method of preparing undamaged DNA in excess of 100 kb, ligating such DNA efficiently to form recombinants of the correct structure, and assuring the uptake of these recombinants into competent *E. coli* recipients. With respect to the ligation step, we are pursuing an approach designed to minimize self-closure of the λSV vector, a reaction that is favored at the expense of the formation of large heterodimers. Cleavage of the vector at two sites (e.g., the unique *Pst*I and *Sal*I sites) to create nonidentical cohesive ends, blocks the self-closure of the vector; although head-to-head homodimeric vector molecules are formed, we find that these are inactive in transforming the specialized λ lysogen N6106 (data not shown). Vectors so treated are expected to accept exogenous DNA fragments terminated with *Pst*I and *Sal*I cohesive ends. Large fragments of these or other sets of nonidentical ends can be generated by partial digestion or by ligation to asymmetric linkers (Bahl *et al.*, 1978).

A third potential application of our system is the recovery of integrated vector DNA from transformed mammalian cells. This application is based on the idea that λSV recombinants can be transferred to mammalian cells, integrated into high-molecular-weight DNA, and stably carried in a form resembling integrated prophage (i.e., flanked by *attL* and *attR* sites). The prophage configuration in mammalian cells might be obtained by transforming these cells with DNA extracted from λSV bacterial lysogens. We have demonstrated that integrated λSV2 can be excised *in vitro* from the bacterial chromosome using partially purified Int, Xis, and host factor (J. Auerbach, M. E. Gottesman, S. Wickner, and B. Howard, unpublished). Rescue of the excised vector by transformation of the λ-sensitive strain N38 occurs with an efficiency of 500–1000 colonies/μgm bacterial DNA. The reaction occurs even in the presence of a large excess of mammalian DNA. This result suggests that it may be possible to rescue integrated recombinants from mammalian cells without resort to restriction-endonuclease digestion, a step that should considerably increase the usefulness of λSV vectors.

Acknowledgments

We thank D. Court and H. Nash for helpful discussions, G. Hobom and F. Blattner for communicating unpublished nucleotide sequences, and R. Weisberg, S. Rogers, and D. Berg for providing strains. We are grateful to Raji Padmanabhan for excellent technical assistance.

References

Alton, N., and Vapnek, D. (1979). *Nature (London)* **282,** 864–869.

Bahl, C. P., Wu, R., Brousseau, R., Sood, A. K., Hsiung, H. M., and Narang, S. A. (1978). *Biochem. Biophys. Res. Commun.* **81,** 695–703.

Birnboim, H. C., and Doly, J. (1979). *Nucleic Acids Res.* **7,** 1513–1523.

Chen, C. W., and Thomas, C. A. (1980). *Anal. Biochem.* **101,** 339–341.

Chia, W., Scott, M. R. D., and Rigby, P. W. J. (1982). *Nucleic Acids Res.* **10,** 2503–2520.

Dove, W. F., Inokuchi, H., and Stevens, W. F. (1971). *In* "The Bacteriophage Lambda" (A. D. Hershey, ed.), pp. 747–771. Cold Spring Harbor Lab., Cold Spring Harbor, New York.

Enquist, L. W., and Skalka, A. (1973). *J. Mol. Biol.* **75,** 185–212.

Fiers, W., Contreras, R., Haegeman, G., Rogiers, R., van de Voorde, A., van Heuverswyn, H., van Herreweghe, J., Volckaert, G., and Ysebaert, M. (1978). *Nature (London)* **273,** 113–120.

Fritsch, E. F., Lawn, R. M., and Maniatis, T. (1980). *Cell* **19,** 959–972.

Gorman, C., Moffat, L., Howard, B. (1982). *Mol. Cell. Biol.* **2,** 1044–1051.

Gottesman, M. E., Adhya, S., and Das, A. (1980). *J. Mol. Biol.* **140,** 57–75.

Howard, B. H., and Gottesman, M. E. (1982). *In* "Eukaryotic Viral Vectors" (Y. Gluzman, ed.), pp. 211–216. Cold Spring Harbor Lab., Cold Spring Harbor, New York.

Hsu, P. L., Ross, W., and Landy, A. (1980). *Nature (London)* **285,** 85–91.

Katz, L., Kingsbury, D. T., and Helinski, D. R. (1973). *J. Bacteriol.* **114,** 577–591.

Kröger, M., and Hobom, G. (1982). *Gene* **20,** 25–38.

Mandel, M., and Higa, A. (1970). *J. Mol. Biol.* **53,** 159–162.

Miller, J. H. (ed.) (1972). "Experiments in Molecular Genetics," p. 433. Cold Spring Harbor Laboratory, Cold Spring Harbor, New York.

Mulligan, R. C., and Berg, P. (1980). *Science (Washington, D.C.)* **209,** 1422–1427.

Mulligan, R. C., and Berg, P. (1981). *Proc. Natl. Acad. Sci. USA* **78,** 2072–2076.

Nash, H. A., and Robertson, C. A. (1981). *J. Biol. Chem.* **256,** 9246–9253.

Queen, C., and Korn, L. J. (1979). *Methods Enzymol.* **65,** 595–609.

Radloff, R., Bauer, W., and Vinograd, J. (1967). *Proc. Natl. Acad. Sci. USA* **57,** 1514–1521.

Reddy, V. G., Thimmappaya, B., Dhar, R., Subramanian, K. N., Zain, B. S., Pan, J., Ghosh, P. K., Celma, M. L., and Weissman, S. M. (1978). *Science (Washington, D.C.)* **200,** 494–502.

Rigby, P. W., Dieckmann, M., Rhodes, C., and Berg, P. (1977). *J. Mol. Biol.* **113,** 237–251.

Seeburg, P. H., Shine, J., Martial, J. A., Baxter, J. D., and Goodman, H. M. (1977). *Nature (London)* **270,** 486–494.

Southern, E. M. (1975). *J. Mol. Biol.* **98,** 503–517.

Sutcliffe, J. G. (1979). *Cold Spring Harbor Symp. Quant. Biol.* **43,** 77–90.

CHAPTER 8

Construction of Highly Transmissible Mammalian Cloning Vehicles Derived from Murine Retroviruses

RICHARD C. MULLIGAN

Center for Cancer Research and Department of Biology
Massachusetts Institute of Technology
Cambridge, Massachusetts

I. Introduction

Since the late 1970s, a considerable amount of progress has been made toward the development of general methods for introducing cloned DNA sequences into mammalian cells. Out of this work has evolved two basic approaches to gene transfer that are in widespread use today:

Experimental Manipulation
of Gene Expression

ISBN 0-12-372380-9

1. Transient expression assays (Banerji *et al.*, 1981; Mellon *et al.*, 1981), in which gene expression is assessed shortly after introducing the cloned sequences into cells (usually 12–72 hr after transfection or microinjection)
2. Stable transformation (Wigler *et al.*, 1977; Mulligan and Berg, 1981a), in which the DNA sequence of interest is stably introduced into the recipient cell, most often through integration of the exogenous sequence into chromosomal DNA of the recipient.

These systems have already become extremely valuable in the definition of common nucleotide sequences required for the expression of most, if not all, eukaryotic genes (McKnight *et al.*, 1981; Benoist and Chambon, 1981; Fromm and Berg, 1983), and they are beginning to shed light on the nature of DNA sequences involved in the regulated expression of specific genes (Huang *et al.*, 1981; Lee *et al.*, 1981; Mellon *et al.*, 1983).

Unfortunately, a major obstacle to this latter work may prove to be the extremely variable transformation frequencies obtained with different cells. Although it appears that many common cell lines used in the laboratory can be readily transformed using a number of standard transformation protocols (Graham and Van der Eb, 1973; Parker and Stark, 1979) and available selectable markers and selective schemes (Wigler *et al.*, 1977; Mulligan and Berg, 1981a; Southern and Berg, 1982), transformation frequencies for a number of specialized cell lines and primary tissues can be dramatically lower (some cells appear totally refractory to DNA-mediated transformation). This variable efficiency has severely restricted the range of recipients available for experiments that require a specific host cell in order to demonstrate the regulated expression of a cloned gene (Mellon *et al.*, 1983). Furthermore, it bodes ill both for the prospect of introducing sequences into a minor component of a heterogeneous cell population and for introducing whole populations of cloned sequences into cells.

In an attempt to overcome these limitations, we set out to develop a more powerful system for gene transfer in mammalian cells. We chose to begin genetic manipulations to convert the murine retroviruses to highly transmissible agents for the transfer of exogenous genes, principally because a number of features of the retrovirus life cycle (Weiss, 1982) appeared uniquely suited for gene transfer both *in vitro* and *in vivo*:

1. Although the viral genome is RNA, during the viral life cycle a DNA-intermediate is formed that is extremely efficiently integrated into the chromosomal DNA of the infected cells (in an invariant fashion with respect to the viral genome).
2. Mammalian cells are not generally killed by productive infection by retroviruses.

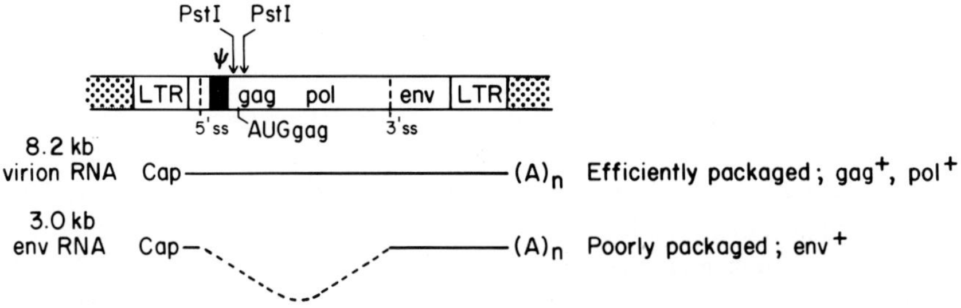

Fig. 1 DNA and RNA representations of a typical murine leukemia virus (WT MuLV) genome. The location of a number of features of the proviral DNA are indicated, including (1) the long terminal repeats (LTRs, see text), (2) the sequence implicated in the encapsidation of viral RNA (R. Mann, R. C. Mulligan, and D. Baltimore, unpublished), (3) the approximate positions of the viral protein-coding regions, and (4) the donor and acceptor sequences for the subgenomic *env* mRNA. The location of the AUG codon, which initiates translation of the viral *gag–pol* protein, is shown with respect to the two unique *Pst*I cleavage sites in MuLV.

3. A wide spectrum of both pluripotent and differentiated cells (*in vitro* and *in vivo*) are susceptible to retrovirus infection.

In this chapter we describe in detail the construction of a prototype retroviral vector, several novel methods for the generation of recombinant virus from cloned DNA sequences, the general characteristics of retrovirus-mediated transformation, and the structure of several useful derivative vectors for introducing both cDNA and genomic sequences into cells. Finally, we discuss some of the prospects for gene transfer using the system.

II. General Strategy

Rather than simply utilize elements of the retroviral genome in the construction of vectors designed for DNA-mediated transformation assays, our aim was to exploit the inherently high efficiency of gene transfer afforded by the infection of cells with intact virus particles. Therefore, our goal was to generate recombinant genomes that would not only promote the expression of inserted sequences but that would also retain the "cis" features necessary for efficient encapsidation into virus particles, reverse transcription, and integration of the proviral DNA intermediate. The location of some of these cis features on the viral genome are shown in Fig. 1, a diagram of both DNA and RNA representations of a typical murine leukemia virus (MuLV) genome. The minimal essential components of the retroviral transcriptional unit appear to be the 5′ and 3′ long terminal repeat

(LTR) sequences, which serve to promote and polyadenylate virion transcripts, respectively. Adjacent to the 5′ LTR are sequences necessary for reverse transcription of the genome (the tRNA primer binding site) and the efficient encapsidation of viral RNA into particles (the ψ site) (R. Mann, R. C. Mulligan, and D. Baltimore, unpublished). Adjacent to the 3′ LTR are sequences implicated in positive-strand viral DNA synthesis during reverse transcription.

One approach to generating infectious recombinant genomes described by Wei *et al.* (1981) involves the stable introduction of recombinant viral DNA constructs into cells and subsequent infection of the transformants with wild-type helper virus in order to rescue the recombinant transcripts as infectious viruses. This procedure is based on the supposition that the integrated sequences would give rise to RNA of the proper structure for encapsidation in virions and would be packaged in the presence of the necessary viral proteins (supplied by the helper virus). Although this method is suitable for the generation of a particular recombinant viral species, it suffers one of the same fundamental limitations as DNA-mediated transformation itself—although the recombinant genome is eventually mobilized to be highly transmissible, the low efficiency of stable DNA-mediated transformation severely restricts the representation of transfected DNA molecules as virus particles. Furthermore, because the integrity of sequences transferred by DNA-mediated transformation is never insured, only a fraction of the transformants may be competent to yield recombinant genomes of the desired structures. In light of our considerable long-term interest in the ability to introduce complete libraries of cDNA recombinants into cells using retrovirus vectors (Section VII), this method was not suitable for our studies.

To effect a more efficient recovery of virus from cloned DNA, we chose instead to rescue virus shortly after transfection of the recominant vector into cells already productively infected with wild-type virus, prior to stable integration of the sequences. In contrast to stable transformation, in which approximately 0.01% of the transfected cells harbor and express the introduced sequences, during this early period of "transient expression" (between 12 and 72 hr after transfection), more than 10% of the transfected cells can actively express foreign sequences (Parker and Stark, 1979; Banerji *et al.*, 1981).

Our protocol for rescuing virus is diagramed in Fig. 2. The vectors we utilize all contain two basic components:

1. A manipulated viral transcriptional unit, which in most cases contains a functional dominant-acting genetic marker to permit direct titering of the rescued recombinant virus and a unique restriction site within a region of the virus genome engineered to promote expression of an inserted gene
2. A plasmid backbone to permit the propagation of the recombinant molecules in bacteria and to promote amplification of the transcriptional template after introduction into animal cells (see Section III for details about the vectors)

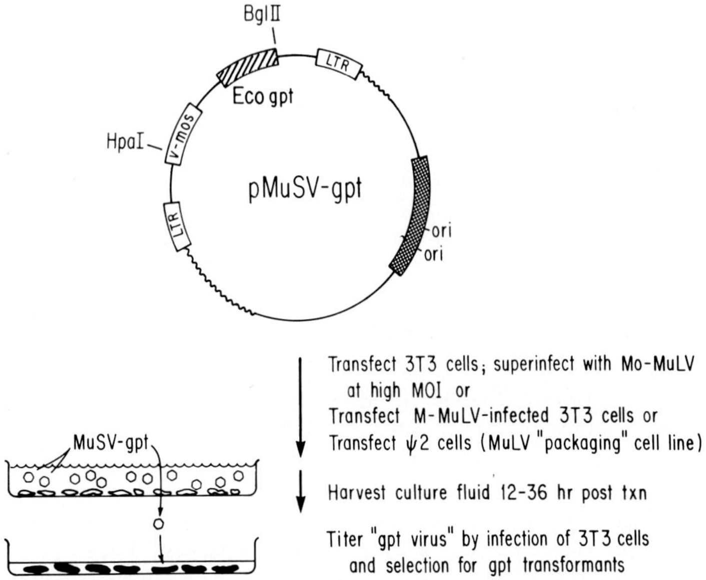

Fig. 2 Protocol for conversion of recombinant retroviral DNA sequences into infectious genomes. txn, Transfection. See text for details.

The vector DNAs are introduced into MuLV-infected NIH/3T3 cells or into a ''packaging'' cell line, which promotes the encapsidation of only recombinant RNA species (R. Mann, R. C. Mulligan, and D. Baltimore, unpublished; and see Section VII). As seen in Section IV, transmissible recombinant virus is then released into the culture fluid in a time-dependent fashion and can be titered by infecting fresh 3T3 cells and subjecting the cultures to the appropriate selective conditions. This ''transiently rescued'' virus can be used directly for infection of the desired recipient or to generate a stable cell line producing a high titer ($>10^6$ colony-forming units/ml culture fluid) and an eternal source of the desired recombinant virus.

III. Construction of a Prototype Retrovirus Vector

As discussed in Section II, our goal in manipulating the viral genome was to generate a hybrid transcriptional unit that upon introduction into the appropriate cell would give rise to a hybrid RNA competent not only for encapsidation into retrovirus particles but also for all subsequent stages in the viral life cycle (e.g., reverse transcription of the viral genome and integration of proviral DNA into genomic DNA of the infected cell). In addition to these constraints, we wanted to begin to explore whether the strong retroviral transcriptional signals could be exploited to promote the expression of exogenous protein-coding sequences inserted in place of normal viral genes.

In our initial constructions, we chose as a matter of convenience to manipulate the genome of the M1 murine sarcoma virus (MuSV), a replication-defective acute transforming retrovirus (Fischinger *et al.*, 1972). This viral genome, by definition, contains all of the sequences necessary for its propagation in the presence of helper virus, in spite of the fact that a considerable amount of its parent genome, Moloney MuLV (M-MuLV), has been replaced with rat-derived cellular sequences. For our work, we utilized a pBR322-MuSV recombinant that contains a 7-kb *Eco*R1 fragment including the entire MuSV proviral DNA as well as some flanking mink-cell DNA sequences (Van de Woude *et al.*, 1979; and see Fig. 3). The first part of the construction involved the generation of a unique restriction site in a region of the MuSV genome expected to promote the expression of inserted sequences. Because in MuSV all of the MuLV *env* coding sequences are missing as well as the putative 3′ splice site for the subgenomic *env* mRNA, we decided to replace the *gag–pol* coding sequences with exogenous DNA. As seen in Fig. 1, the AUG codon initiating translation of the *gag–pol* polyprotein is straddled by two *Pst*I restriction-endonuclease cleavage sites ~180 nucleotides apart (Shinnick *et al.*, 1981). In an attempt to position the foreign coding sequences as close to the normal viral *gag* coding sequence as possible, we introduced a *Bgl*II site at the first of the two *Pst*I sites, thus rendering the site of insertion ~57 nucleotides from the normal start of *gag–pol* translation. Coupled with that manipulation was the introduction of a modified pBR322 backbone, which lacks the "poison sequences" known to inhibit the replication of pBR322-based transient expression vectors in animal cells (Lusky and Botchan, 1981).

The exogenous coding sequence we chose to insert into the viral backbone at the *Bgl*II site was a 0.9-kb segment of *Escherichia coli* DNA encoding the bacterial xanthine-guanine phosphoribosyl transferase, a purine salvage enzyme. Exquisitely sensitive enzymatic assays exist for the detection of this protein in cells (Mulligan and Berg, 1980), and more importantly, the gene can be used as a dominant-acting genetic marker in most mammalian cells (Mulligan and Berg, 1981a). The gene is therefore an ideal choice not only for quantifying the level of translational expression that can be be obtained by insertion of sequences into various sites in the viral genome, but also for providing a direct means of titering rescued recombinant virus (using the *Eco gpt* transformation assay).

The final aspect of the construction involved the introduction of a segment of the polyoma virus genome into the plasmid backbone. Polyoma virus is a papovavirus that productively infects mouse cells in culture (Tooze, 1973). The

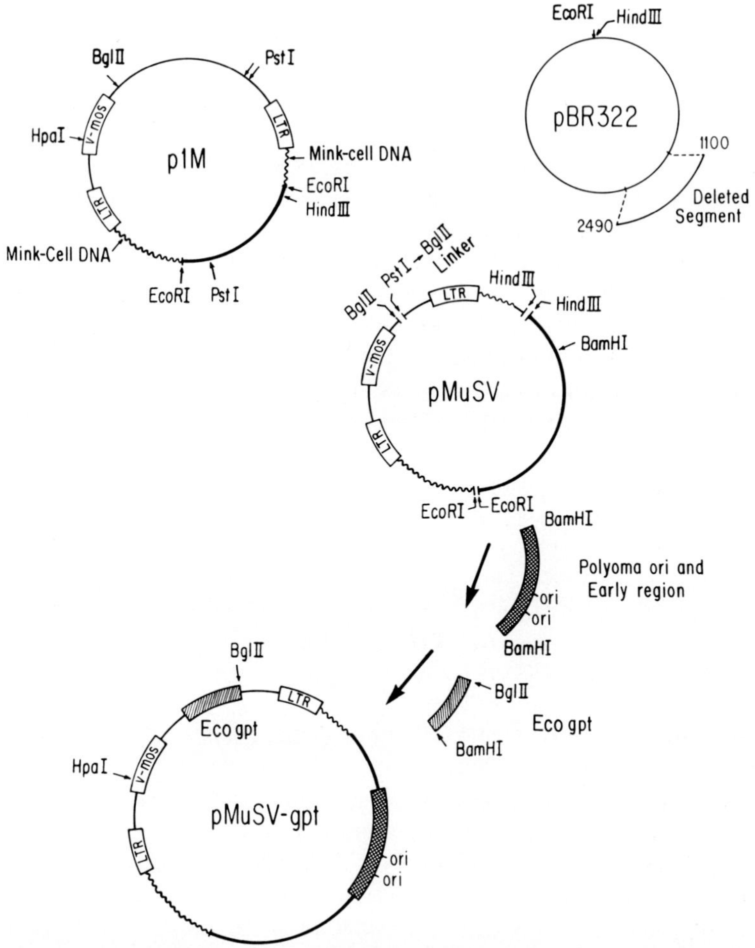

Fig. 3 Construction of pMuSV-*gpt*. The starting materials for this construction are shown at the top of the figure: p1M, a hybrid plasmid DNA containing a mink cell DNA-derived integrated copy of the M1 M-MuSV viral genome (see Van de Woude *et al.*, 1979) and a modified pBR322 genome that lacks ''poison'' sequences, which have been shown to inhibit replication of plasmid recombinants in mammalian cells (Lusky and Botchan, 1981). Through the series of enzymatic manipulations outlined, the final product, pMuSV-*gpt,* was constructed. In pMuSV-*gpt,* the direction of viral transcription is counterclockwise, as is the direction of transcription of the polyoma virus early region. The boxed long terminal repeat (LTR) and v-*mos* sequences denote the long terminal redundancies found in retrovirus proviral DNA and MSV transforming sequences, respectively. The *Pst*I site in the pMuSV molecule is located ~50 nucleotides to the 5′ side of the AUG codon initiating *gag–pol* translation (Shinnick *et al.*, 1981). The actual details of the construction are presented elsewhere (R. C. Mulligan, unpublished).

"early" region of the viral genome (that expressed shortly after infection) has been shown by a number of workers to encode a gene product (termed large-T antigen) that serves to drive the replication of the viral genome in a productive infection through interactions with the viral origin of replication (sequences located at one end of the large-T gene). When excised from the viral genome and attached to a bacterial replicon, this early region promotes the transient replication of the recombinant plasmid when introduced into mouse cells by transfection. Such recombinant plasmids have been used by a number of workers (particularly using the analogous SV40 sequences and monkey cells) to increase the transient expression of sequences through the transient amplification of transcriptional templates (Mellon *et al.*, 1981; Mulligan and Berg, 1981b). For our studies, a 3.8-kb segment of a mutant polyoma genome (Magnusson and Nilsson, 1978) bounded by *Bam*HI restriction sites was used that contains the entire large-T antigen transcriptional unit and two viral origins of replication. This segment was introduced into the vector at the *Bam*HI site within pBR322 sequences.

IV. Rescue of Recombinant Genomes as Infectious Virus

A. Use of MuLV-Infected NIH/3T3 Cells as Recipients

To test for the recovery of transmissible *gpt* virus, NIH/3T3 cells productively infected with M-MuLV were transfected with 10 μgm of pMuSV-*gpt* DNA using a modification (Parker and Stark, 1979) of the method of Graham and Van der Eb (1973). After the glycerol shock, harvests of culture fluid from the transfected cells were made every 8–10 hr for several days and frozen at −80°C. The putative recombinant virus stocks were then titered for the presence of both helper and recombinant virus. For M-MuLV titers, standard XC plaque assays were performed and indicated a titer of $>10^6$ XC plaques/ml for each harvest period. To determine the titers of *gpt* virus, dilutions of culture fluid were used to infect fresh 3T3 cells in the presence of 8 μgm/ml polybrene (Aldrich Chemical Co.) for 2.5 hr. After culturing the cells under nonselective conditions for 3 days, they were split into mycophenolic acid-containing selective media (Mulligan and Berg, 1981a) and fed every 3 days. The results for the *gpt* titering are shown in Fig. 4. At the peak of *Eco gpt* virus production, between 8 and 18 hr after transfection, $>10^5$ *gpt* colony-forming units/ml (gm cfu/ml) were produced in the transfected cell cultures. The characterization of these transformants is discussed in Section V. This is a surprisingly high titer of virus production in

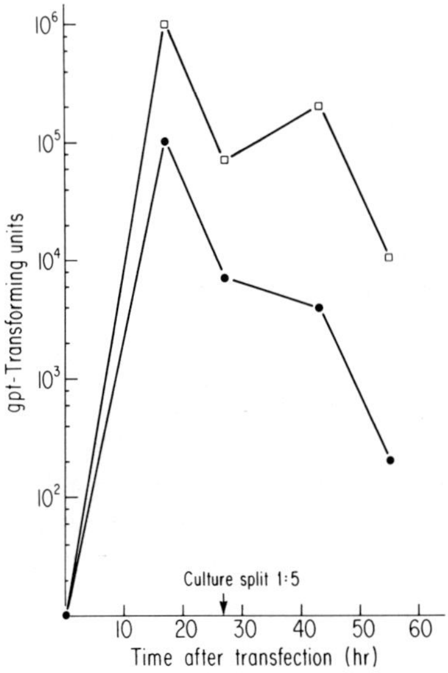

Fig. 4 Time-course titering of transiently rescued *gpt* virus. MuLV-infected 3T3 cells (1–5 × 10^6 cells) were transfected with 10 μgm of *gpt* DNA, as described in the text. Culture fluid was harvested at the indicated times, and dilutions were used to infect 3T3 cells. After each harvest, the transfected culture was replaced with fresh media. At 27 hr after transfection, the cells approached confluence and were therefore split 1:5 into fresh dishes. ●, *gpt* colonies/ml; □, total number of *gpt* colonies.

light of the fact that equivalent titers of acute transforming virus are often obtained from cultures in which 100% of the cells harbor the transforming genome. Furthermore, the high *gpt* titer is at least suggestive that the process of representing cloned recombinant DNAs as virus is indeed extraordinarily efficient. In particular, the data imply that the system would permit the recovery of a recombinant molecule constituting less than 10^{-3}% of the transfected DNA population. This efficiency approaches that anticipated as necessary to insure the introduction of single-copy cDNAs present in a complete cDNA library into cells.

B. Use of ψ2, an MuLV "Packaging" Cell Line, as Recipient

Although for many applications the presence of the MuLV helper virus genome in transformants is irrelevant, there exist a variety of situations in which this may not be desirable. This is a particular concern in experiments in which the vectors are used to introduce new sequences into the intact animal (Section VII). In anticipation of these situations, we constructed a mutant MuLV genome in which the region of the genome implicated in the encapsidation of viral RNA into virions was deleted (the ψ sequence in Fig. 1). By stably introducing this genome into NIH/3T3 by DNA cotransformation, we were able to isolate stable transformants that produced all of the viral proteins used for encapsidation, yet that budded noninfectious particles (R. Mann, R. C. Mulligan, and D. Baltimore, unpublished). We further demonstrated that this cell line could be used as a recipient for transient-rescue experiments, resulting in the efficient production of helper-free stocks of recombinant virus (once a replication-defective genome is encapsidated, the particle contains the proteins necessary to permit infection and efficient integration).

We routinely use the packaging cell line ψ2 for transient-rescue experiments. The kinetics of virus rescue appear very similar to those obtained using MuLV-invected cells, and the titers of virus are only slightly depressed (<10-fold). As an alternative to transient rescue as a means of generating virus, any recombinant viral DNA construct can also be stably introduced into the ψ2 line by standard DNA cotransformation procedures, resulting in the generation of stable cell lines that produce high titers (>10^6 cfu/ml) of the desired recombinant virus, free of any detectable helper.

V. Characteristics of Retrovirus-Mediated Transformation

From the wealth of information available regarding the retrovirus life cycle, a number of straightforward predictions could be made about the characteristics of retrovirus-mediated transformation. The first prediction was that most of the *gpt*-positive colonies obtained after the transient-rescue protocol outlined in Section III,A would in fact be high-titer producers of both recombinant *gpt* virus and helper virus. This seemed likely because of the relative excess of helper titer during each harvest period. To test this idea, a number of the *gpt*-selected colonies were expanded and culture fluids harvested and again titered for *gpt* transforming activity. Every colony tested was indeed a producer, with the titers ranging from 10^5 to 10^6 cfu/ml. Nonproducer *gpt* transformants (transformants harboring only a recombinant genome) could also be obtained using high dilutions of the mixed stocks, and viruses could be subsequently rescued from those cells after their superinfection with wild-type virus. Transformants generated by transient rescue through ψ2 cells in no case were viral producers, although *gpt* virus could also be rescued by superinfection of the cells with wild-type virus.

Another expectation was that the recombinant viral genomes would be integrated into cellular DNA in the same way as wild-type retrovirus genomes; integration would occur so as to preserve the entire viral genome, and duplication of the LTRs would occur as a consequence of reverse transcription. Furthermore,

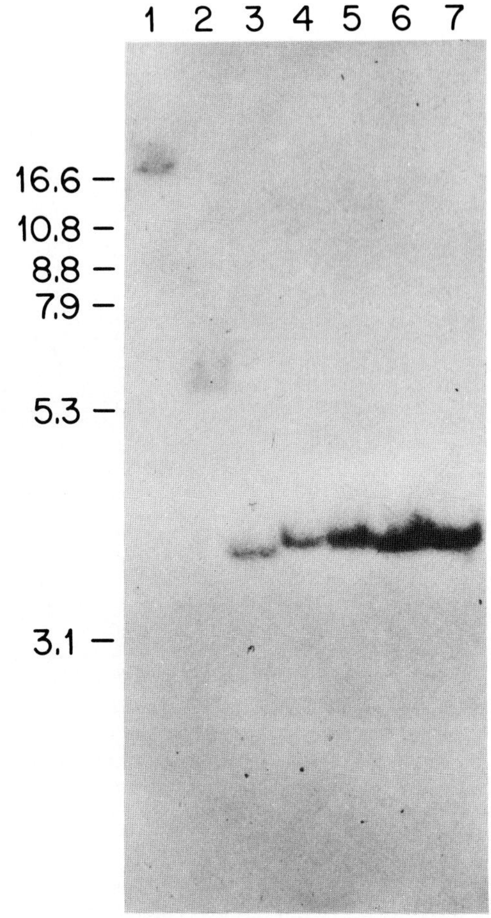

Fig. 5 Southern analysis of transformed cell DNAs. High-molecular-weight DNA was isolated from four independently derived *gpt* transformants as described by Wigler *et al.* (1977). Then, 10 μgm of transformant-cell DNA were cleaved with either *Xba*I or *Eco*R1, electrophoresed on a 1.2% agarose gel, and transferred to nitrocellulose (Southern, 1975). The ''blot'' was hydridized with $\sim 10^7$ cpm of ^{32}P-labeled nick-translated *gpt* probe (10^8 cpm/μgm), washed, and exposed for 2 days. (Tracks 1 and 2) *Eco*R1-cleaved DNA from two transformants. (Track 3) 10 pgm MuSV *gpt* DNA and 10 μgm 3T3 DNA. (Tracks 4–7) *Xba*I-cleaved DNA from four transformants. Size standards (kb) are indicated to the left of the figure.

because the transformation events were mediated by virus-particle infections, only a few copies of the viral genome should have been integrated, in contrast to the results often obtained using DNA-mediated transformation. To examine these points, transformed-cell DNA was analyzed by the method of Southern (1975; Fig. 5). First, to assess the integrity of the transferred recombinant genomes, cell DNA from a number of transformants was cleaved with the re-

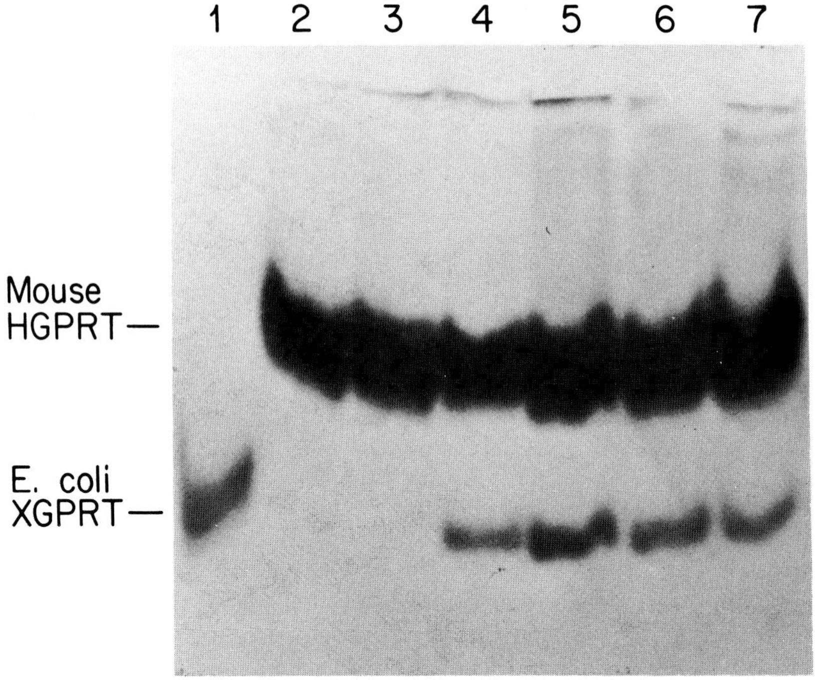

Fig. 6 *In situ gpt* enzyme assay of transformant cell extracts. Protein extracts from transformants were prepared and assayed as described by Mulligan and Berg (1980). Extract from $\sim 10^5$ cells was loaded in each track. The positions of the mammalian and bacterial enzymes are indicated. (Track 1) *Escherichia coli* extract. (Track 2) pSV2 *gpt*-transformed cell extract. (Track 3) 3T3 cell extract. (Tracks 4–7) Retrovirus-mediated transformant extracts.

striction endonuclease *Xba*I (which cleaves the input viral genome once in each LTR), electrophoresed on a 1.2% agarose gel, and transferred to nitrocellulose as described by Southern (1975). As marker, 10 pgm of the input pMuSV *gpt* DNA were mixed with 3T3-cell DNA and cleaved with *Xba*I as well. The blot was then hybridized with a ^{32}P-labeled *gpt* probe and processed in the standard way. Tracks 4–7 (Fig. 5) indicate that no gross rearrangement of input-vector sequences has occurred during gene transfer. Furthermore, the generation of unique hybridizing bands after *Xba*I digestion indicate generation of the expected proviral structure after reverse transcription of the recombinant template. To determine copy number in the transformants, two of the cell DNAs were cleaved with *Eco*R1, which does not cleave the viral genome. Tracks 1 and 2 show that, as expected, relatively few copies of the viral genome were transferred. To date, we have examined a wide variety of retrovirus-mediated transformants extensively by both northern and Southern analysis to assess the integrity of transferred vector sequences and copy number, and in almost all cases the results confirm those presented here.

Another rather obvious characteristic of the transformants is that they synthesize the bacterial *Eco gpt* enzyme, because *gpt* selection was the basis for their isolation. A question of interest, however, was whether or not the placement of the *Eco gpt* coding sequence within the viral genome would lead to high-level expression of the enzyme, as predicted by the unusual strength of the retrovirus promoter (Weiss, 1982). Figure 6 shows the results of an *in situ* enzymatic assay for the *E. coli* XGPRT performed on cell-transformant protein extracts. Indeed, every retrovirus-mediated transformant synthesized the bacterial gene product (tracks 4–7). More interesting, however, is the fact that the level of enzyme expression in the retrovirus-mediated transformants is significantly higher than in a typical transformant generated by DNA-mediated transformant using an SV40-based vector [the enzyme activity in the pSV2 *gpt* transformation shown in track 2 (Fig. 6) is below detection after reproduction]. In a variety of constructs with a variety of different genes, we have confirmed this strong translational expression.

VI. Useful Derivative Vectors

A. Vectors for Promoting cDNA Expression

To date, we have examined in detail two different strategies for engineering the expression of cDNA sequences using retrovirus vectors. The first involves use of the retrovirus transcriptional signals themselves to drive the expression of both a selectable marker and the cDNA. As diagrammed in Fig. 1, MuSV gene products are normally expressed by virtue of the formation of two distinct viral RNA species. The *gag–pol* protein is translated from the genomic length RNA, and the *env* gene product from the subgenomic, spliced message (Weiss, 1982). Although work with the pMuSV *gpt* vector demonstrated that insertions could effectively be made within the *gag–pol* region, we could not assess the feasibility of introducing similar insertions in the *env* coding region using the same vector, because sequences involved in the biogenesis of the *env* mRNA, define the 3′ splice site, (3′ss) are not present in the particular MSV backbone utilized. To address this question, we have constructed derivatives of the MuSV-based vectors in which a segment of the MLV genome containing the *env* 3′ss has been added in such a way that unique restriction sites exist for insertion in both the *gag–pol* and *env* regions (J. Hellerman, H. Kronenberg, and R. C. Mulligan, unpublished). By introducing either the *Eco gpt* or Tn5 phosphotransferase sequences within the *env* region and a variety of cDNA sequences into the *gag–pol* region, we have been able to demonstrate the successful dual expression of selectable and nonselectable genes (see Fig. 7 for an example of such a construct).

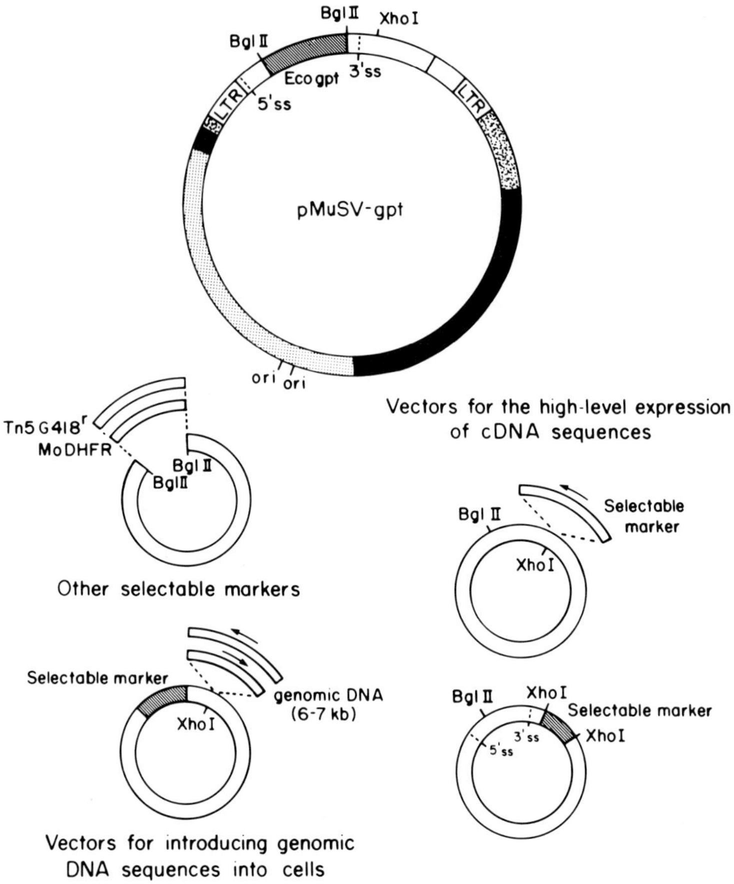

Fig. 7 Useful derivative retrovirus vectors (see text for explanation). Important restriction-endonuclease cleavage sites are shown (the *Bgl*II and *Xho*I sites are located in the same positions as the *Bgl*II and *Hpa*I sites in Fig. 1, respectively). The arrows indicate transcriptional orientation of inserted sequences. The retroviral transcriptional orientation is clockwise in each diagram.

A variation on this theme is the use of the *gag–pol* site for the insertion of cDNAs and an exogenous chimeric gene for expression of the selectable marker. A prototype vector of this design that we have constructed and utilized successfully makes use of a chimeric *gpt* gene consisting of promoter sequences derived from the SV40 genome, the *Eco gpt* segment, and the large intervening sequence and polyadenylation site of the human β-globin gene. This type of vector has the advantage that marker expression is not dependent on efficient production of an analog of the *env* message. This may be a particularly useful feature of our vectors; in recombinants that do require production of an analog of *env* message, marker titer appears to be influenced by the nature of insertions in the *gag–pol* site, presumably because aberrant splicing into the inserted sequence (Mulligan and Berg, 1981b) can in some cases greatly decrease the amount of normally spliced marker message and therefore limit expression of the marker gene product (J. Hellerman, H. Kronenberg, and R. C. Mulligan, unpublished).

Because in a number of instances it would be desirable to modulate the level of cDNA expression rather than obtain constitutive high-level expression, we have introduced totally exogenous transcriptional units into retrovirus vectors for the expression of cDNAs as well. In these vectors, selectable marker expression is mediated by the retrovirus transcriptional signals (Fig. 7).

B. Vectors for Introducing Genomic DNA Sequences into Cells

The backbone of vectors of this type contain (1) a selectable marker, such as the *Eco gpt* or Tn5 phosphotransferase genes, the expression of which is driven either by retrovirus transcriptional signals or by totally exogenous signals (as described in Section VI,A) and (2) a unique restriction site for the insertion of DNA sequences (Fig. 7). In the design of these vectors, particular attention has been paid to minimizing the amount of viral sequences present in the recombinant genome. Although we do not yet know the maximum genome size that can be efficiently encapsidated into virions, the largest murine genome isolated approaches 12kb in size (S. P. Goff, personal communication). If this does reflect the upper limit of packagability, then our standard vectors can accommodate ~7 kb of exogenous sequence. This size limit may seriously limit the usefulness of retrovirus vectors for introducing genomic sequences containing intervening sequences into cells.

In the introduction of sequences into the vectors, careful consideration must be given to the optimal orientation of the segment. One of the prerequisites for the high transmissibility of recombinant genomes is that virion transcription not be perturbed. The reason for this is that the terminally repeated sequences present at both the 5′ and 3′ ends of the virion RNA are necessary for the genome's reverse transcription. Any insertion of sequence that prematurely terminates transcription is expected to drastically reduce virus transmissibility. If a genomic sequence is introduced into a vector such that the transcriptional direction of the gene is the same as that of viral transcription (the *sense* direction), the genomic sequence's natural polyadenylation site may be used in preference to the viral site within the 3′ LTR. [Although the rules governing the usage of tandem polyadenylation sites have not yet been formulated, it is clear that some sites are used exclusively when placed in competition with others (M. Horowitz and P. A. Sharp, unpublished).] In light of this potential problem, a considerable amount of variability in viral titers may be expected from construct to construct. Another difficulty with inserting sequences into vectors in the sense orientation is that the strong retroviral signals for transcription may confuse analysis of any gene expression driven by the gene's own transcriptional signals.

Therefore, it appears unlikely that useful information will generally be obtained through the sense orientation of genomic sequences in retrovirus vectors. A practical application of such constructs, however, may be the recovery of truncated copies of genomic DNAs that have lost their introns during the retrovirus life cycle. Temin's laboratory has reported an example of such "intron processing" in studies in which a segment of the human β-globin gene was introduced into an avian retrovirus transcriptional unit (Shimotohno and Temin, 1982). If this phenomenon generally holds true, it may be possible to generate cDNAs from any genomic sequence by insertion of the genomic DNA into a retrovirus vector.

Orientation of genomic DNA sequences in the antisense direction presents fewer clear-cut problems. Nonetheless, in this situation there is concern over the potential effect of opposing transcriptional units on the expression of either virion RNA or the gene's transcript. At this point we have information on two constructions of this type that suggests no unique problem exists. We have constructed many more recombinants of this type and are characterizing expression of the inserted sequences.

VII. Conclusions and Prospects

The retrovirus vector system described here appears to possess a number of distinct advantages over currently available methods for introducing DNA sequences into mammalian cells. Perhaps the single most powerful feature of the system is the extraordinarily efficient means of gene transfer afforded by the infection of recipient cells with intact virus particles. In sharp contrast to other techniques such as DNA-mediated transformation and protoplast fusion (Schaffner, 1980), the efficiency of transforming cells using retroviral vectors can be 100%. This property may be particularly useful in cases in which it is necessary to transform an entire population of cells (e.g., when attempting to introduce sequences into a minor component of a heterogeneous cell population, such as the stem cells in murine bone marrow). A related property of the system is the ability to represent a large number of independent cloned DNA molecules as infectious viruses. This makes it feasible to envisage the widespread use of mammalian gene transfer as a means of gene isolation (in particular, the introduction of entire cDNA-vector libraries into cells in a functional form).

Another potential advantage of using intact retrovirus particles for gene transfer derives from the well-recognized ability of retroviruses to interact with other classes of enveloped viruses in the formation of viral pseudotypes with altered host ranges (Weiss *et al.*, 1975). Although the system as described permits the introduction of exogenous sequences only into murine cells, it appears likely that the host range of the vectors could be significantly extended through this type of approach. We have obtained support for this notion through experiments in which several recombinant genomes have been successfully encapsidated in amphotropic rather than ecotropic ''coats'' (Mulligan *et al.*, unpublished). These amphotropic pseudotypes infect a wide variety of mammalian cells (Hartley and Rowe, 1976).

From a paractical standpoint, the usefulness of the type of vector described here depends upon the availability of derivatives of our prototype vectors that promote the expression of both selectable and nonselectable genes. The design of these types of vectors (briefly outlined in Fig. 7) is a major effort in our laboratory and one that is evolving as we understand better the strategies of normal retroviral gene expression. Areas of current study that bear on this effort include

1. The mechanism that regulates the production of genomic versus subgenomic viral RNA
2. The consequences of opposing, overlapping transcriptional units.
3. The nature of viral sequences necessary for integration and/or reverse transcription
4. The constraints on genomic size for encapsidation

One important advance in vector design is the development of derivative vectors that permit extremely rapid recovery of recombinant genomes propagated in mammalian cells as molecular clones (C. Cepko and R. C. Mulligan, unpublished). This feature will be particularly useful in attempts to isolate genes on the basis of their expression in mammalian cells.

A final feature of the system that is worthy of note is the use of ψ2, an MuLV-packaging cell line, to rescue pure stocks of recombinant virus free of helper. As mentioned in Section IV,B, although the generation of mixed stocks of MuLV and recombinant virus is irrelevant for most gene-transfer experiments, a number of situations exist in which the presence of helper virus could cause problems. One example would be an attempt to introduce multiple copies of recombinant genomes in cells, either to amplify the level of a gene's expression or to attempt to mutate a locus by insertional mutagenesis. One feature of the retrovirus life cycle that makes this difficult is that a cell infected with wild-type virus becomes resistant to superinfection by viruses with the same envelope properties, because of interactions between the viral envelope protein expressed by the resident genome and the actual receptor for viral uptake (Rubin, 1960). Although cells can be synchronously infected with high multiplicities of virus, one cannot easily obtain repeated infection. The ability to produce stocks of recombinant virus free of helper overcomes this limitation, because no helper-virus genomes are ever expressed in transformants generated with ψ2 virus.

We suspect that the use of helper-free recombinant virus studies may also be important for *in vivo* gene-transfer experiments. One major focus in our laboratory is to attempt to introduce recombinant genomes into the mouse, either by the infection of embryos (Jaenisch, 1976) or bone marrow tissue. In both of these situations, the presence of helper virus may have profound consequences, in that the recipient is capable of actively replicating wild-type virus. The use of ψ2-generated virus should overcome these difficulties.

Acknowledgments

I acknowledge the support of John T. Potts, Jr. in the early phases of this work; David Baltimore and Richard Mann, my collaborators in the ψ2 experiments; and the current members of my laboratory. The work was supported by grants from the NCI, the Whitaker Foundation, and the McArthur Foundation.

References

Banerji, J., Rusconi, S., and Schaffner, W. (1981). *Cell* **27,** 299–308.

Benoist, C., and Chambon, P. (1981). *Nature (London)* **290,** 304–310.

Fan, H. (1977). *Cell* **11,** 297–305.

Fischinger, P. J., Nomura, S., Peebles, P. T., Haapala, D. K., and Bassin, R. H. (1972). *Science (Washington, D.C.)* **176,** 1033–1035.

Fromm, M., and Berg, P. (1983). *J. Mol. Appl. Genet.* **1,** 457–481.

Graham, I. L., and Van der Eb, A. J. (1973). *Virology* **52,** 456–467.

Hartley, J. W., and Rowe, W. P. (1976). *J. Virol.* **19,** 19–25.

Huang, A. L., Ostrowski, M. C., Benard, D., and Hager, G. (1981). *Cell* **27,** 245–255.

Lee, F., Mulligan, R. C., Berg, P., and Ringold, G. (1981). *Nature (London)* **294,** 228–232.

Lusky, M., and Botchan, M. (1981). *Nature (London)* **293,** 79–81.

Magnusson, G., and Nilsson, M. G. (1978). *J. Virol.* **22,** 646–653.

McKnight, S. L., Gavis, E. R., Kingsburg, R., and Axel, R. (1981). *Cell* **25,** 385–398.

Mellon, P., Parker, V., Gluzman, Y., and Maniatis, T. (1981). *Cell* **27,** 279–288.

Mulligan, R. C., and Berg, P. (1980). *Science (Washington, D.C.)* **209,** 1422–1427.

Mulligan, R. C., and Berg, P. (1981a). *Proc. Natl. Acad. Sci. USA* **78,** 2072–2076.

Mulligan, R. C., and Berg, P. (1981b). *Mol. Cell. Biol.* **1,** 449–459.

Parker, B., and Stark, G. S. (1979). *J. Virol.* **31,** 360–369.

Rubin, H. (1960). *Proc. Natl. Acad. Sci. USA* **46,** 1105–1119.

Schaffner, W. (1980). *Proc. Natl. Acad. Sci. USA* **77,** 2163–2167.

Shimotohno, K., and Temin, H. M. (1982). *Nature (London)* **299,** 265–268.

Shinnick, T. M., Lerner, R. A., and Sutcliffe, J. G. (1981). *Nature (London)* **293,** 543–548.

Southern, E. M. (1975). *J. Mol. Biol.* **98,** 503–517.

Southern, P. J., and Berg, P. (1982). *J. Mol. Appl. Genet.* **1,** 327–341.

Tooze, J. (ed.) (1973). ''Molecular Biology of Tumor Viruses.'' Cold Spring Harbor Lab., Cold Spring Harbor, New York.

Van de Woude, G. F., Oskarsson, M., McClements, W. L., Enquist, L. W., Blair, D. G., Fischinger, P. J., Maizel, J. V., and Sullivan, M. (1979). *Cold Spring Harbor Symp. Quant. Biol.* **44,** 735–745.

Wei, C. M., Gibson, M., Spear, P. G., and Scolnick, E. M. (1981). *J. Virol.* **39,** 935–944.

Weiss, R. A. (1982). ''The Molecular Biology of RNA Tumor Viruses.'' Cold Spring Harbor Lab., Cold Spring Harbor, New York.

Weiss, R. A., Boettiger, D., and Love, D. N. (1975). *Cold Spring Harbor Symp. Quant. Bio.* **39,** 913–918.

Wigler, M., Silverstein, S., Lee, L. S., Pellicer, A., Cheng, Y. C., and Axel, R. (1977). *Cell* **11,** 223–232.

CHAPTER 9

Use of Retrovirus-Derived Vectors to Introduce and Express Genes in Mammalian Cells

ELI GILBOA

Department of Biochemical Sciences
Princeton University
Princeton, New Jersey

I. Introduction

There has been a surge in the development of experimental systems to introduce and express genes in mammalian cells. The main feature of such transduction systems is a *vector*, a defined DNA fragment carrying various properties affecting the propagation and/or expression of the transduced gene in the recipient cells. The first and most widely used mammalian transduction system was

Experimental Manipulation
of Gene Expression

ISBN 0-12-372380-9

derived from the genome of SV40, a DNA virus of monkey origin (Hamer and Leder, 1979; Mulligan *et al.,* 1979, Gruss and Khoury, 1981).

In this chapter I describe a mammalian transduction system derived from the genome of a murine retrovirus, Moloney murine leukemia virus (M-MuLV). This experimental system is designed to introduce genes into the chromosome of mammalian cells at the rate of one or a few copies per genome and to permit expression of the transduced genetic information in a stable fashion, enabling the isolation and indefinite propagation of cells carrying the newly acquired gene. Several retrovirus-derived vectors have been constructed and used successfully to transfer genes into recipient cells (Blair *et al.,* 1981; Huang *et al.,* 1981; Lee *et al.,* 1981; Wei *et al.,* 1981; Shimotono and Temin, 1981; Doehmer *et al.,* 1982; Joyner *et al.,* 1982; Tabin *et al.,* 1982).

The use of retrovirus-derived vectors offers a general approach to introduce and express genes in mammalian cells. Several properties of this class of viruses suggest that a limited number of vectors can be constructed to transduce genes into many mammalian species as well as a wide variety of *in vitro* or *in vivo* cell types. A special property of the M-MuLV vectors constructed in our lab, and the topic of this chapter, is their general ability to introduce *any* desired DNA fragment into mammalian cells, regardless of whether it is expressed or not. Moreover, these vectors enable the *expression* of genes containing no more than a complete, uninterrupted coding sequence (e.g., reverse transcripts of mRNA or bacterial, plant, and invertebrate genes).

Probably the most exciting and distinctive use of retrovirus-derived vectors, which at present is mere speculation, is the development of efficient DNA transfection procedures, which is the main stumbling block in the development of gene-therapy procedures. This possibility stems from a unique property of retroviruses, involving their integration into the host chromosome. I elaborate on this topic at the end of this chapter.

II. Organization of the M-MuLV Genome

In general, the genome organization of retroviruses, of which M-MuLV is a prototype, is simple (as long as it is not investigated in detail). The main features, in a somewhat simplified manner, are shown in Fig. 1. The viral DNA, integrated into the host genome, is 8858 nucleotides long and is flanked by directly repeated sequences 594 nucleotides long called long terminal repeats (LTRs; Gilboa *et al.,* 1979a; Shinnick *et al.,* 1981). One set of genes, the *gag* and *pol* genes, is expressed from the larger of the two identified mRNAs, the 35-S RNA species, which appears to be an unspliced transcript of the viral genome

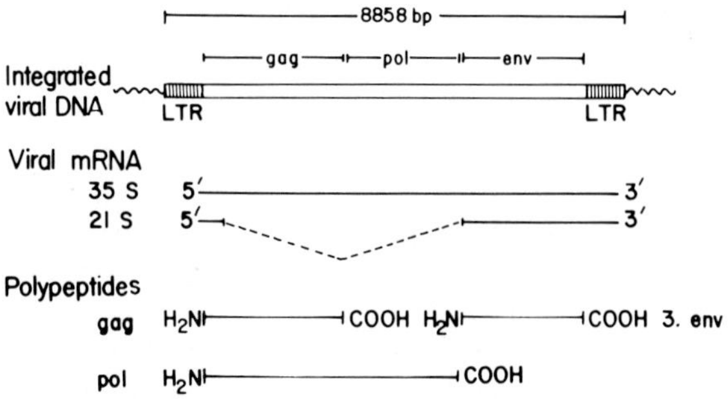

Fig. 1 Organization of the retrovirus genome.

(Fig. 1). Its 5′ end, which is capped, maps within the left LTR. The initiation codon for the *gag* gene product has been tentatively mapped to a position 476 nucleotides outside the left LTR (Oroszlan *et al.*, 1978; Shinnick *et al.*, 1981). The 3′ end of the 35-S mRNA, which is polyadenylated, lies within the right LTR (Fig. 1). The *pol* gene product is believed to be a read-through product of the *gag* gene (Philipson *et al.*, 1978; Shinnick *et al.*, 1981). The second gene, the envelope gene, is expressed from a 21-S mRNA species from which most of the *gag* and *pol* gene sequences have been spliced out. The 21-S mRNA species has termini identical to the unspliced 35-S mRNA. The 5′ splice junction (''donor'' splice sequence) has been tentatively mapped to a sequence 61 nucleotides outside the left LTR. The 3′ splice junction is located 200–300 nucleotides upstream from the envelope-coding sequences (Rothenberg *et al.*, 1978; Shinnick *et al.*, 1981).

I shall describe very briefly the replication cycle of M-MuLVs, emphasizing the aspects relevant to their functions as mammalian expression vectors. The virion particle contains two identical copies of the single-stranded RNA species that constitute the retrovirus genome. In the first or early phase of the viral replication cycle, the virion particle is introduced into the cell cytoplasm, where its genomic RNA is reverse transcribed into a linear, double-stranded DNA molecule. (The end of the DNA molecule is terminally redundant, 594 nucleotides long in M-MuLV, and is the LTR.) For a detailed discussion of this step see Gilboa *et al.* (1979b) and Varmus and Swanstrom (1982). In the nucleus of the infected cell two circular forms of the viral DNA appear (formed as a result of circularization of the linear form?) that have one or two copies of the LTR. The viral DNA integrates into the cell chromosome (one or both circular forms serving as the precursor?), and the structure of the integrated viral form (*provirus*) is collinear with the free cytoplasmic linear form (i.e., it is flanked by two LTRs; Fig. 1). The integrated provirus serves as the template for the expression of the viral genes and the progeny genomic RNA, which is encapsidated into a virion particle and released into the surroundings (the medium in tissue-culture-grown cells). Two points should be emphasized. First, it appears (though it has not been rigorously proven) that the integration of the incoming viral DNA is an obligatory step in the replication cycle of the virus. Second, with a few exceptions *no deleterious effects* are associated with retrovirus infection; the integrated viral genetic information is continuously expressed into mature virions without apparently affecting cell viability. Thus one can isolate and propagate retrovirus-infected cells that continuously produce and release progeny viruses into the surrounding medium.

In a simple way one can view a mammalian gene as composed of two parts: a structural part, consisting of the coding sequences for the gene product, and a regulatory part, coding for cis-acting functions required for the expression of the structural part (e.g., RNA promoter, splice junctions, and polyadenylation signals). The M-MuLV DNA sequences comprised of such regulatory sequences are present on three regions of the genome (Fig. 1):

1. A region at the 5′ end including the left LTR and additional sequences code for cis-acting functions (e.g., RNA polymerase promoter) required for the expression of both the *gag–pol* and *env* genes
2. A sequence upstream from the envelope-coding sequences required for the expression of the envelope gene (e.g., the acceptor splice junction)
3. A sequence at the 3′ end of the genome specifying the polyadenylation of both the 35- and 21-S mRNAs

A detailed description of the structure, replication, and biology of retroviruses can be found in the book edited by Weiss *et al.* (1982).

III. Use of Retrovirus Vectors to Study the Mechanism of Gene Expression of the M-MuLV Genome

A. Experimental Strategy

Which are the *cis-acting elements* required for the expression of a mammalian gene? Three such elements have been identified as functionally required in at least some cases: (1) a promoter sequence for RNA initiation of transcription, (2) a polyadenylation signal for RNA "termination," and (3) splice junctions, required for circumventing a coding sequence or, alternatively, increasing stability or efficiency of transport of the processed RNA transcript. Are there additional cis-acting elements present on all or some of the mammalian genes affecting the qualitative or quanitative expression of its coding sequences? For example, do (some?) mammalian genes contain a ribosome-binding site, which had been thought to be absent in mammalian genes (Kozak, 1978, 1981)?

The main line of our research, only briefly touched on in this chapter, is the construction of retrovirus vectors for systematic analysis of the cis-acting functions that govern the expression of the M-MuLV genome. The M-MuLV genome serves as a model system for mammalian (murine) genes and is particularly suited for such studies because it is naturally expressed from within the cell chromosome by the cellular machinery, thus closely resembling a cellular gene.

The experimental strategy for identifying and characterizing the cis-acting sequences that regulate the expression of the viral genes is to identify a region of the M-MuLV genome that codes for the regulatory elements necessary to express an adjacent coding sequence and, subsequently, to use *in vitro* mutagenesis to identify sequences involved in the various steps of its expression. To do this, DNA fragments are generated from various regions of the M-MuLV genome where regulatory sequences are likely to reside and are tested for their ability to direct the expression of an adjacent coding sequence. The following experimental protocol has been employed. The M-MuLV–derived DNA fragments are covalently linked to DNA fragments carrying a coding sequence the expression of which is monitored. The covalent joining is achieved by enzymatic ligation followed by purification and amplification of the desired ligation product through molecular cloning in pBR322. The hybrid structure is introduced into tissue-culture-grown mouse cells using DNA transfection procedures (Graham and van der Eb, 1973; Wigler *et al.*, 1977; Maitland and McDougall, 1977), followed by identification of cells that express the virus-linked gene. Because DNA transfection is an inefficient procedure (i.e., only a small fraction of cells incorporate and express a newly introduced gene), the gene product must have an associated selection method. We use two selectable genes in our studies. One is the *Herpes* virus thymidine kinase gene (HSV *tk*) as a *Bgl*II–*Eco*RI fragment. This DNA fragment lacks its own promoter and will not be expressed unless linked to an appropriate vector (McKnight, 1980; McKnight *et al.*, 1981; Wagner *et al.*, 1981). Cells expressing the *tk* gene are selected by the HAT procedure (Szybalska and Szybalski, 1962). The second selectable gene used in our studies is the bacterial Neo^R gene carried on Tn5 (we use a *Bgl*II–*Bam*HI fragment; Jorgenson *et al.*, 1979). Cells expressing this gene can be selected by using a compound termed G418, which is an analog of gentamycin (Jimenez and Davies, 1980). Again, this gene cannot be expressed in mouse cells unless it is properly coupled to an expression vector that provides the functions missing in the bacterial gene.

The M-MuLV DNA-dependent expression of the HSV *tk* and bacterial Neo^R genes is demonstrated by the appearance of HAT- or G418-resistant colonies, respectively. Such a viral DNA fragment carrying the necessary functions to

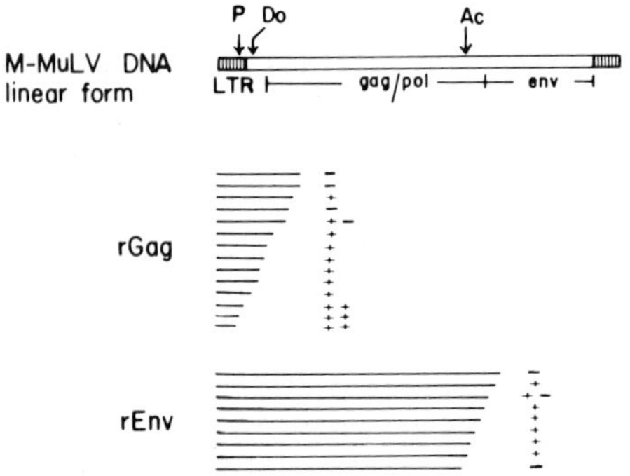

Fig. 2 M-MuLV–derived rGag and rEnv vectors. The actual substrate for the generation of the DNA fragments was the small circular DNA from cloned in pBR322. DNA fragments from the rGag series were generated by isolation of a restirction fragment (*Bam*HI–*Pst*I, and *Bgl*I–*Xho*I) followed by *Bal*31 digestion and addition of *Eco*RI linkers. Likewise, rEnv vectors were generated by digestion with *Hpa*I followed by *Bal*31 digestion and addition of *Eco*RI linkers. Thus the left ends of the rGag and rEnv fragments actually contain a sequence derived from the 3′ end of the viral genome. P, Promotor; Do, "donor" splice sequence; Ac, "acceptor" splice sequence; LTR, long terminal repeat; + and −, ability or inability of the respective DNA fragment to express a covalently linked HSV *tk* or bacterial Neo^R gene, as measured by the appearance of drug-resistant colonies.

express adjacent genes is by definition an *expression vector*. One important aspect of this transduction system should be emphasized here. The transduced DNA (i.e., the *tk* or Neo^R gene) is integrated in the chromosome of the recipient cells in one or more copies, from which it is *stably* expressed in a constitutive manner. Thus cell lines can be isolated and propagated indefinitely, carrying and expressing the newly introduced DNA.

B. rGag and rEnv Vectors

Figure 2 is a summary of our method for analyzing two of the regulatory regions involved in the expression of the M-MuLV genes: the left end of the viral genome responsible for the expression of both the *gag–pol* and the *env* genes and the "middle" region involved in the expression of the envelope gene. As shown in Fig. 2, we have tested various DNA fragments obtained from the left end of the viral genome for their ability to express the HSV *tk* or bacterial Neo^R genes. Most, but not all, DNA fragments tested (shown in Fig. 2) were found to express the covalently linked selectable genes, as evidenced by the appearance of drug-resistant colonies. We have termed the vectors derived from this region of M-MuLV *rGag vectors* because they carry the functions required for the expression of the adjacent *gag–pol* genes. In a similar approach, to study the expression of the viral envelope gene we have derived a set of DNA fragments termed *rEnv vectors* (Fig. 2), from which the coding sequences of the envelope gene as well as varying amounts of additional upstream sequences have been removed. The ability of these DNA fragments to express the *tk* and Neo^R genes has been similarly tested. Note that the hybrid DNA structures shown schematically in Figs. 2, 3, and 4 and constructed throughout these studies are cloned in pBR322 in a step of purification and amplification. The borders between the various parts of these constructs are restriction sites that are either naturally present or artifically added as DNA linkers. In addition, in all hybrid constructs the *tk* and Neo^R genes are expressed only if their transcriptional orientation is parallel to that of the M-MuLV vector. The *tk*- and *neo*^{r}-containing DNA fragments are not expressed as a biologically active gene product if their transcriptional orientation is opposite to that of the M-MuLV vector (i.e., in head-to-head or tail-to-tail orientation).

Molecular biologists seem to have defined a functional expression vector as a DNA fragment that directs the expression of an adjacent gene and have monitored it by using a selection procedure that results in the appearance of drug-resistant cells. This interpretation deserves a word of caution. The appearance of, say, a HAT-resistant cell means that an amount of thymidine kinase above a minimum threshold is synthesized, rendering the cell resistant to the selection procedure. Thus two expression vectors varying 100-fold in their efficiency of expressing the *tk* gene may still record the same number of HAT-resistant colonies. Therefore, the suitability of a DNA fragment to serve as an expression vector has to be measured in more direct ways (i.e., by measuring the levels of mRNAs and/or the corresponding gene products). These quantitative studies have to be performed in cells preferably carrying *one copy* of the newly introduced hybrid DNA.

Obviously, to study the expression the M-MuLV genes using the strategy outlined in Fig. 2, one has to compare quantitatively the efficiency by which vectors from the rGag and Env series express adjacent genes. This line of study is being pursued in our laboratory.

IV. A General Transduction System Derived from the M-MuLV Genome

A. Double Expression Vectors

In this section I describe the construction of a mammalian expression vector from the genome of M-MuLV that can be used to introduce any DNA fragment into the chromosome of mammalian cells and to express stably its genetic information through functions provided by the viral vector.

The rationale for the construction of such a vector from M-MuLV is very

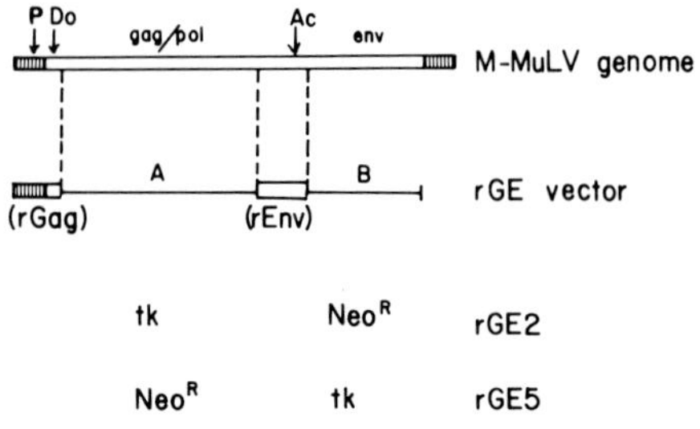

Fig. 3 Structure of M-MuLV–derived double expression vectors.

simple. As shown in Fig. 3, the genome of M-MuLV consists of essentially two "expression units," the *gag–pol* and the *env* genes. A general transduction vector is constructed by removing (in both the physical and functional senses) the two major viral coding sequences (*gag–pol* and *env*) and replacing them with new DNA fragments carrying their own coding sequences (A and B in Fig. 3). This is a *double expression vector,* in which both inserts are expressed simultaneously through functions present in the vector that are normally used to express the viral genes. In analogy to the drug-resistance genes of pBR322, by introducing a *selectable gene* into one of the two positions one can introduce *any DNA fragment* into the second position and use the associated selection procedure to identify cells that have incorporated this vector into their chromosome (see Southern and Berg, 1982; Roberts and Axel, 1982). If that DNA fragment contains a complete coding sequence, it will be expressed to a biologically active product through the functions present on the vector that are normally used to express its own genes, which have been removed and replaced with the new transduced genes.

The actual construction of such a double expression vector from the genome of M-MuLV has been achieved by combining a previously identified rGag vector with the relevant portion of an rEnv vector (Fig. 2). These composite vectors are accordingly termed *rGE vectors.* To demonstrate the biological activity of these vectors we have used the two selectable genes previously described (the HSV *tk* and the bacterial Neo^R genes) to insert and express through an rGE vector. As shown in Fig. 3, two hybrid vectors have been constructed that carry both selectable genes in the two possible combinations rGE2 and rGE5. The left end of rGE2 or rGE5 is a DNA fragment derived from the left end of the M-MuLV genome that was previously identified as a functional rGag vector. In this specific example the right end (3′) boundary of the rGag DNA fragment is the viral *Bal*I site (converted to an *Eco*RI site by DNA linkers) that lies 69 nucleotides beyond the left LTR and that also includes the putative donor splice sequence required for the expression of the viral envelope gene. The first insert (*tk* in rGE2 and the Neo^R gene in rGE5) is expressed through functions present in this part of the vector. The second part of the vector as shown in Fig. 3 is derived from a functional rEnv vector. (The left boundary of this fragment is an arbitrarily chosen *Hin*dIII site that is likely to be outside the region required for the expression of the envelope gene.) One obvious function this part is expected to contain is an acceptor splice sequence for the envelope mRNA (Ac in Fig. 3). Thus the second insert (Neo^R gene in rGE2 and *tk* in rGE5) is expressed through functions present on both M-MuLV fragments.

To demonstrate the simultaneous expression of both inserts in rGE2 and rGE5, tk^- containing mouse L cells were transfected with the two constructs, and the expression of both inserts to a biologically active product was easily demonstrated using either of the two associated selection procedures. Moreover, one can select cells expressing one insert and show that the second, *unselected insert* is expressed as well (using the associated selection procedure). All of these experiments have been performed with both vectors rGE2 and rGE5 in either combination (first and second selection) with appropriate controls.

In summary, we have used the two selectable markers to show that the double expression vectors constructed from the genome of M-MuLV can indeed direct the expression of two inserts incapable of being expressed independently in the absence of cis elements donated by the vector.

One very important piece of information, though, is missing from the description of these vectors, that is, How *efficiently* are the two inserts expressed? Efficiency of expression is measured by quantification of the gene product or the corresponding mRNA in cells harboring one copy of the hybrid vector. To construct an *efficient* double expression vector (and for other reasons as well), we screen the rGag and rEnv vectors we have identified (Fig. 2) for vectors that express their adjacent genes most efficiently. The best rGag and rEnv vector will, accordingly, be chosen to construct an efficient double expression vector. Moreover, one can envisage uses of double expression vectors in which one insert is expressed at high level and the second insert at low level, for which an appropriate combination of rGag and rEnv will be chosen.

Finally, I wish to point out that I have not forgotten to consider the need (?) for polyadenylation signals (Fitzgerald and Shenk, 1981) in construction of the vectors described here. Preliminary experience with polyadenylation signals shows that they appear not to be required! Two lines of evidence point to this. First, rEnv vectors linked to the HSV *tk* or the bacterial Neo^R gene have the same transfection efficiency (i.e., number of resistant colonies/μgm DNA). The HSV *tk* gene has its own polyadenylation signal, whereas the Neo^R gene is of bacterial origin. Second, the transfection efficiency of a vector carrying the Neo^R gene with or without an added polyadenylation signal (derived from M-MuLV) is the same. Again, the catch is that we are looking at the appearance of drug-resistant cells and not at how efficiently the genes are expressed. A detailed study of this observation is now under way.

B. Transmission Vectors

The retrovirus-derived vectors described (rGag, rEnv, and rGE vectors) can be termed *integration vectors,* that is, the genes transduced by these vectors are integrated into the chromosome of the recipient cells, becoming a part of the cellular genome. A major shortcoming associated with this type of vector results from the fact that the hybrid vector constructed *in vitro* carrying the transduced gene is introduced into the recipient cells by *DNA transfection procedures*. DNA transfection is an inefficient method of introducing genes into cells; at best, 0.1–1% of cells acquire the newly introduced DNA (in a genetically stable manner), and it also is limited to a few established cell lines. Thus the efficiency of transfection of most cells *in vitro* and, especially, *in vivo* is very low. In this section I describe the construction of vectors derived from the previously discussed integration vectors that can be transmitted as retrovirus virions. These vectors, termed *transmission vectors* (Wei *et al.*, 1981; Shimotono and Temin, 1981; Tabin *et al.*, 1982), allow the efficient introduction of genes into mammalian cells, and taking advantage of the biology of retroviruses greatly increases the host range of gene transfer using the same vector system.

A retrovirus-derived transmission vector functions in the following way. The desired DNA fragment(s) is linked to the vector *in vitro,* is purified by cloning in pBR322, and is introduced by DNA transfection into a recipient cell, where it integrates into the cell chromosome. The cell harboring the integrated hybrid vector DNA is isolated. Next, the transmission vector (or more precisely the RNA transcribed from the integrated DNA) carrying the transduced gene is incorporated (or pseudotyped) into a retrovirus virion, which is released into the medium. The virion components missing from the vector are supplied by a helper, replication-competent retrovirus. Thus a population of retrovirus virions is obtained that carries the transduced gene as a hybrid RNA structure linked to the transmission vector. This transduced gene can be *efficiently* introduced into susceptible cells by viral *infection* (a very efficient process compared to inefficient DNA transfection), which is the raison d'être of this whole exercise.

A retrovirus-derived transmission vector requires two structural elements acting in cis to complete a normal replication cycle (the rest of the components are complemented in trans by a helper virus). One set of elements are the two LTRs present in the same orientation at the left and right ends of the vector. This requirement stems from the mechanism of reverse transcription (Gilboa *et al.*, 1979b; Varmus and Swanstrom, 1982). [To be accurate, the tRNA-binding site flanking the left LTR and the (+)ss DNA primer region flanking the right LTR also are required.] The second cis-acting element required is involved in the packaging of the viral RNA into virions (Goldberg *et al.*, 1976). Analysis of a mutant isolated by Shank and Linial (1980) suggests that this "packaging signal" is located outside the left LTR at the 5′ end of the viral genome. In other words, RNA transcribed from a DNA region flanked by two LTRs and contain-

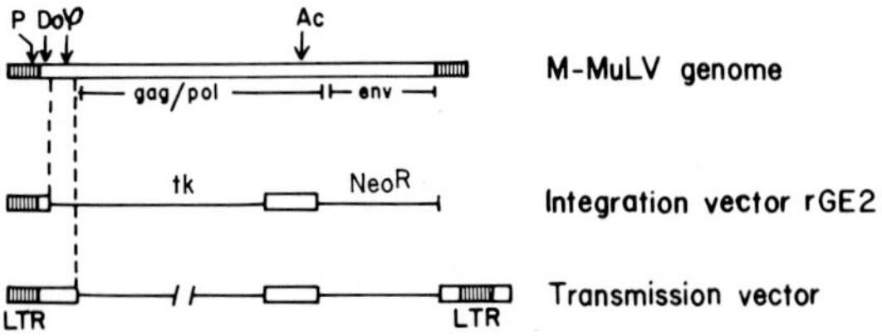

Fig. 4 Structure of M-MuLV–derived transmission vectors. φ, Approximate location of the viral packaging signal.

ing the viral packaging signal will be packaged into retrovirus virion particles with all other structural components complemented through a helper virus. Such "recombinant virions," like normal virions, are secreted to the medium and can infect new cells (susceptible to the virus used as a helper).

We have constructed a transmission vector along these lines, from a double expression vector carrying the HSV *tk* and the bacterial Neo^R genes (Fig. 4). This vector differs from rGE-2, an integration vector, by containing an additional LTR at the 3′ end and another, longer rGag moiety derived from the 5′ end of M-MuLV that contains the viral packaging signal (in Fig. 4). We have shown that this construct can transmit its genetic information through virions by a protocol involving two cycles of transmission through the medium (Fig. 5). This experimental protocol demonstrates both the retrovirus virion-mediated transmission of transduced genes and the special properties of double expression vectors, in which one expressed gene can be used as the selectable marker (the Neo^R gene selected with G418) to transmit the second, nonselected gene (*tk* gene; Fig. 5).

One shortcoming of the use of retrovirus transmission vectors in this protocol is that the medium containing the recombinant virions, which carry the transduced gene(s), also contains infectious replication-competent helper retrovirus. R. Mann, R. Mulligan and D. Baltimore (personal communication) have eliminated the presence of helper virus by constructing a cell line containing a packaging-defective retrovirus that expresses all of the genes necessary to complement the packaging of a transmission vector into virions (Mulligan, Chapter 8).

To summarize the use of retrovirus transmission vectors, their obvious advantage is their high efficiency of gene introduction into mammalian cells, because the vector-linked genes are introduced into cells through virion infection, which is an efficient process. Their shortcoming lies in the fact that obtaining the "recombinant" virions carrying the transduced gene is a time-consuming process. First, the hybrid vector is constructed *in vitro* and is purified by cloning in pBR322. Next, it is introduced into tissue-culture-grown cells by transfection, productively transfected cells are isolated by a selection procedure, and finally, virions carrying the recombinant information are rescued into virions (secreted into the medium) by superinfection with a helper retrovirus. No superinfection is required if cell lines of the type derived by Mann, Mulligan, and Baltimore (see

Transfect NIH *tk*− cells

↓

isolate colonies expressing the Neo^R gene (G418)

↓

superinfect with M-MuLV

↓

collect supernatant

↓

infect NIH *tk*− cells with supernatant

↓

First passage: select HAT- (G418)-resistant colonies

↓

collect supernatant

↓

infect NIH *tk*− cells

↓

Second passage: select HAT- (G418)-resistant colonies

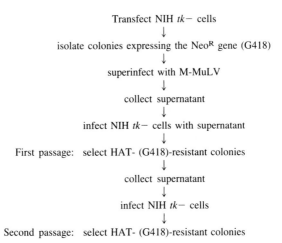

Fig. 5 Assay of transmission vectors carrying the HSV *tk* gene and the bacterial Neo^R gene.

Chapter 8) are used for the transfection, though this approach limits the host range of the recombinant virions. This potentially wide host range is probably the most exciting feature of the retrovirus-derived transmission vectors for gene transfer in mammalian systems and merits some elaboration.

The host range of retrovirus-derived transmission vectors is determined by two factors. One is the host range of the retroviruses and the other is the extent to which packaging signals are recognized by different retroviruses. Starting with the second aspect, existing evidence suggests that the retrovirus packaging signal is universal (Goldberg *et al.*, 1976). This is rather surprising in view of the otherwise large structural differences existing between different members of this group of viruses. Regarding the *host range* of this group of viruses, retroviruses are ubiquitious among mammalian species. Although most retroviruses are species specific, some groups have a broader host range. Examples include the mouse xenotropic isolates that cannot grow in the mouse but that grow in other mammalian species (e.g., cat, mink, dog, and monkey) or the mink-cell focus-forming (MCF) viruses that can grow in (some) mouse cells as well as in cells from other mammalian species. In addition, most retroviruses can infect a wide variety of cell types, both *in vitro* and *in vivo*.

The host range of *retrovirus transmission vectors* (i.e., the capacity to introduce a desired gene into a wide variety of cells using a *universal vector*) is determined by the host range of retroviruses and the universality of the retrovirus packaging signal and is greatly amplified by a combination of the two features. A limited number of transmission vectors can be constructed, introduced into a tissue-culture-grown cell line by DNA transfection, and "rescued" into virions by heterologous packaging, using a helper retrovirus with a desired host range.

V. Summary and Prospects

The development of mammalian expression vectors from the genome of retroviruses is in its infancy, and it is far too early to appreciate the full potential of this new experimental tool (as well as its limitations). It is likely that retrovirus-derived vectors will become the vector system most widely used to insert defined DNA fragments into the chromosomes of mammalian cells, either by DNA transfection procedures (using integration vectors; Fig. 3) or by viral infection (using transmission vectors; Fig. 4). The first approach is simpler but is limited by low transfection efficiency and the narrow host range of susceptible cells. The second approach is more time consuming but results in a high efficiency of gene transduction in the recipient cells and a wider host range. The potentially wide host range of retrovirus vectors is a very exciting feature of this transduction system. This will allow us to use *one vector system* (e.g., the M-MuLV vectors described here) to transfer genes into a wide variety of cell types *in vitro* and *in vivo* in a wide variety of mammalian species. In this chapter, I have described the construction of double expression vectors from the M-MuLV genome (rGE vectors) that enable the transduction (and expression) of any DNA fragment desired. This feature adds another degree of versatility to the use of this type of vectors.

Another practical aspect to be considered briefly is the cloning capacity of this system—the size limitation on the DNA fragments used in the transduction experiment. This question is relevant when transmission vectors are used because it is likely that a limitation in the size of RNA to be packaged exists. Here again, retrovirus-derived vectors prove to be very advantageous because there is great flexibility in the size of RNA that can be packaged in the virions, the upper limit being 9–10 kb (J. Sorge and S. Hughes, personal communication) and the lower 2–3 kb or less (Goldfarb and Weinberg, 1981). Thus, using the same M-MuLV vector and including a selectable marker DNA, inserts of 0–6 kb can be accommodated.

I conclude this chapter with a somewhat speculative discussion of what may prove to be a most important feature, unique to retrovirus-derived vectors. It is possible to envisage the development of retrovirus-derived mammalian expression vectors that can be *efficiently* introduced into the chromosome of cells by *DNA transfection procedures* in which *most or all* cells acquire the newly introduced DNA stably integrated into the cell chromosome. This suggestion is based on the following reasoning. The newly synthesized free retrovirus DNA reverse transcribed from the incoming virion RNA *integrates* into the cell chromosome, from which the viral genes are expressed. This integration step, unique to retroviruses, is mediated by a process related to bacterial transposition (Shoemaker *et al.*, 1980; Shimotono *et al.*, 1980), and although no quantitative data are available, it is reasonably to assume that it is an efficient process. If this integration function can be introduced into a retrovirus expression vector, once in the nucleus, it will integrate efficiently into the cell chromosome. Indeed, evidence is accumulating that the low efficiency of (stable) DNA transfection is at the step of integration, whereas the introduction of the DNA into the nucleus is an efficient process (Banerji *et al.*, 1981; Mellon *et al.*, 1981).

Of course, the prerequisite (and the difficult part) in developing such a ''super vector'' is to identify biochemically the function(s) involved in retrovirus integration. It is reasonable to assume, although not necessarily correct, that this specific function is coded by the virus, which will simplify its identification, but probably not too much because it will likely be a posttranslational cleavage product from the *gag* or *pol* region. [One of the most promising candidates is the endonuclease described by Golomb *et al.* (1981).]

Anyone using DNA transfection procedures in mammalian cells will surely appreciate a high-efficiency transfection procedure. A most important application, though, of such a technique may be in developing effective *gene-therapy* procedures. Successful gene therapy requires the *quantitative* introduction of a new gene into the tissues of an individual. The introduction of DNA by current DNA transfection procedures is very inefficient and thus not suitable for this purpose. One approach for increasing the proportion of cells productively transfected is use of a selection procedure, *in vitro* or *in vivo,* which is quite unthinkable practicably (Cline *et al.*, 1980; Mercola *et al.*, 1980). An alternative approach, discussed in this chapter, is to use retrovirus transmission vectors in which the introduction of genes occurs via viral infection, which is an efficient process. This is a realistic approach, though reluctance in using this approach is anticipated because of the use of retroviruses, some members of which are tumor viruses. Therefore, the third alternative, an efficient transfection procedure, may be an attractive approach, but like gene therapy it is barely more than sheer fantasy.

References

Banerji, J., Rusconi, S., and Shattner, M. (1981). *Cell* **27,** 299–308.

Blair, D. G., Oskarsson, M., Wood, T. G., McClements, W. L., Fishinger, D. J., and Van de Woude, G. G. (1981). *Science (Washington, D.C.)* **212,** 941–943.

Cline, M. J., Stang, H., Mercola, K., Morse, L., Ruprecht, R., Browne, J., and Salser, W. (1980). *Nature (London)* **284,** 422–425.

Doehmer, J., Barinuga, M., Vale, W., Rosenfeld, M. G., Verma, I. M., and Evans, R. M. (1982). *Proc. Natl. Acad. Sci. USA* **79,** 2268–2272.

Fitzgerald, M., and Shenk, T. (1981). *Cell* **24,** 251–260.

Gilboa, E., Goff, S., Shields, A., Yoshimura, F., Mitra, S., and Baltimore, D. (1979a). *Cell* **16,** 863.

Gilboa, E., Mitra, S., Goff, S., and Baltimore, D. (1979b). *Cell* **18,** 93.

Goldberg, R. J., Levin, R., Parks, W. P., and Scolnick, E. M. (1976). *J. Virol.* **17,** 43–50.

Goldfarb, M. P., and Weinberg, R. A. (1981). *J. Virol.* **38,** 136–150.

Golomb, M., Grangenett, D. P., and Mason, W. (1981). *J. Virol.* **38,** 548–555.

Graham, F. L., and van der Eb, A. J. (1973). *Virology* **52,** 456–467.

Gruss, P., and Khoury, G. (1981). *Proc. Natl. Acad. Sci. USA* **78,** 133–137.

Hamer, D., and Leder, P. (1979). *Nature (London)* **281,** 35–40.

Huang, A. L., Ostrowski, M. C., Berard, D., and Hager, G. L. (1981). *Cell* **27,** 245–255.

Jimenez, A., and Davies, J. (1980). *Nature (London)* **287,** 869–871.

Jorgensen, R. A., Rothstein, S., and Reznikoff, W. S. (1979). *Mol. Gen. Genet.* **177,** 65–72.

Joyner, A., Yamamoto, Y., and Bernstein, A. (1982). *Proc. Natl. Acad. Sci. USA* **79,** 1573–1577.

Kozak, M. (1978). *Cell* **15,** 1109–1123.

Kozak, M. (1981). *Nucleic Acids Res.* **9,** 5233–5252.

Lee, F., Mulligan, R., Berg, P., and Ringold, G. (1981). *Nature (London)* **299,** 228–232.

Maitland, N., and McDougall, J. K. (1977). *Cell* **11,** 233–241.

McKnight, S. L. (1980). *Nucleic Acids Res.* **8,** 5949–5964.

McKnight, S. L., Gavis, E. R., Kingsbury, R., and Axel, R. (1981). *Cell* **25,** 385–398.

Mellon, P., Parker, V., Gluzman, Y., and Maniatis, T. (1981). *Cell* **27,** 279–288.

Mercola, K., Stang, M., Broome, J., Salser, W., and Cline, M. (1980). *Science (Washington, D.C.)* **208,** 1033–1035.

Mulligan, R., Howard, B., and Berg, P. (1979). *Nature (London)* **277,** 108–114.

Oroszlan, S., Henderson, L. E., Stephenson, J. R., Copeland, T. D., Lang, C. W., Ihle, J. N., and Gilden, R. V. (1978). *Proc. Natl. Acad. Sci. USA* **75,** 1404–1408.

Philipson, L., Anderson, P., Olshevsky, U., Weinberg, R., Baltimore, D., and Gesteland, R. (1978). *Cell* **13,** 189.

Roberts, J. M., and Axel, R. (1982). *Cell* **29,** 109–119.

Rothenberg, E., Donoghue, D. J., and Baltimore, D. (1978). *Cell* **13,** 435.

Shank, P., and Linial, M. (1980). *J. Virol.* **36,** 450–456.

Shimotono, B. K., and Temin, H. (1981). *Cell* **26,** 67–77.

Shimotono, B. K., Mizutani, S., and Temin, H. (1980). *Nature (London)* **285,** 550–554.

Shinnick, T. M., Lerner, R. A., and Sutcliffe, J. G. (1981). *Nature (London)* **293,** 543–548.

Shoemaker, C., Goff, S., Gilboa, E., Paskind, M., Mitra, S., and Baltimore, D. (1980). *Proc. Natl. Acad. Sci. USA* **77,** 3932–3930.

Southern, P., and Berg, P. (1982). *In* "Eucaryotic Viral Vectors" (J. Gluzman, ed.), pp. 41–45. Cold Spring Harbor Lab., Cold Spring Harbor, New York.

Szybalska, E. H., and Szybalski, W. (1962). *Proc. Natl. Acad. Sci. USA* **48,** 2026–2034.

Tabin, C. J., Hoffman, J. W., Goff, S. P., and Weinberg, R. A. (1982). *Mol. Cell. Biol.* **2,** 426–436.

Varmus, M., and Swanstrom, L. (1982). *In* "RNA Tumor Viruses" (R. Weiss, N. Teich, H. Varmus, and J. Cottin, eds.), pp. 369–512. Cold Spring Harbor Lab., Cold Spring Harbor, New York.

Wagner, M. J., Sharp, J. A., and Summers, W. C. (1981). *Proc. Natl. Acad. Sci. USA* **78,** 1441–1445.

Wei, C. M., Gibson, M., Spear, P. G., and Scolnick, E. M. (1981). *J. Virol.* **39,** 935–944.

Weiss, R., Teich, N., Varmus, H., and Cottin, J. (eds.) (1982). "RNA Tumor Viruses." Cold Spring Harbor Lab., Cold Spring Harbor, New York.

Wigler, M., Silverstein, S., Lee, L. S., Pellicer, A., Cheng, Y. C., and Axel, R. (1977). *Cell* **11,** 223–232.

CHAPTER 10

Production of Posttranslationally Modified Proteins in the SV40–Monkey Cell System

DEAN H. HAMER

Laboratory of Biochemistry
National Cancer Institute
National Institutes of Health
Bethesda, Maryland

I. Introduction

An important goal of recombinant DNA research is to develop methods for overproducing useful gene products. Most work in this field has concentrated on microbial hosts, such as the enterobacteria and yeasts, because they can be

Experimental Manipulation
of Gene Expression

ISBN 0-12-372380-9

grown in large quantity at low cost. But such hosts are not appropriate for producing certain higher eukaryotic proteins that must be posttranslationally modified to achieve their biological activity (e.g., proteins that require glycosylation or assembly into particles). For this purpose it is necessary to employ a eukaryotic host containing the enzymes and cellular factors required for such modifications.

The SV40–monkey cell system has proven especially useful for this purpose. The detailed information available on the molecular biology of SV40 allows the construction of vectors in which cloned cDNAs are efficiently expressed under the control of viral regulatory sequences. Moreover, because the signals for eukaryotic RNA transcription, polyadenylation, and splicing are recognized in this system, it is possible to work directly with genomic DNA clones. Finally, the recombinant molecules can be packaged into virions and efficiently introduced into cells so that reasonably high yields of product are obtained. In this chapter we concentrate on the use of SV40 vectors to produce two posttranslationally modified gene products: human growth hormone (hGH) and hepatitis B virus surface antigen (HBsAg). These experiments have been described in detail by Pavlakis *et al.* (1981) and Moriarty *et al.* (1981). The use of SV40 as a cloning vector has been reviewed by Hamer (1980), Elder *et al.* (1981), and Rigby (1982). More general information about SV40 can be found in the volume edited by Tooze (1980).

II. SV40 Late-Replacement Vectors

Lytic infection with late-replacement recombinants is the preferred route for obtaining high-level expression of inserted sequences. In this system the DNA is efficiently introduced into the cells and replicates to high copy number. Moreover, large quantities of mRNA can be obtained by inserting the foreign sequences under the control of the strong SV40 late-region promoter.

The construction and propagation of such recombinants involves 3 steps:

1. A fragment of DNA from the SV40 late region is replaced with a foreign DNA fragment of equal or smaller size.
2. The recombinant DNA is introduced into cultured monkey kidney cells, the permissive host for SV40, together with DNA from a temperature-sensitive early gene mutant (*tsA* mutant) as helper. The transfected cells are maintained at the nonpermissive temperature for the helper (40°C) so that virions will be produced only in those cells infected with both the recombinant and the helper. The recombinant provides the functional early gene product (T antigen) required

for DNA replication, whereas the helper provides the late gene products (VP1, 2, and 3) necessary for encapsidation.

3. The resulting viral stock, containing a mixture of recombinant and helper particles, is used to infect a fresh culture of monkey cells at high multiplicity. The expression of the inserted gene is usually assayed at 2–3 days after infection, well into the late portion of the lytic cycle but before the cells actually lyse.

In the present work we have used vectors prepared by cleavage with *Bam*HI (nucleotide 2533) and either *Hpa*II (nucleotide 346) or *Hha*I (nucleotide 833). Both vectors retain the late promoter region and polyadenylation site. The *Bam*HI–*Hha*I vector also retains the splice junction for late 19-S mRNA and the AUG for coat protein VP2. In contrast, the *Bam*HI–*Hpa*II vector contains no known intact splice junction and retains only the "agnogene" AUG. Both vectors have been cloned in plasmid pBR322 so that the recombinant molecules can be constructed and propagated in *Escherichia coli.*

III. Human Growth Hormone

Human growth hormone (hGH) is produced by the acidophil cells of the anterior pituitary. Like all know polypeptide hormones, it is synthesized as a prehormone containing a hydrophibic N-terminal signal sequence that is removed during secretion. Thus hGH provides us with an opportunity to examine two interesting posttranslational events: signal sequence cleavage and secretion.

Mature hGH exhibits multiple biological effects *in vivo,* including diabetogenic, insulin-like, lactogenic, and growth-promoting activities (reviewed by Raiti, 1973). Most hGH preparations can be separated into multiple bands by high-resolution isoelectric focusing or gel electrophoresis (Cheever and Lewis, 1969; Bala *et al.*, 1973; Chrambach *et al.*, 1973). Some of these appear to result from posttranslational modification (Lewis and Cheever, 1965; Lewis *et al.*, 1970), and others represent primary sequence variants. The best studied of these is the M_r 20,000 ("20 K") variant, which is a single polypeptide chain identical to the major form of hGH except that it lacks amino acid residues 32–46 (Lewis *et al.*, 1978, 1980). It is also known that hGH exists in heterogeneous forms in human plasma with respect to molecular size and biological properties (Goodman *et al.*, 1972; Gorden *et al.*, 1973, 1976).

DNA cloning expriments have shown that the human genome contains at least seven different hGH-related genes (Fiddes *et al.*, 1979; Goodman *et al.*, 1980). One of these, designated here as the hGH1 gene, has been shown by sequence analysis to be capable of encoding the predominant form of pituitary hGH (De-

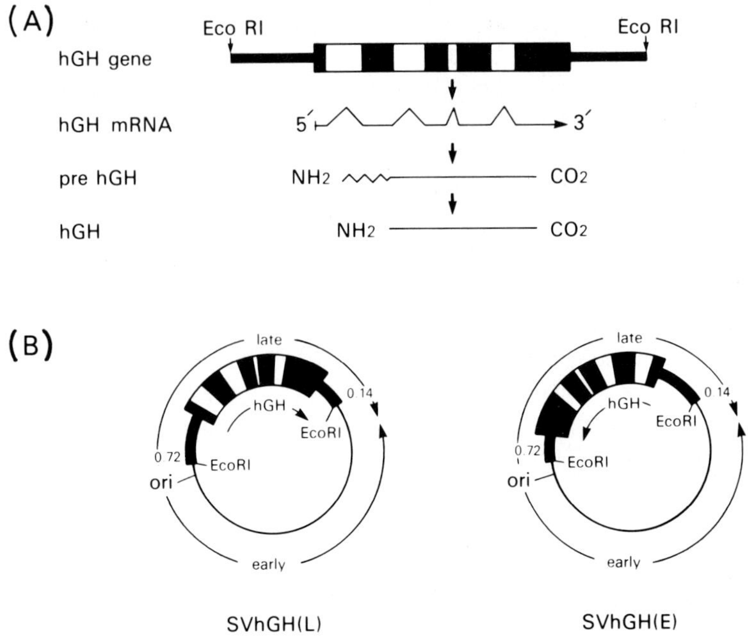

Fig. 1 SV40–hGH recombinants. (A) Both hGH genes were cloned on 2.7-kb *Eco*RI fragments containing five structural sequences (■) and four intervening sequences (□) together with 500 bp of 5′-flanking sequences and 550 bp of 3′-flanking sequences (thick segments). Translation of hGH mRNA yields pre-hGH containing an N-terminal signal sequence of 23 amino acids. This is processed to generate mature hGH. (B) The two hGH genes were inserted, in both possible orientations, into an SV40/*Bam*HI-plus-*Hpa*II vector in which both sites were converted to *Eco*RI sites using oligonucleotide linkers. *ori,* Origin of replication. From Pavlakis *et al.* (1981).

Noto *et al.,* 1981). A second gene, referred to as the hGH2 gene, is highly homologous to the hGH1 gene but contains 14 point mutations that are expected to lead to amino acid substitutions in the mature hormone (Seeburg, 1982). This suggests that some of the heterogeneity of hGH might be due to the expression of multiple closely related but nonidentical genes. We have used SV40 recombinants to compare the synthesis, secretion, and processing of hGH1 and hGH2 and to prepare sufficient amounts of these proteins for initial characterization.

A. SV40–hGH Recombinants

The hGH1 and hGH2 genes were originally cloned in phage λ as 2.7-kb *Eco*RI fragments of human placental DNA (Fiddes *et al.,* 1979). The two genes are approximately 95% homologous but can be readily distinguished from one another by the fact that gene 1 contains one *Bam*HI site whereas gene 2 contains two sites. DNA sequence analysis shows that both gene-1 and gene-2 *Eco*RI fragments contain ~500 bp of 5′-flanking sequences and ~550 bp of 3′-flanking sequences as well as five hGH structural sequences (exons) separated by four intervening sequences (introns) (Fig. 1A). The two different 2.7-kb hGH gene fragments were inserted, in both possible orientations, into the *Bam*HI–*Hpa*II SV40 vector (Fig. 1B). In the SVhGH(L) recombinants, the hGH1 and hGH2 genes are in the same orientation as SV40 late-gene transcription, whereas in the SVhGH(E) recombinants they are in the opposite or early orientation. These recombinant molecules were constructed by cloning in *E. coli* and then propagated in monkey kidney cells as described in Section II.

B. Synthesis, Processing, and Secretion of hGH in Infected Monkey Cells

The ability of these recombinants to direct the synthesis of hGH was tested by labeling infected cells with [^{3}H]leucine and analyzing the cellular proteins by immunoprecipitation and acrylamide/NaDodSO_4-gel electrophoresis. As shown in Fig. 2A, cells infected with each of the four recombinants synthesized a protein that comigrated with authentic pituitary hGH and that was absent from uninfected and wild-type SV40-infected controls. The amount of this protein synthesized was similar for SVhGH1(L) compared to SVhGH2(L) and for SVhGH1(E) compared to SVhGH2(E); in both cases about threefold more polypeptide was made in the late (L) than the early (E) orientation. Thus the two genes function equally well, but the level of expression depends on their orientation relative to the SV40 late-region promoter.

To determine whether the hGH was secreted from monkey cells, we repeated the immunoprecipitation and gel analysis on the media from the control and recombinant-infected cells. Figure 2B shows that both the hGH1 and hGH2 proteins are present in the media. Quantification of the gel lanes by microdensitometry showed that the secreted material represents approximately 80% of the total hGH synthesized in a 3-hr pulse with [^{3}H]leucine. To exclude the possibility of cell leakiness or lysis, as compared to active transport, we also analyzed the total intracellular and media proteins without immunoprecipitation. The cell extract (Fig. 2C) showed a complex array of bands, as expected from the fact that SV40 does not shut down host-cell synthesis, and it was not possible to resolve hGH from the background of cellular proteins. In contrast, the media (Fig. 2D) contained a much more discrete set of proteins, and hGH could readily be visualized as a predominant band in the recombinant-infected samples, but not in the controls. No SV40 capsid proteins were detected in the media, indicating that cell lysis was not occurring at the time of labeling (40 hr after infection).

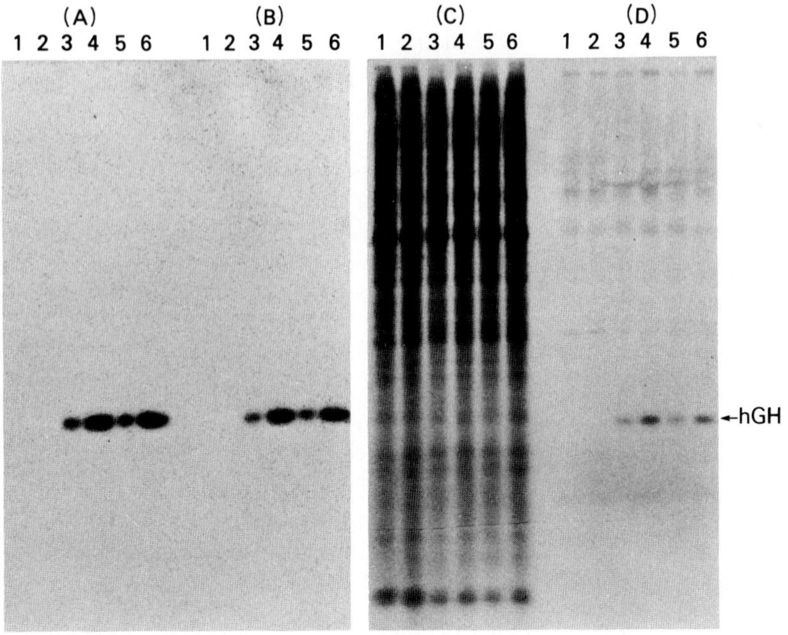

Fig. 2 Synthesis of hGH in infected monkey cells. Monolayers of 2×10^7 monkey kidney cells were labeled with [^{3}H]leucine (200 μci/ml) at 40–43 hr after infection. Cellular and media proteins were analyzed by immunoprecipitation with excess anti-hGH antibody and 20% acrylamide/$NaDodSO_4$ electrophoresis. (A) Immunoprecipitates of cell extracts from 5×10^6 cells. (B) Immunoprecipitates of media from 10^6 cells. (C) Total cellular extracts from 10^5 cells. (D) Total media from 10^5 cells. In each panel the lanes represent cells infected with: 1, no virus; 2, wild-type SV40; 3, SVhGH1(E) plus helper; 4, SVhGH1(L) plus helper; 5, SVhGH2(E) plus helper; 6, SVhGH2(L) plus helper. The symbols (E) and (L) refer to the orientation of the genes relative to the late promoter of the SV40 vector. From Pavlakis *et al.* (1981).

These results show that both hGH1 and hGH2 are specifically and efficiently secreted from infected monkey cells.

The structures of the secreted hGH1 and hGH2 were analyzed by partial chymotrypsin digestion of the $NaDodSO_4$-denatured proteins followed by acrylamide/$NaDodSO_4$-gel electrophoresis. Figure 3 shows that both proteins give rise to identical [^{3}H]leucine-containing chymotryptic peptides and that these comigrate with the peptides obtained from unlabeled pituitary hGH. These data, in conjunction with the fact that the intact proteins comigrate with pituitary hGH on $NaDodSO_4$ gels, suggest that the N-terminal signal sequences have been appropriately removed. This was expected because the necessary enzymes are known to be present in a variety of species and cell types (Blobel *et al.*, 1979). Also, Gruss and Khoury (1981) have presented evidence that preproinsulin is processed to proinsulin in monkey cells infected with an SV40–insulin gene

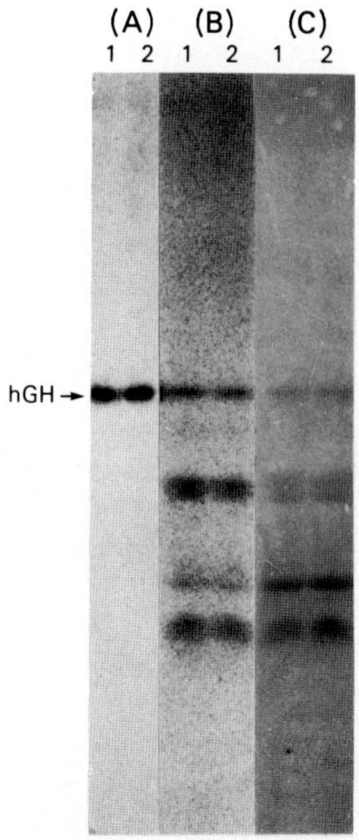

Fig. 3 Partial chymotrypsin digests. [^{3}H]Leucine-labeled hGH1 and hGH2 were purified by immunoprecipitation of the media from cells infected with SVhGH1(L) or SvhGH2(L). The precipitated proteins were eluted by boiling in $NaDodSO_4$, mixing with 20 μgm of unlabeled pituitary hGH, partially digesting with chymotrypsin, and were analyzed by 20% acrylamide/$NaDodSO_4$-gel electrophoresis. (A) Fluorogram of undigested samples. (B) Fluorogram of chymotrypsin-digested samples. (C) Panel (B) stained with Coomassie Brilliant Blue. In each panel lane 1 represents hGH1 and lane 2 hGH2. From Pavalkis *et al.* (1981).

recombinant; however, the mature hormone was not produced owing to failure to remove the internal C peptide. In pituitary cells hGH is sequestered into secretory granules prior to release into the bloodstream. The specific and efficient secretion of hGH from monkey kidney cells, which lack these granules, suggests that they are not essential for transport across the cell membrane.

C. Binding of hGH1 and hGH2 to Antibodies and Receptors

Figure 4 shows double-antibody radioimmunoassays of the media from SVhGH1- and SVhGH2-infected cells using either rabbit (Fig. 4A) or guinea pig (Fig. 4B) anti-hGH antibody. In order to ensure that we compared equal quantities of hGH1 and hGH2 polypeptides, the media used in this experiment were

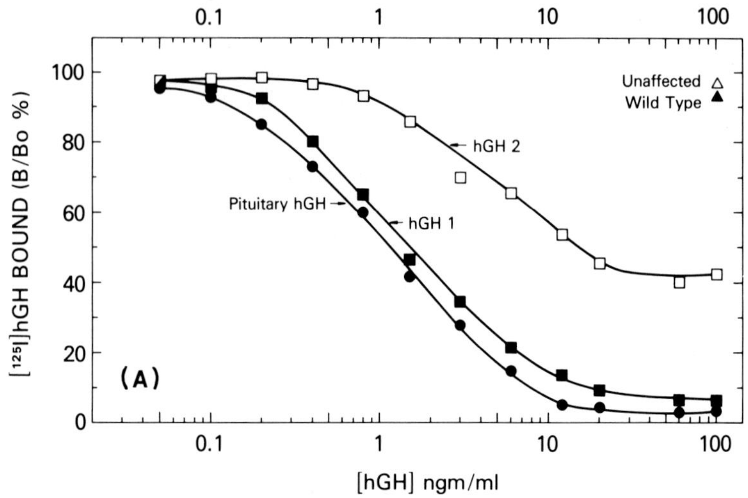

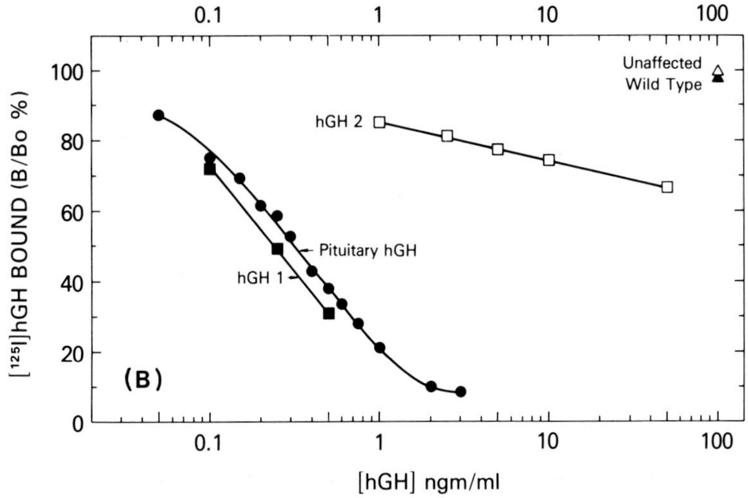

Fig. 4 Radioimmunoassay of hGH preparations. (A) Rabbit anti-hGH serum used. (B) Guinea pig anti-hGH serum used. For the assay shown in (B) the standard ^{125}I-labeled hGH (^{125}I-hGH, specific activity ~30 μci/μgm) was the same used in the radioreceptor assays shown in Fig. 5. The media used in this experiment were collected from ~2×10^7 cells that had been infected with SVhGH1(L) or SVhGH2(L) and labeled in 20 ml of medium containing [^{3}H]leucine at 20 μci/ml from 24 to 48 hr after infection. Analysis of these samples by acrylamide/NaDodSO$_4$-gel electrophoresis and fluorography showed that they contained approximately equal amounts of [^{3}H]leucine-labeled hGH. The lower scale refers to the concentration of pituitary hGH standard added to the assay, and the upper scale refers to the amount of infected cell medium added. The total sample volume was 100 μl in all assays. Controls of medium from cells incubated without virus (''unaffected'') and cells infected with wild-type SV40 are shown. B, [^{125}I]hGH bound; B$_0$, [^{125}I]hGH bound in the absence of competitor. From Pavlakis *et al.* (1981).

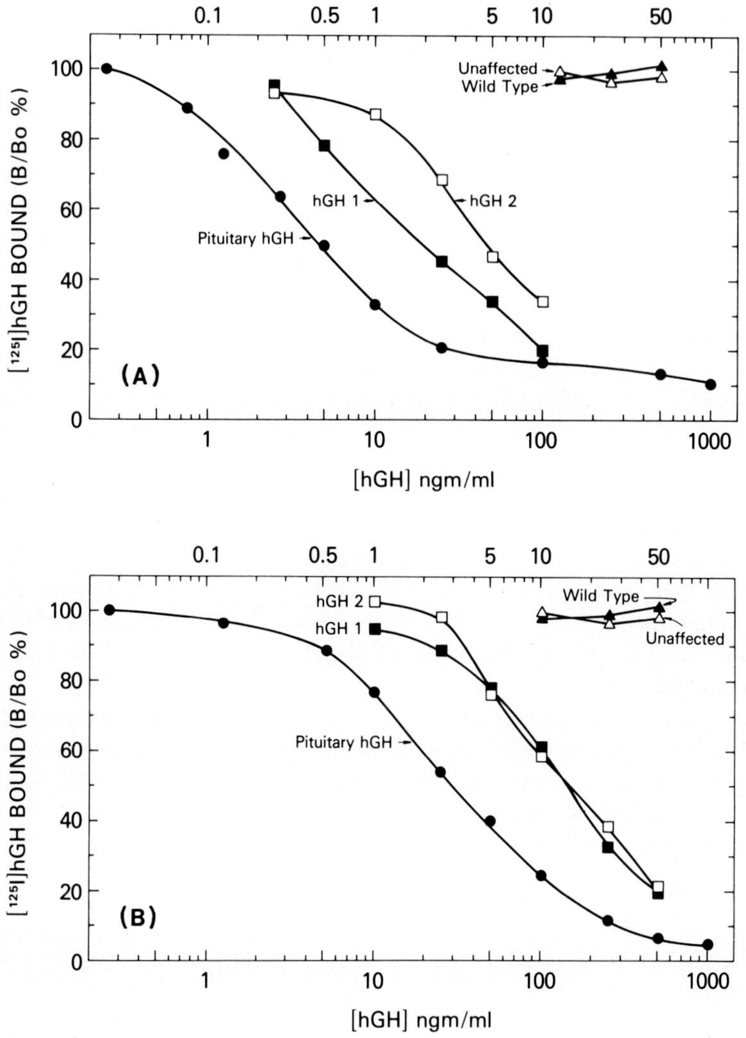

Fig. 5 Radioreceptor assay of hGH preparations. (A) Data from the IM-9 cultured human lymphocyte assay. (B) Data from the pregnant rabbit liver membranes assay. In each instance the assay contained the receptor preparation, [^{125}I]hGH, and various amounts of the unknown hGH preparation. These ingredients were diluted in buffer to final volume of 500 μl, and the incubation and separation of the bound and free components were carried out as described by Tsushima and Friesen (1973) and Eastman *et al.* (1979). The media were the same samples as used in Fig. 4. Note that the relationship of the lower and upper scales is different from that in Fig. 4. From Pavlakis *et al.* (1981).

collected from cells uniformly labeled with $[^3H]$leucine. Scans of the NaDod-SO_4-gel fluorograms showed that the SVhGH1 and SVhGH2 samples contained equal amounts of hGH polypeptide. The dose–response curves of hGH1 were parallel to the pituitary hGH standard in both assays. In contrast, hGH2 gave nonparallel curves with both antisera, and it reacted less with the guinea pig antibody than with the rabbit antibody. Because of the nonidentical cross-reactivity, the most dilute samples of hGH2 give the highest values in radioimmunoassay. Under these conditions, hGH2 had approximately 10% of the immunoactivity of hGH1 with the rabbit antibody and less than 5% with the guinea pig antibody.

The ability of hGH1 and hGH2 to bind to hGH cell-surface receptors was tested by radioreceptor assays using either the human lymphocyte line IM-9 (Eastman *et al.*, 1979) or pregnant rabbit liver membranes (Tsushima and Friesen, 1973) as the receptor sources. In both systems hGH1 was indistinguishable from pituitary hGH (Fig. 5). Surprisingly, hGH2 was 50% as active as hGH1 in the lymphocyte assay and 100% as active in the liver-membrane assay and gave dose–response curves parallel to the standard in both systems. Thus the ratio of receptor to immunoassay activity is approximately 1 for hGH1, whereas it is 10 or greater for hGH2. The identity of the hGH1 and hGH2 preparations in the liver assay is in contrast to the behavior of the M_r 20,000 hGH variant, which is less potent than pituitary hGH in this system (Sigel *et al.*, 1981; Hizuka *et al.*, 1982).

Because cell-surface receptors are thought to mediate the biological activities of hGH, it is conceivable that the hGH2 protein represents a previously unrecognized growth factor or an hGH agonist. Such factors could play a role in acromegalia or variant short stature. To test these ideas it will be necessary to conduct animal tests on hGH2 and to search for this protein in human tissues.

IV. Hepatitis B Surface Antigen

Hepatitis B virus (HBV) is the causitive agent for serum hepatitis. At least half of the world population shows evidence for past or recent infection by this virus, and the $\sim 200 \times 10^6$ carriers in the world are at serious risk of liver disease and, possibly, primary liver cancer. The classic marker for HBV infection is the surface antigen, HBsAg, which circulates in the serum in three forms: 22-nm spherical particles, 22-nm filaments of various lengths, and the 42-nm spherical virions (Dane particles) that contain the viral genome. The 22-nm particles and filaments are noninfectious, subviral forms that circulate in the sera of chronic carriers at concentrations as high as 100–200 μgm/ml. They contain two predominant polypeptides, with apparent molecular weights of ~23,000 and ~29,000, that carry both the group (a) and subtype (d/y) antigenic determinants of HBsAg. Protein- and DNA-sequencing experiments indicate that polypeptide P2 is identical to P1 except that it is glycosylated (reviewed in Gerin and Shin, 1978). Thus this system offers the opportunity to study two further aspects of posttranslational modification, namely, glycosylation and assembly.

Characterization of the life cycle and biology of HBV has been hampered by its narrow host range (it is restricted to humans and a few other primates) and its inability to grow in cultured cells. However, several groups have succeeded in cloning the viral genome in *E. coli* phage λ and plasmid vectors and in determining its primary structure (Valenzuela *et al.*, 1979; Pasek *et al.*, 1979; Gailbert *et al.*, 1979). This has allowed the identification of a continuous 892-bp sequence that could encode surface antigen as well as a 549-bp sequence that may specify the core antigen and several additional open sequences of unknown function. Although the DNA sequence provides crucial structural information, it is clearly not sufficient to establish how the HBV gene products are assembled or how these products interact during infection of the target cell. As a first step in understanding these problems, we have constructed an SV40 recombinant carrying the HBsAg gene. Monkey kidney cells infected with the recombinant synthesize surface antigen that is both glycosylated and assembled. Similar results have been obtained by introducing the HBsAg gene into mouse L cells by cotransformation (Dubois *et al.*, 1980) and into monkey cells by acute transfection with an SV40 plasmid vector (Liu and Levinson, 1982). Moreover, we have demonstrated that the 22-nm particles produced by this technique are immunogenic in chimpanzees (Section IV,C).

A. Cloning of the HBsAg Gene and Insertion into SV40

The SV40–HBV recombinant used in this work carries a 1350-bp fragment of HBV DNA, representing about 40% of the HBV genome, inserted into the late-gene region of SV40. The first step in the construction of this recombinant was to amplify the HBV genome by cloning it in an *E. coli* plasmid vector (step A in Fig. 6). Dane particles were purified from the serum of a chronic HBsAg carrier, subtype adw, and the partially single-stranded viral genome was repaired by an endogenous DNA polymerase reaction. Two fragments, 1350 and 1850 bp, were obtained after cleavage of this DNA with *Bam*HI. Partial digestion with *Bam*HI generated a full HBV genome that was ligated to *Bam*HI-cleaved plasmid pBR322 DNA and cloned in *E. coli*.

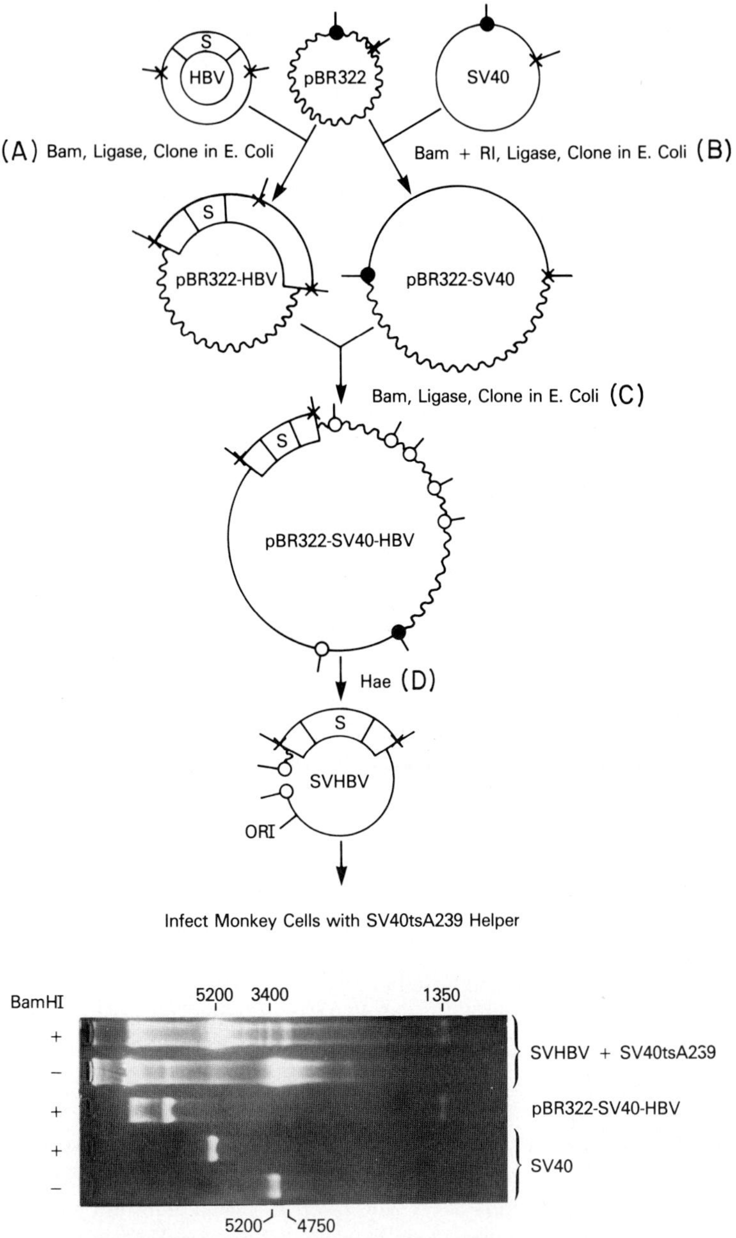

Fig. 6 Construction and analysis of the SV40–HBV recombinant. (Above) The HBV genome was cloned in pBR322, and the 1350-bp fragment containing the HBsAg gene was subcloned in a pBR322-SV40 vector. Following removal of the plasmid sequences, the SV40–HBV recombinant

From the published sequence data we anticipated with the HBsAg coding sequence would be located within the 1350-bp *Bam*HI fragment. This fragment was purified by electrophoresis, subcloned in pBR322, isolated, ligated to a *Bam*HI-cleaved pBR322–SV40 vector plasmid, and recloned in *E. coli* (steps B and C in Fig. 6). *Hae*II digestion of the resultant pBR322–SV40–HBV "double recombinant" plasmid removed all but 143 bp of the pBR322 DNA and yielded a homogenous preparation of 4950-bp SV40–HBV linear recombinant molecules, containing the HBsAg gene inserted between the *Hae*II and *Bam*HI sites of SV40, that were propagated as virions as described in Section II (step D in Fig. 6). Analysis of the intracellular viral DNA showed that the stock contained approximately 75% helper genomes and 5% SV40–HBV recombinant genomes retaining the complete 1350-bp HBV fragment (Fig. 6, bottom). The remaining 20% of the DNA was found in a heterogeneous collection of genomes with lengths ranging from ~3000 to ~4900 bp; although not examined in detail, these molecules contained no more HBV genetic information than that present in the original subcloned 1350-bp fragment. We refer to this complex mixture of virus as SVHBV.

B. Assembly and Glycosylation of HBsAg

Specific immunological assays showed that monkey kidney cells infected with SVHBV synthesize HBsAg but no other established HBV antigens. Immunofluorescence analysis revealed that approximately 45% of the cells infected with SVHBV expressed cytoplasmic HBsAg by 72 hr after infection, whereas uninfected cells and cells infected with wild-type SV40 were negative. Quantitative radioimmunoassays showed that a culture of 2×10^7 cells produced a total of 2.5 μgm of HBsAg, of which 40% was found in the medium and 60% was released from the cells by freeze-thawing and sonication. Subtype analysis

---

was propagated as virus in monkey cells. Cleavage sites; X, *Bam*HI; ●, *Eco*RI; ○, *Hae*II. (Below) Ethidium bromide-stained 1% agarose gel containing intracellular viral DNA from cells infected with the SVHBV stock (two upper lanes), plasmid pBR322–SV40–HBV DNA (middle lane), and purified SV40 DNA (two lower lanes). The numbers above the gel represent the lengths of linear molecules; the numbers below refer to the lengths of covalently closed circular DNAs. The uncleaved SVHBV DNA (second lane from top) shows a predominant band of helper DNA at 5200 bp, a band of SVHBV DNA at 4750 bp, and a heterogeneous collection of shorter DNAs; the more slowly migrating bands represent nicked circular and host-cell DNA. *Bam*HI cleavage of this DNA (top lane) generated a predominant 5200-bp band of helper DNA, a 3400-bp band of SV40 vector DNA, a 1350-bp band of inserted HBV DNA that comigrated with its authentic counterpart from *Bam*HI-cleaved pBR322–SV40–HBV DNA (middle lane), and several other bands of unknown origin. From Moriarty *et al.* (1981).

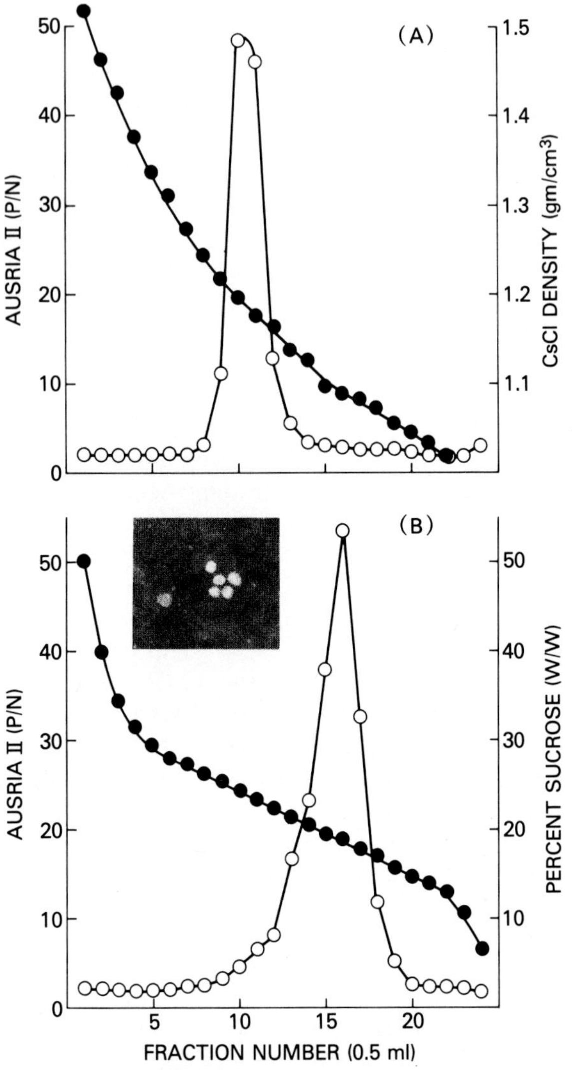

Fig. 7 Biophysical properties and appearance of HBsAg from SVHBV-infected monkey kidney cells. (A) The medium of monkey kidney cells, harvested 72 hr after infection with SVHBV, was banded in a CsCl step gradient. Fractions were assayed for HBsAg by a commercial radioimmunoassay (○; Ausria II); results are expressed as the ratio of ^{125}I cpm in the sample (*P*) to the negative control (*N*; 93 cpm). CsCl density (●) was determined by refractometry. (B) HBsAg purified through the CsCl gradient was sedimented in a sucrose gradient (●) and fractions were assayed for HBsAg (○) as in (A). (Inset) Electron micrograph of particles stained with 1% phosphotungstic acid; these particles are from fractions 15–17 (B) and are 20–24 nm in diameter (magnification 100,000×). From Moriarty *et al.* (1981).

showed that SVHBV–HBsAg has the same antigenic composition (d+, y−) as the antigen from the original donor of the HBV DNA. SVHBV-infected cells were negative for HBcAg and δ antigen by immunofluorescence, and the culture medium was negative for HBcAg, HBeAg, and δ antigen by solid-phase radioimmunoassay.

The HBsAg secreted from infected monkey cells is found in 22-nm particles and filaments indistinguishable from those found in human serum. As shown in Fig. 7A, isopycnic banding of the SVHBV–HBsAg revealed a buoyant density (1.2 gm/cm^3) identical to that of purified serum HBsAg run in parallel. Furthermore, most of the SVHBV-specific antigen sedimented at the same rate as human 22-nm particles in a rate-zonal centrifugation experiment (Fig. 7B). Examination of the antigen by electron microscopy of the pooled, concentrated peak fractions from the sucrose gradient (Fig. 7B, inset) revealed 22-nm spherical particles with the same appearance as the predominant form of HBsAg from human sera. Examination of the less highly purified material from the CsCl gradient also showed 22-nm filaments of variable lengths (results not shown). No Dane particles were observed.

The polypeptide composition of the excreted SVHBV–HBsAg was analyzed by immunoprecipitation and gel electrophoresis of both chemically and biosynthetically labeled antigen. SVHBV–HBsAg from the rate-sedimentation step was radioiodinated, resedimented, and immunoprecipitated by anti-HBsAg/ad antiserum and antibodies against the group (a) and d-subtype specific determinants. More than 60% of the iodinated antigen was precipitated by anti-HBs and by both of the monospecific antibodies. $NaDodSO_4$-Gel electrophoresis of the denatured and reduced precipitates revealed two predominant polypeptides (P1 and P2) with the same mobilities as those isolated from human serum (Fig. 8A), as well as small amounts of a more slowly migrating species desginated P5. Similar results were obtained with antigen labeled by incubating the infected cells with [^{35}S]methionine for 4 hr late in the lytic cycle, although in this case no specifically precipitated higher molecular weight forms were observed (Fig. 8C). It was also possible to visualize P1 and P2 in the complete, nonprecipitated medium from SVHBV-infected cells though not from mock or wild-type SV40-infected cells (Fig. 8B). This clearly demonstrates that HBsAg is actively excreted rather than merely released by cell lysis, because only a subpopulation of the total cell proteins was found in the medium. The ratio of P1 to P2 in these experiments was similar if not identical to that found for HBsAg from human serum. Because P2 is simply the glycosylated form(s) of P1, this suggests that the surface antigen is glycosylated with normal efficiency in infected monkey cells. Glycosylation has also been observed for influenza virus hemagglutinin produced in monkey cells infected with an SV40-influenza recombinant (Gething and Sambrook, 1981).

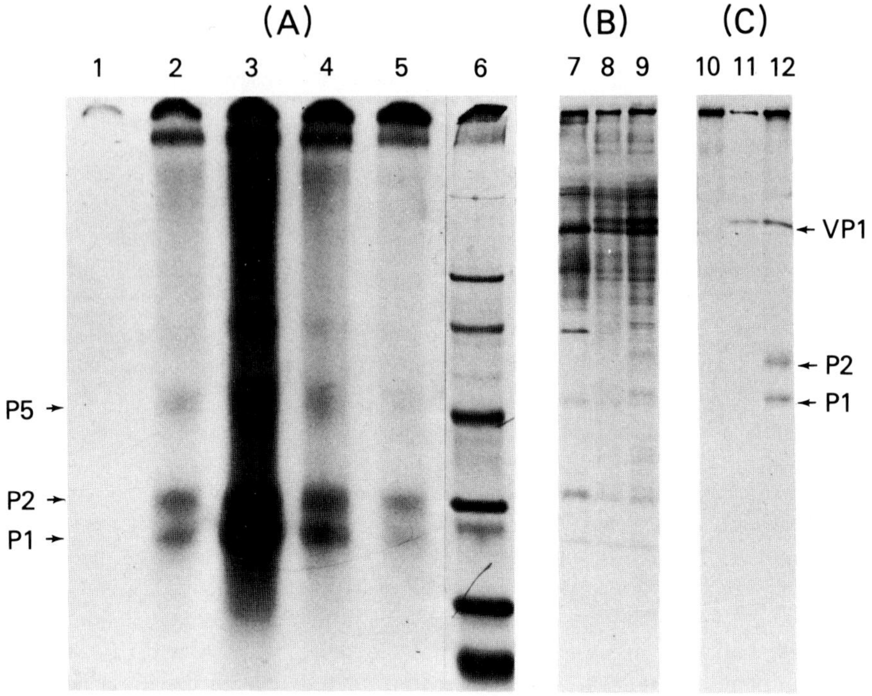

Fig. 8 Electrophoretic analysis of SVHBV–HBsAg polypeptides. (A) Immunoprecipitation of ^{125}I-labeled SVHBV–HBsAg and human HBsAg/ad. The immunoprecipitated antigens were electrophorsed through a $NaDodSO_4$/7.5% acrylamide gel. Lanes: 1, SVHBV–HBsAg precipitated with preserum; 2, SVHBV–HBsAg precipitated with anti-ad; 3, HBsAg precipitated with anti-ad; 4, SVHBV–HBsAg precipitated with anti-d; 5, SVHBV–HBsAg precipitated with anti-a; 6, ^{14}C-labeled protein standards (MW: myosin, 20,000; phosphorylase b, 92,500; bovine serum albumin, 68,000; ovalbumin, 45,000; chymotrypsinogen A 25,700; β-lactoglobulin, 18,400; cytochrome c, 11,700). (B) Total medium from [^{35}S]methionine-labeled cells. Confluent monolayers of $\sim 2 \times 10^7$ monkey kidney cells incubated with SVHBV (lane 9), with wild-type SV40 (lane 8), or without virus as controls (lane 7). At 68 hr after infection the medium was replaced with 5 ml of methionine-free medium containing 200 μci of [^{35}S]methionine per milliliter. After 5 hr the medium was removed, and an aliquot was electrophoresed through a $NaDodSO_4$/20% polyacrylamide gel. (C) Immunoprecipitation of [^{35}S]methionine-labeled peptides. The medium from (B) was treated with anti-ad antiserum followed by formalin-fixed *Staphylococcus aureus* Cowan I bacteria, and the precipitates were solubilized and electrophoresed through the same $NaDodSO_4$/20% polyacrylamide gel as in (B). Lanes: 10, uninfected; 11, wild-type SV40; 12, SVHBV. Arrows point to the HBsAg polypeptides P1, P2, and P5 and the SV40 major coat protein VP1. From Moriarty *et al.* (1981).

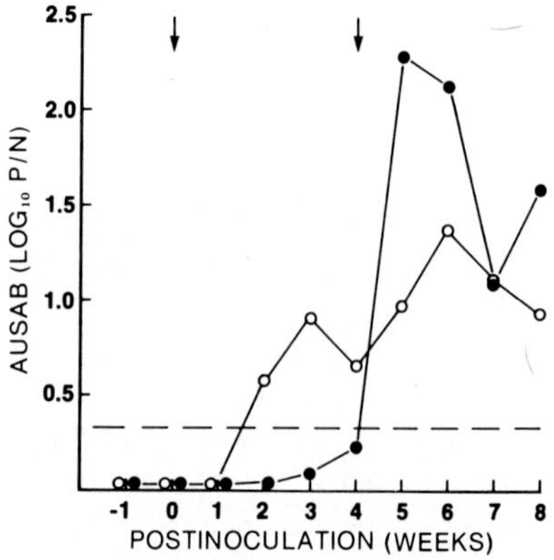

Fig. 9 Time course of anti-HBs response in two chimpanzees inoculated with HBsAg particles purified from the medium of SVHBV-infected monkey cells. Each animal was immunized intramuscularly at 0 and 4 weeks (arrows) with 5 μgm of HBsAg absorbed on alum. Serum was tested weekly for anti-HBs by the commercial Ausab test.

C. Immunogenicity in Chimpanzees

Hepatitis B is a major remaining "unsolved" problem in infectious diseases. Because of the narrow host range of the virus, and its apparent inability to grow in cultured cells, it has been necessary to resort to producing vaccine from the sera of chronic carriers. This is a costly and time-consuming process. Clearly, it would be desirable to use recombinant DNA techniques to develop a more economical and widely available vaccine.

As a first step in exploring this possibility we have collaborated with J. Gerin and R. Purcell to test the immunogenicity of the SV40-derived antigen in chimpanzees, the best studied animal system for HBV. Two seronegative animals were injected intramuscularly with 5 μgm of formalin-inactivated, alum-absorbed SVHBV–HBsAg and boosted with the same inoculum at 4 weeks. As shown in Fig. 9, both chimpanzees developed anti-surface-antigen antibody within 5 weeks. The kinetics of the response and final levels of antibody were as good, if not better, than in chimpanzees treated in an identical fashion with particles purified from human serum. The efficacy of the immunization was tested by a challenge with 10^3 CID_{50} units of infectious HBV at 10 weeks. The one animal available for follow-up survived the challenge and showed no overt signs of hepatitis. This animal continued to show high levels of antibody more than 9 months after immunization.

V. Conclusions and Prospects

We have used the hGH and HBsAg genes as model systems to study post-translational events in monkey cells infected with SV40 recombinants. Our results demonstrate that signal-sequence cleavage, glycosylation, secretion, and particle assembly occur in this system and that the final products are indistinguishable from their authentic counterparts by a variety of biochemical, physical, and immunological tests. The success of these initial experiments opens several interesting experimental opportunities.

First, it becomes possible to study the effects of amino acid sequence alterations on the posttranslational and biological activity of animal-cell proteins. In the case of hGH this was achieved by using a naturally occurring variant of the normal gene. Similar studies could be conducted using genes that have been more specifically altered by *in vitro* mutagenesis.

Second, SV40 vectors may be useful for characterizing proteins for which the genes have been cloned but the functions are unknown. This should be especially true for secreted proteins because, at least for hGH and HBsAg, they can be recovered from the media in highly enriched forms and can be purified without the use of antibodies. It is becoming clear that many secreted proteins are encoded by multigene families; for example, there appear to be at least seven hGH-related genes, whereas for the immunoglobulins there may be thousands or even millions of different genes. The sophistication of recombinant DNA technology makes it easier to purify and characterize the genes than the corresponding proteins. By inserting such genomic clones directly into SV40 vectors it should be possible to isolate sufficient protein for initial characterization and antibody preparation.

Finally, these experiments set a precedent for using gene transfer into mammalian cells as a means for producing medically important proteins. In particular, our demonstration that the SV40-derived HBsAg particles are as immunogenic in chimpanzees as particles from human serum raises the possibility of using this or similar material as a vaccine. Obviously, there are many practical difficulties, both in terms of expense and safety, in producing a human therapeutic from SV40-infected tissue-culture cells. Eventually it will probably be necessary to use artificially or naturally incapacitated vectors and to either increase the yields of product or to decrease the cost of growing cells. But given the importance of hepatitis as a world health problem and the failure of all attempts to produce native HBsAg polypeptide or particles in *E. coli,* it seems to be a project well worth pursuing.

Acknowledgments

I think my many colleagues who collaborated on these experiments and Gail Taff for excellent preparation of the manuscript.

References

Bala, R. M., Ferguson, K. A., and Beck, J. C. (1973). *In* "Advances in Human Growth Hormone Research" (S. Raiti, ed.), Publ. No. 74–612, pp. 494–516. Dept. Health, Educ. and Welfare, Washington, D.C.

Blobel, G., Walter, P., Chang, C. N., Goodman, B., Erickson, A. H., and Lingappa, V. R. (1979). *Symp. Soc. Exp. Biol.* **33,** 9–36.

Cheever, E. V., and Lewis, U. J. (1969). *Endocrinology (Baltimore)* **85,** 465–470.

Chrambach, A., Yadley, R. A., Ben-David, M., and Rodbard, D. (1973). *Endocrinology (Baltimore)* **93,** 848–857.

DeNoto, F. M., Moore, D. D., and Goodman, H. M. (1981). *Nucleic Acids Res.* **9,** 3719–3730.

Dubois, M.-F., Pourcel, C., Rousset, S., Chany, C., and Tiollais, P. (1980). *Proc. Natl. Acad. Sci. USA* **77,** 4549–4553.

Eastman, R. C., Lesniak, M. A., Roth, J., DeMeyts, P., and Gorden, P. J. (1979). *J. Clin. Endocrinol. Metab.* **49,** 262–268.

Elder, J. T., Spritz, R. A., and Weissman, S. M. (1981). *Annu. Rev. Genet.*, pp. 295–340.

Fiddes, J. C., Seeburg, P. H., DeNoto, F. M., Hallewell, R. A., Baxter, J. D., and Goodman, H. M. (1979). *Proc. Natl. Acad. Sci. USA* **76,** 4294–4298.

Galibert, F., Mandart, E., Fitoussi, F., Tiollais, P., and Charnay, P. (1979). *Nature (London)* **281,** 646–650.

Gerin, J. L., and Shih, J.W.-K. (1978). *In* "Viral Hepatitis" (G. N. Vyas, S. N. Cohen, and R. Schmid, eds.), pp. 147–153. Franklin Inst., Philadelphia, Pennsylvania.

Gething, M. J., and Sambrook. J. (1981). *Nature (London)* **293,** 620–625.

Goodman, A. D., Tanenbaum, R., and Rabinowitz, D. (1972). *J. Clin. Endocrinol. Metab.* **36,** 178–184.

Goodman, H. M., DeNoto, F., Fiddes, J. C., Hallewell, R. A., Page, G. S., Smith, S., and Tischer, E. (1980). *In* "Mobilization and Reassembly of Genetic Information" (W. A. Scott, R. Werner, D. R. Joseph, and J. Schultz, eds.), pp. 155–179. Academic Press, New York.

Gorden, P., Hendricks, C. M., and Roth, J. (1973). *J. Clin. Endocrinol. Metab.* **36,** 178–184.

Gorden, P., Lesniak, M. A., Eastman, R., Hendricks, C. M., and Roth, J. (1976). *J. Clin. Endocrinol. Metab.* **43,** 364–373.

Gruss, P., and Khoury, G. (1981). *Proc. Natl. Acad. Sci. USA* **78,** 133–137.

Hamer, D. H. (1980). *Genet. Eng.* **2,** 83–101.

Hizuka, N., Hendricks, C., Pavlakis, G. N., Hamer, D. H., and Gorden, P. (1982). *J. Clin. Endocrinol. Metab.* **55,** 545–550.

Lewis, U. J., and Cheever, E. V. (1965). *J. Biol. Chem.* **240,** 247–250.

Lewis, U. J.,, Cheever, E. V., and Hopkins, W. C. (1970). *Biochim. Biophys. Acta* **214,** 498–508.

Lewis, U. J., Dunn, J. T., Bonewald, L. F., Seavey, B. K., and VanderLaan, W. P. (1978). *J. Biol. Chem.* **253,** 2679–2687.

Lewis, U. H., Bonewald, L. F., and Lewis, L. J. (1980). *Biochem. Biophys. Res. Commun.* **92,** 511–516.

Liu, C. C., and Levinson, A. D. (1982). *In* "Eukaryotic Viral Vectors" pp. 55–60. Cold Spring Harbor Lab., Cold Spring Harbor, New York.

Moriarty, A. M., Hoyer, B. H., Shih, J. W.-K., Gerin, J. L., and Hamer, D. H. (1981). *Proc. Natl. Acad. Sci. USA* **78,** 2606–2610.

Pasek, M., Gato, F., Gilbert, W., Zink, B., Schaller, H., Mackay, P., Leadbetter, G., and Murray, K. (1979). *Nature (London)* **282,** 575–579.

Pavlakis, G. N., Hizuka, N., Gorden, P., Seeburg, P., and Hamer, D. H. (1981). *Proc. Natl. Acad. Sci. USA* **78,** 7398–7402.

Raiti, S. (ed.) (1973). "Advances in Human Growth Hormone Research." Dept. Health, Educ., and Welfare, Washington, D.C.

Rigby, P. W. J. (1982). *Genet. Eng.* **3,** 84–141.

Seeburg, P. H. (1982). *DNA* **1,** 239–250.

Sigel, M. B., Thorpe, N. A., Kobrin, M. S., Lewis, U. J., and Vanderlaan, W. P. (1981). *Endocrimonology (Baltimore)* **108,** 1600–1603.

Tooze, J. (ed.) (1980). "Molecular Biology of Tumor Viruses," Part 2. Cold Spring Harbor Lab., Cold Spring Harbor, New York.

Tsushima, T., and Friesen, H. G. (1973). *J. Clin. Endrocrinol. Metab.* **37,** 334–337.

Valenzuela, P., Gray, P., Quiroga, M., Zaldivar, J., Goodman, H. M., and Rutter, W. G. (1979). *Nature (London)* **280,** 815–819.

CHAPTER 11

Adenovirus Type 5 Region-E1A Transcriptional Control Sequences

PATRICK HEARING
THOMAS SHENK

Department of Microbiology
Health Sciences Center
State University of New York at Stony Brook
Stony Brook, New York

I. Introduction

The adenovirus genome is a linear, double-stranded DNA molecule containing ~36,000 bp. Region E1A is located at the far left end of the adenovirus genome,

Experimental Manipulation
of Gene Expression

ISBN 0-12-372380-9

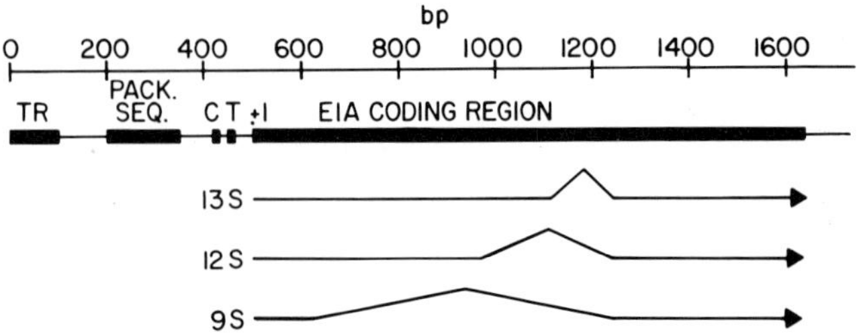

Fig. 1 Schematic of the left end of the Ad5 genome. The left end of the adenovirus type 5 genome including region E1A and its 5′-flanking sequences is diagrammed. The E1A cap site is at nucleotide 499 (according to the sequence published by van Ormondt *et al.*, 1978) and is designated +1. Transcription of region E1A proceeds in a rightward direction, yielding three polyadenylated, cytoplasmic mRNAs designated 13, 12, and 9 S. T, "TATA" homology at −30; C, "CAT" box at −70; PACK. SEQ., Ad5 packing sequence between −170 and −305; TR, terminal repeat.

mapping between 1.4 and 4.5 map units. The complete nucleotide sequence of the left 4.5% of the Ad5 genome has been reported (van Ormondt *et al.*, 1978). Figure 1 is a schematic of the left end of the adenovirus type 5 (Ad5) genome including region E1A and its 5′-flanking sequences. The unique cap site (+1) for E1A transcription is located at nucleotide 499 from the left end of the genome (Baker and Ziff, 1981). Three mRNAs are synthesized from region E1A during the course of Ad5 infection; they are 5′ and 3′ coterminal and differ by their splicing patterns (Berk and Sharp, 1978; Spector *et al.*, 1978; Chow *et al.*, 1979; Baker and Ziff, 1981). Two mRNAs (13 and 12 S) are observed in the cytoplasm early after infection, and an additional transcript (9 S) is detected late in infection (Spector *et al.*, 1978; Chow *et al.*, 1979). The 13-S mRNA encodes a product required to activate all of the remaining viral transcription units (Berk *et al.*, 1979; Jones and Shenk, 1979; Carlock and Jones, 1981; Nevins, 1981; Montell *et al.*, 1982).

Here we describe a mutational analysis of the Ad5 region-E1A transcriptional control region. We compare the effects of deletion mutations in the E1A 5′-flanking sequences on transcription *in vitro,* using the soluble system of Manley *et al.* (1980), and *in vivo,* after reconstruction of the mutations back into virus. The sequences critical for efficient transcription *in vitro* include the "TATA" homology (Goldberg, 1979) at −30 and the region downstream through the E1A cap site (+1). *In vivo* analyses identify two sets of transcriptional control signals. Sequences around the cap site (+1), including the TATA box at −30, determine the site of initiation but do not significantly influence the levels of transcription, and a regulatory region mapping well upstream of these sequences (around −230) modulates the levels of transcription. Thus it appears that only a subset of the signals that operate *in vivo* are recognized in cell-free extracts.

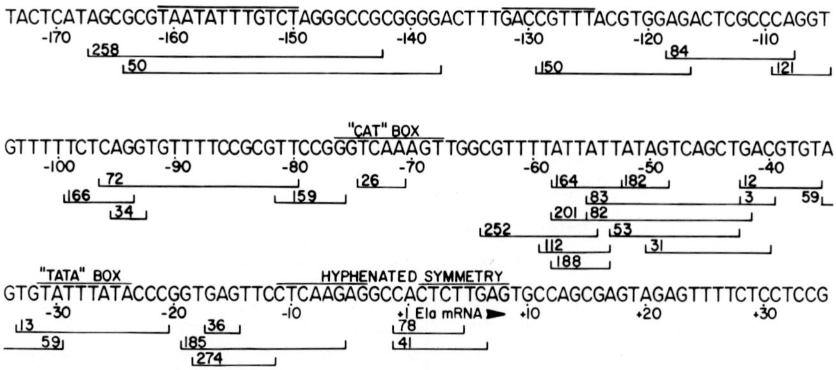

Fig. 2 Deletions generated within the E1A 5′-flanking region by D-loop mutagenesis. The sequence around the E1A cap site (+1) is presented. The boundaries of each deletion are shown by brackets, with the plasmid number indicated. The hyphenated symmetry at the cap site, the TATA homology at −30, and the CAT box at −70 are shown.

II. Deletion Mutations in the 5′-Flanking Sequences of Ad5 Region E1A

We cloned the left end of the Ad5 genome into pBR322 and constructed *in vitro* two sets of deletion mutations that map in the E1A 5′-flanking sequences. The first protocol utilized is a modification of the D-loop mutagenesis procedure described by Shortle *et al.* (1980). The procedure generates small (3–25 bp), random deletions in the area of interest. The mutagenized plasmids were subjected to nucleotide-sequence analysis, and the exact location of each deletion is presented in Fig. 2.

A second set of deletion mutations was constructed that begin farther upstream of the E1A cap (at −392) and progress toward the cap site. The end points of each of these deletions were determined by nucleotide-sequence analysis and are presented in Fig. 3. Mutations in the 5′-flanking sequences of region E1A were reconstructed into intact virus using the procedure of Stow (1981), and mutant viruses were propagated in 293 cells (a permissive human cell line that contains and expresses the left 11% of the Ad5 genome; (Graham *et al.*, 1977). Variants carrying mutations in the region between 194 and 355 bp from the left end of the genome were severely defective for packaging of viral DNA into virions. A cis-acting packaging sequence is known to be near the left end of the viral chromosome (Tibbetts, 1977; Hammarskjold and Winberg, 1980). To overcome this packaging defect, mutations that overlapped the packaging sequence were rebuilt

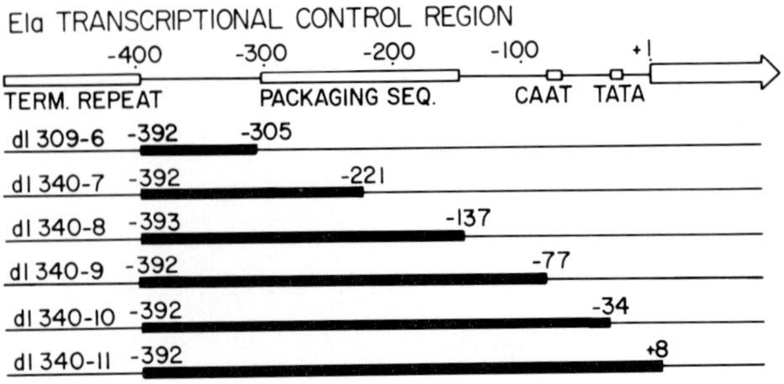

Fig. 3 Large-deletion mutations constructed within the E1A 5′-flanking sequences. Deletion mutations were constructed that begin far upstream of the E1A cap site (at −392) and progress toward the cap site. The end points of these deletions were determined by nucleotide-sequence analysis. The numbers of each mutant represent the plasmid number (Berk *et al.*, 1979; Jones and Shenk, 1979; Manley *et al.*, 1980; Carlock and Jones, 1981; Nevins, 1981; Montell *et al.*, 1982) and the virus background into which the deletion was rebuilt (*dl*309 or *in*340; see text for details).

into an Ad5 variant, *in*340, which contains a copy of the packaging sequence at the right end of the viral genome. The variant *in*340 grows as well as wild-type Ad5.

III. Analysis of Mutagenized Templates in Cell-Free Transcription Extracts

Plasmids containing deletions in the E1A 5′-flanking sequences were linearized with *Xba*I and used to program the *in vitro* transcription system described by Manley *et al.* (1980). An example of the results obtained by these analyses is shown in Fig. 4. When no DNA was added to the extract, two endogenous bands of ~1800 and ~750 nucleotides were labeled. When the wild-type plasmid (pE1A-WT) was used to program the *in vitro* reaction, several new transcripts were synthesized. The 840-nucleotide product initiates at the E1A cap site (containing nucleotides 499 to 1339). A 5′-S1 analysis was used to confirm that this transcript had the same 5′ end as mature cytoplasmic E1A mRNA found after infection of HeLa cells. Several other larger transcripts were also produced using the wild-type plasmid. The 1320-nucleotide product is actually a doublet representing a transcript initiating upstream of the E1A cap site with a 5′ end in the vicinity of the Ad5 terminal repeat, along with a separate transcript initiated in pBR322 sequences and transcribed from the opposite stand of the E1A transcript. The transcript at 2300 nucleotides was also initiated in pBR322 sequences from the opposite stand of the E1A transcripts. Transcription of all of these products

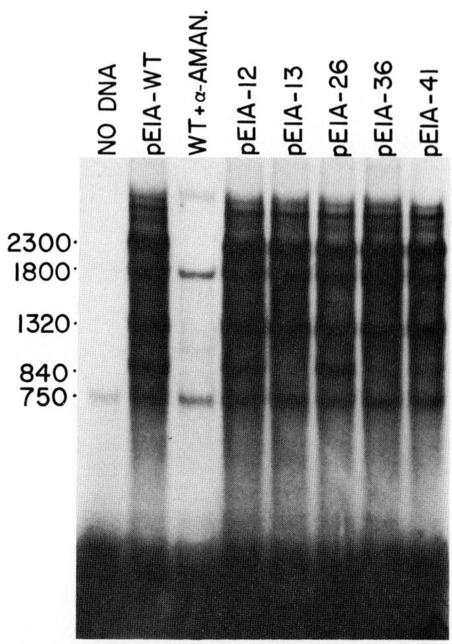

Fig. 4 Analysis of mutagenized templates in cell-free transcription extracts. Plasmids were linearized with *Xba*I and used to program *in vitro* transcription reactions as described by Handa *et al.* (1981). Products from the *in vitro* transcription reactions were glyoxylated and analyzed on a 1.8% agarose gel (McMaster and Carmichael, 1977). The length of each runoff product is indicated in nucleotides. The template DNA that was added to each reaction is shown at the top of each lane. WT + α-AMAN, Products from a reaction with pE1A-WT plus 1 μgm/ml α-amanitin.

was sensitive to low concentrations of α-amanitin (1 μgm/ml), demonstrating that they are products of RNA polymerase II (Roeder, 1976). Several other minor species were observed, but they were not further investigated.

When plasmids with deletions generated by D-loop mutagenesis were used as templates in the *in vitro* transcription system, a block of nucleotides ~40 bp long was found to be critical for efficient E1A transcription. These sequences included the TATA box at −30, the hyphenated symmetry at the site of initiation of transcription, and the sequences between these two regions (pE1A-13, -36, and -41; Fig. 4). The results from analyses of all of the D-loop mutants *in vitro* are presented quantitatively in Table I. Deletion of the TATA box (pE1A-13) abolished *in vitro* initiation at the E1A cap site. Plasmids with deletions downstream of the TATA box (e.g., pE1A-36 and -41), showed reduced efficiency of E1A transcription *in vitro*. The level of reduction (Table I) depended on the particular mutation and ranged from 2- to 10-fold. Deletion of sequences upstream of the TATA box had no effect on the efficiency of E1A transcription *in vitro*. Similar results were obtained using the second set of mutations, which

Table I Transcriptional Activity of Mutant Templates

In vitro		*In vivo*	
Plasmid	% pE1A-WT[a]	Virus	% *dl*309[b]
pE1A-12	88	*dl*309-12	112
pE1A-13	0	*dl*309-13	62
pE1A-26	95	*dl*309-26	96
pE1A-36	10	*dl*309-36	80
pE1A-41	30	*dl*309-41	69
pE1A-59	16	*dl*309-59	81
pE1A-78	31	*dl*309-78	102
pE1A-82	96	*dl*309-82	99
pE1A-185	49	*dl*309-185	102
pE1A-274	24	*dl*309-274	92

[a]Band intensity was quantified by densitometer tracing.
[b]Bands at the 12- and 13-S locations were excised from the gel, and the radioactivity was quantified. The percentage indicated is the average of three independent experiments.

contained larger deletions in the E1A 5′-flanking sequences (Fig. 3). Again, a critical requirement for the TATA homology was observed, whereas deletions upstream of this sequence did not reduce E1A transcription *in vitro*.

IV. Analysis of Cytoplasmic E1A mRNAs Found in Vivo after Infection with Deletion Mutants

HeLa cells were infected with reconstructed viruses to analyze the effects of the various 5′-flanking mutations on the efficiency of E1A transcription *in vivo*. Cytoplasmic, polyadenylated RNA was isolated from HeLa cells at 5 hr after infection and subjected to Northern-type analysis. The results from a number of the viruses containing deletions originally generated by D-loop mutagenesis (i.e., deletions spanning the first 170 bp of 5′-flanking sequence) are shown in Fig. 5. As previously stated, polyadenylated E1a mRNAs are detected (12 and 13 S) early after infection with Ad5. As is apparent, none of the deletion mutations upstream of the E1A cap site to −170 significantly affected the quantities of steady-state cytoplasmic E1A RNAs synthesized after infection. The actual levels of the E1A mRNAs were quantified by excising the bands from the Northern blot and counting the filters in a liquid scintillation counter (Table I). Virus *dl*309-13, which lacks the TATA box at −30, consistently produced E1A

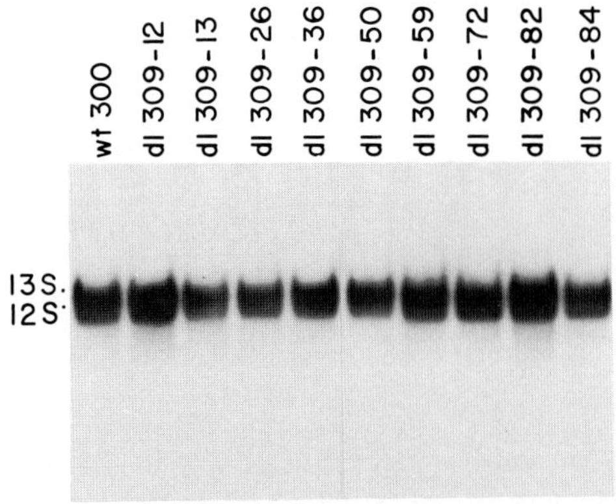

Fig. 5 Northern-type analysis of E1A mRNAs found *in vivo* after infection of HeLa cells with D-loop deletion mutants. Spinner cultures of HeLa cells were infected with wild-type Ad5 and the indicated deletion mutants at a multiplicity of 25 plaque-forming units/cell for 30 min at 37°C. Media containing 25 μgm/ml araC were added, and cytoplasmic, polyadenylated RNA was isolated (McGrogan *et al.*, 1979) at 5 hr after infection. RNAs were subjected to Northern-type analysis (Alwine *et al.*, 1977), using pE1A-WT as a probe. The 13- and 12-S early E1A mRNAs are indicated.

mRNAs at about 60% of the wild-type level. Also, certain mutations downstream of the TATA box reduced the mRNA level by 20–30%. Results from several experiments have shown that minor variations detectable with the other mutants (i.e., upstream of the TATA box) are not significant. Thus the sequences critical for efficient initiation of E1A transcription *in vitro* are not required for efficient transcription *in vivo*. Further, deletion of the "CAT" box or −68 homology (Benoist *et al.*, 1980; Efstratiadis *et al.*, 1980) had no detectable effect on E1A transcription *in vivo*.

Northern analyses of cytoplasmic, polyadenylated RNA synthesized after infection of HeLa cells with mutants carrying deletions farther upstream of the E1A cap site are shown in Fig. 6. Virus *dl*309-6, with a deletion from −392 to −305, produced wild-type levels of E1A transcripts. Deletions that progressed farther toward the E1A cap site, however, resulted in reduced levels of E1A transcription. The deletion present in *dl*340-7 resulted in a 3-fold decrease in cytoplasmic E1A transcripts, whereas *dl*340-8 produced 20-fold less E1A mRNAs than the wild-type virus. These two mutations define a region critical for regulating the levels of E1A transcription that maps between −305 and −144. It is interesting to note that *dl*340-9 and -10 produced slightly elevated levels of E1A transcripts relative to *dl*340-8. We have no definitive explanation for these results. Virus *dl*340-11 again showed a 20-fold reduction in the level of E1A transcripts relative to wild-type virus.

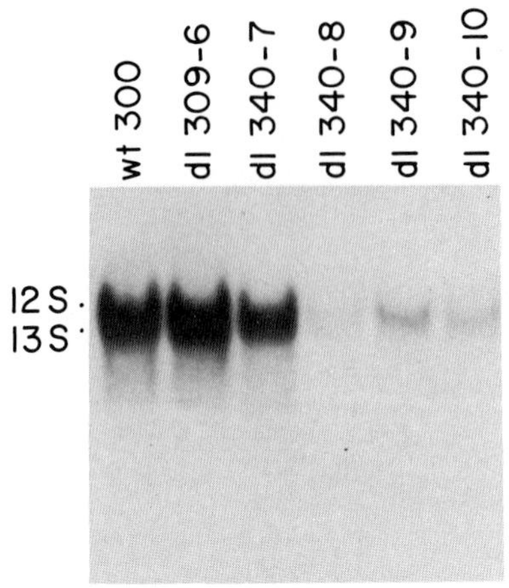

Fig. 6 Northern-type analysis of E1A mRNAs found *in vivo* after infection with mutants containing large deletions in the E1A 5′-flanking sequences. Infections, isolation of early RNAs, and Northern hybridization analyses were performed as described in the legend to Fig. 5.

V. 5′-End Analyses of E1A mRNAs Synthesized in Vivo *after Infection with Deletion Mutants*

The 5′ ends of the E1A transcripts produced after infection of HeLa cells with the various mutants were analyzed to determine what effect the deletions had on the initiation site (Fig. 7). Using cytoplasmic, polyadenylated RNA isolated from *wt*300-infected cells, a predominant band was observed by 5′-S1 analysis that correlated precisely to a 5′ end at +1 (nucleotide 499). The minor band evident at +5 was not consistently observed. It is presumably an artifact of the 5′-S1 mapping procedure, because the cap analyses of Ad5 RNAs reported by Baker and Ziff (1980) only demonstrated one E1A cap. Deletion of all or part of the TATA homology at −30 (*dl*309-13 and -59) resulted in heterogeneity of the 5′ ends of the E1A transcripts. With *dl*309-13, a predominant 5′ end was detected at +23, but a number of minor 5′ends were also evident that map between +23 and −20. A similar result was found with *dl*309-59, although approximately 50% of the transcripts contained a wild-type 5′ end. Mutants with deletions downstream of the TATA box all synthesized E1A transcripts with 5′ ends displaced downstream by approximately the size of the deletion. For example, *dl*309-274 contains a 7-bp deletion (−12 to −18; Fig. 2) and synthesizes E1A RNAs with 5′ ends mapping at +6 and +7. Mutants that contain deletions upstream of the TATA homology synthesized E1A RNAs with wild-type 5′ ends. Results obtained with the larger-deletion mutants (*dl*309-6 and *dl*340-7 to -11; Fig. 3) confirmed these results.

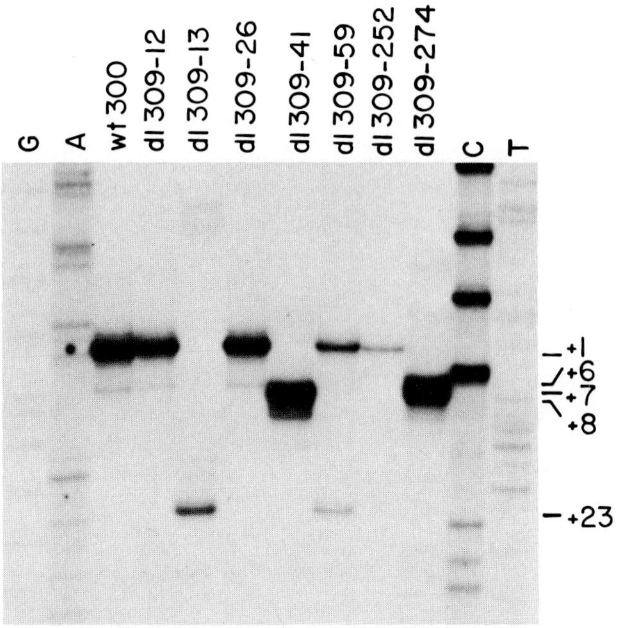

Fig. 7 5′-S1 analyses of E1A mRNAs found *in vivo* after infection with D-loop deletion mutants. RNAs from the experiment described in the legend to Fig. 5 were subjected to 5′-S1 analysis (Weaver and Weissmann, 1979). The wild-type E1A 5′ end is indicated as +1. The positions of altered 5′-ends with the various mutants are shown. G, A, C, and T designate sequencing reactions using the procedure of Sanger *et al.* (1977); they were employed as size makers.

VI. E1A Transcriptional Control Region and Comparison to Other Eukaryotic Control Regions

Our results define sequences that are essential for E1A transcription in cell-free extracts and within virus-infected cells. Only a limited block of nucleotides are required for efficient transcription of region E1A *in vitro*. These sequences map from approximately −35 to +10 relative to the E1A cap site (+1) and include the TATA box at −30, the cap site at +1 (contained in a hyphenated symmetry), and the sequences located between these two regions. Templates lacking the TATA box appeared completely inactive, as has been found for a great variety of eukaryotic polymerase II control regions (Corden *et al.*, 1980; Wasylyk *et al.*, 1980; Hu and Manley, 1981; Mathis and Chambon, 1981;

Sassone-Corsi *et al.*, 1981; Tsai *et al.*, 1981; Tsujimoto *et al.*, 1981; Wasylyk and Chambon, 1981; see reviews in Breathnach and Chambon, 1981; Shenk, 1981). Mutations downstream of the TATA homology also affected the efficiency of E1A transcription *in vitro*. The reduction (2- to 10-fold) was less dramatic than that caused by deletion of the TATA box and depended on the particular mutation assayed. This is also true for a number of other transcription units. Deletion of the cap sites of the Ad2 major late promoter (Corden *et al.*, 1980), chicken conalbumin (Sassone-Corsi *et al.*, 1981), and silk fibroin gene (Tsujimoto *et al.*, 1981) results in approximately a 10-fold reduction in the transcription of these genes. In the studies reported by Talkington and Leder (1982) on the α_1- and α_4-globin gene promoters and by Hu and Manley (1981) on the Ad2 major late promoter, only slight reductions in transcription were apparent after deletion of the cap site (about 2-fold). Deletion of sequences upstream of the E1A TATA box did not affect the efficiency of E1A transcription *in vitro*.

Our data define a more extensive set of signals operating to regulate E1A transcription *in vivo*. Deletion of the TATA homology only modestly reduced E1A transcription *in vivo*. Deletion of all or part of this sequence resulted in heterogeneity of the 5′ ends of cytoplasmic E1A transcripts. This finding is consistent with results reported for the sea urchin histone H2A gene (Grosschedl and Birnstiel, 1980a), the SV40 early transcription unit (Benoist and Chambon, 1981), the Ad2 E1A transcription unit (Osborne *et al.*, 1982), and the rabbit β-globin gene (Grosveld *et al.*, 1982). Some, but not all, TATA box mutations have been reported to reduce the levels of *in vivo* transcription as much as 5- to 10-fold (sea urchin H2A, Grosschedl and Birnstiel, 1980a; Ad2 E1A, Osborne *et al.*, 1982; rabbit β-globin, Grosveld *et al.*, 1982). It is likely that the phenotype of TATA mutations depends on the precise nature and extent of the alteration, the particular gene under study, and the specific assay utilized.

Although the 5′ ends of the E1A transcripts synthesized by the TATA box mutants (*dl*309-13 and -59) show heterogeneity, the cap sites are still maintained within ~25 bp of the normal wild-type site. This result implies that there are additional signals other than the TATA box that are recognized by polymerase II and that act in selection of the site of initiation of transcription. Consistent with this possibility is the fact that there are numerous TATA box-like sequences present upstream of the E1A cap site, but only one homology appears to be utilized *in vivo*. That a predominant 5′ end is observed with the TATA box mutants at +23 (Fig. 7) suggests that an alternative signal has been unmasked in these mutants that directs the selection of an E1A cap site. The hyphenated symmetry at the wild-type initiation site is centered ~25 nucleotides upstream of this new cap site. Perhaps this symmetry plays a role, at least in the E1A TATA-box mutant and possibly in the wild-type virus as well, in the selection of the initiation site for E1A transcription.

Deletion of sequences downstream of the TATA box resulted in displacement of the 5′ end of the E1A transcripts downstream by approximately the size of the deletion. These results are consistent with analyses by several groups (Grosschedl and Birnstiel, 1980a; Benoist and Chambon, 1981; Ghosh *et al.*, 1981; Grosveld *et al.*, 1982), and suggest the polymerase uses the TATA box as a reference point and initiates transcription ~30 nucleotides downstream of this sequence.

We have also identified a region located 144–305 bp upstream of the E1A cap site that modulates the levels of E1A transcription *in vivo*. Deletion of the distal half of this region (*dl*309-7, −392 to −229) resulted in a 3-fold decrease in the level of E1A transcripts found after infection of HeLa cells. Deletion of the entire region (*dl*340-8, −392 to −144) reduced the levels of E1A transcripts 20-fold relative to wild-type Ad5. This region lies far upstream of the CAT box or −68 homology, which could be important for optimal transcription of the human α-globin gene (critical sequence −87 to −55; Mellon *et al.*, 1981), the rabbit β-globin gene (−100 to −34; Grosveld *et al.*, 1982), and the HSV *tk* gene (−105 to −50; Ghosh *et al.*, 1981; McKnight *et al.*, 1981; McKnight and Kingsbury, 1982). In each case, when sequences surrounding and containing this homology were deleted, transcription was reduced 10- to 20-fold. Deletion of the E1A CAT box (*dl*309-26) as well as the region surrounding this sequence (*dl*309-72, -159, and -252) did not affect the levels of E1A transcription. In fact, the proximal boundary of the critical sequence for regulation of E1A transcription lies at least 90 nucleotides upstream of this region.

Far-upstream regulatory sites have been identified for the SV40 early transcription unit (−260 to −116; Gruss *et al.*, 1981), the polyoma virus early transcription unit (−420 to −180; Tyndall *et al.*, 1981), and the sea urchin H2A gene (−450 to −110; Grosschedl and Birnstiel, 1980b). There are no striking similarities or homologies between the sequences in these and the E1A control regions. It is not surprising that the control regions of the various transcription units differ in their location relative to the cap site and in nucleotide sequence. Each of the transcription units studied is unrelated to the others and would not be expected to be regulated in an analogous manner (again with the exception of the globin genes, in which the control regions appear similar in location and sequence; Mellon *et al.*, 1981; Grosveld *et al.*, 1982). On the other hand, a regulatory element such as the TATA box, which in each case serves the common purpose of localizing the site of initiation of transcription, would be expected to be more highly conserved.

The region that was found to be critical for regulation of E1A transcription *in vivo* was found to be dispensable for efficient E1A transcription *in vitro*. Clearly, the *in vitro* transcription systems only recognize a subset of the regulatory sequences that are operational *in vivo*. It is unclear whether these results represent differences in templates *in vivo* versus *in vitro* (i.e., chromatin structure) or the loss of positive or negative transcription factors during the preparation of the *in vitro* extracts.

Acknowledgments

We acknowledge the competent technical assistance of Martha Marlow and the secretarial skills of Mary Fils-Aime. This work was supported by a grant from the American Cancer Society (MV-45). Hearing is a Fellow of the Jane Coffin Childs Memorial Fund for Medical Research and Shenk is an Established Investigator of the American Heart Association.

References

Alwine, J. C., Kemp, D. J., and Stark, G. R. (1977). *Proc. Natl. Acad. Sci. USA* **74,** 5350–5354.

Baker, C., and Ziff, E. (1981). *J. Mol. Biol.* **149,** 189–221.

Benoist, C., and Chambon, P. (1981). *Nature (London)* **290,** 304–310.

Benoist, C., O'Hare, K., Breathnach, R., and Chambon, P. (1980). *Nucleic Acids Res.* **8,** 127–142.

Berk, A. J., and Sharp, P. A. (1978). *Cell* **14,** 695–711.

Berk, A. J., Lee, F., Harrison, T., Williams, J., and Sharp, P. A. (1979). *Cell* **17,** 935–944.

Breathnach, R., and Chambon, P. (1981). *Annu. Rev. Biochem.* **50,** 349–383.

Carlock, L. R., and Jones, N. C. (1981). *J. Virol.* **40,** 657–664.

Chow, L. T., Broker, T. R., and Lewis, J. B. (1979). *J. Mol. Biol.* **134,** 265–303.

Corden, J., Wasylyk, B., Buchwalder, A., Sassone-Corsi, P., Kedinger, C., and Chambon P. (1980). *Science (Washington, D.C.)* **209,** 1406–1414.

Efstratiadis, A., Posakony, J., Maniatis, T., Lawn, R., O'Connell, C., Spritz, R., DeRiel, J., Forget, B., Weissman, S., Slightom, J., Blechl, A., Smithies, O., Baralle, F., Shoulders, C., and Proudfoot, N. (1980). *Cell* **21,** 653–668.

Ghosh, P., Lebowitz, P., Frisque, R. J., and Gluzman, Y. (1981). *Proc. Natl. Acad. Sci. USA* **78,** 100–104.

Goldberg, M. (1979). Ph.D. Thesis, Stanford Univ., Stanford, California.

Graham, F. L., Smiley, J., Russell, W. C., and Nairu, R. (1977). *J. Gen. Virol.* **36,** 59–72.

Grosschedl, R., and Birnstiel, M. L. (1980a). *Proc. Natl. Acad. Sci. USA* **77,** 1432–1436.

Grosschedl, R., and Birnstiel, M. L. (1980b). *Proc. Natl. Acad. Sci. USA* **77,** 7102–7106.

Grosveld, G. C., deBoer, E., Shewmaker, C. K., and Flavel, R. A. (1982). *Nature (London)* **295,** 120–126.

Gruss, P., Dhar, R., and Khoury, G. (1981). *Proc. Natl. Acad. Sci. USA* **78,** 943–947.

Hammarskjold, M.-L., and Winberg, G. (1980). *Cell* **20,** 787–795.

Handa, H., Kaufman, R. I., Manley, J., Gefter, M., and Sharp, P. A. (1981). *J. Biol. Chem.* **256,** 478–482.

Hu, S. L., and Manley, J. (1981). *Proc. Natl. Acad. Sci. USA* **78,** 820–824.

Jones, N., and Shenk, T. (1979). *Proc. Natl. Acad. Sci. USA* **76,** 3665–3669.

Manley, J. L., Fire, A., Sharp, P. A., and Gefter, M. L. (1980). *Proc. Natl. Acad. Sci. USA* **77,** 3855–3859.

Mathis, D., and Chambon, P. (1981). *Nature (London)* **290,** 310–315.

McGrogan, M., Spector, D. J., Goldenberg, C., Halbert, D. N., and Raskas, H. J. (1979). *Nucleic Acids Res.* **6,** 593–608.

McKnight, S. L., and Kingsbury, R. (1982). *Science (Washington, D.C.)* **217,** 316–324.

McKnight, S. L., Gavis, E. R., Kingsbury, R., and Axel, R. (1981). *Cell* **25,** 385–398.

McMaster, G. K., and Carmichael, G. G. (1977). *Proc. Natl. Acad. Sci. USA* **74,** 4835–4838.

Mellon, P., Parker, V., Gluzman, Y., and Maniatis, T. (1981). *Cell* **27,** 279–288.

Montell, C., Fisher, E., Caruthers, M., and Berk, A. J. (1982). *Nature (London)* **295,** 380–384.

Nevins, J. R. (1981). *Cell* **26,** 213–220.

Osborne, T. F., Gaynor, R. B., and Berk, A. J. (1982). *Cell* **29,** 139–148.

Roeder, R. G. (1976). *In* "RNA Polymerase" (R. Losick, and M. Chamberlin, eds.), pp. 285–329. Cold Spring Harbor Lab., Cold Spring Harbor, New York.

Sanger, F., Nicklen, S., and Coulson, A. R. (1977). *Proc. Natl. Acad. Sci. USA* **74,** 5463–5467.

Sassone-Corsi, P., Corden, J., Kedinger, C., and Chambon, P. (1981). *Nucleic Acids Res.* **9,** 3941–3958.

Shenk, T. (1981). *Curr. Top. Microbiol. Immunol.,* pp. 25–46.

Shortle, D., Koshland, D., Weinstock, G. M., and Botstein, D. (1980). *Proc. Natl. Acad. Sci. USA* **77,** 5375–5379.

Spector, D. J., McGrogan, M., and Raskas, H. J. (1978). *J. Mol. Biol.* **126,** 395–414.

Stow, N. (1981). *J. Virol.* **37,** 171–180.

Talkington, C. A., and Leder, P. (1982). *Nature (London)* **298,** 192–195.

Tibbetts, C. (1977). *Cell* **12,** 243–249.

Tsai, S. Y., Tsai, M. J., and O'Malley, B. W. (1981). *Proc. Natl. Acad. Sci. USA* **78,** 879–883.

Tsujimoto, Y., Hirose, S., Tsuda, M., and Suzuki, Y. (1981). *Proc. Natl. Acad. Sci. USA* **78,** 4838–4842.

Tyndall, C., LaMantia, G., Thacker, C. M., Favaloro, J., and Kamen, R. (1981). *Nucleic Acids Res.* **9,** 6321–6250.

van Ormondt, H., Maat, J., deWaard, A., and van der Eb., A. (1978). *Gene* **4,** 309–328.

Wasylyk, B., and Chambon, P. (1981). *Nucleic Acids Res.* **9,** 1813–1824.

Wasylyk, B., Derbyshire, R., Guy, A., Molko, D., Roget, A., Teoule, R., and Chambon, P. (1980). *Proc. Natl. Acad. Sci. USA* **77,** 7024–7028.

Weaver, R. F., and Weissmann, C. (1979). *Nucleic Acids Res.* **7,** 1175–1193.

CHAPTER 12

Expression of Proteins on the Cell Surface Using Mammalian Vectors

JOE SAMBROOK
MARY-JANE GETHING

Cold Spring Harbor Laboratory
Cold Spring Harbor, New York

Experimental Manipulation
of Gene Expression

ISBN 0-12-372380-9

I. How Proteins Are Normally Expressed on Mammalian Cell Surfaces

The surface of the mammalian cell is populated by a large number of protein molecules that display a great diversity of physical properties, carry out a large number of different functions, and collectively endow upon the cell many of its more specialized and interesting properties. Although there is no accurate accounting of the total number of protein molecules on the cell's surface, nor of the number of different molecules that are found there, it is clear from a large number of studies that such proteins are not merely a random cross-section of the total proteins of the cell, but rather are a unique subset. It therefore follows that the cell has devised a mechanism by which proteins destined to be displayed on the surface are segregated from those that are to occupy intracellular locations. Thus, despite their ultimate diversity of structure and function, all of these membrane proteins are tagged with a common marker that causes them to be synthesized along biosynthetic pathways leading to the cell surface; in addition, they carry sequences that interact with the lipid bilayer and that cause them to be held securely in place in the plasma membrane.

An important insight into how proteins are directed to the surface membrane of the cell came from the discovery in 1972 that secreted proteins are synthesized as higher molecular weight precursors (Milstein *et al.*, 1972). When messenger RNA encoding a secretory protein (in this case the light chain of immunoglobulin) is translated *in vitro,* the polypeptide synthesized is 20 amino acids longer than the product secreted by plasma cells. However, light-chain polypeptide synthesized *in vitro* in the presence of rough microsomal membranes lacks the additional N-terminal peptide. Dozens of secretory and integral membrane proteins have been studied since 1972 and have been found to behave in a similar manner. Invariably, the additional sequences have been found to be hydrophobic in nature; almost always they carry at the N-terminus of the nascent polypeptide chain. These findings have provided strong experimental support for the proposal (Blobel and Dobberstein, 1975) that the leader peptide acts as a signal to direct the growing polypeptide chain to sites in the membrane of the rough endoplasmic reticulum serving as portals to the biosynthetic pathway that leads to the cell surface (for reviews see Blobel *et al.*, 1979; Davis and Tai, 1980; Sabatini *et al.*, 1982).

The initial stages in this process appear to be as follows. Free ribosomal subunits in the cytosol attach to an mRNA coding for a secretory or integral membrane protein, and translation begins in the usual manner. After some 40 amino acids have been polymerized, the hydrophobic N-terminus of the growing polypeptide chain begins to be extruded from the ribosome (Blobel and Sabatini, 1970) and binds strongly to a large protein complex, the signal-recognition protein, which is a normal component of microsomal membranes (Walter and Blobel, 1982; Walter *et al.*, 1982; Meyer and Dobberstein, 1980). Further translation of the secretory protein ceases shortly thereafter unless the ribosome–signal-recognition protein complex "docks" at a suitable site in the rough endoplasmic reticulum (Kreibich *et al.*, 1978). Here, another protein relieves the translational block and allows the nascent polypeptide chain to thread its way through the membrane of the rough endoplasmic reticulum as it is synthesized (Meyer *et al.*, 1982). Before synthesis of the entire polypeptide chain has been completed, the signal sequence is cleaved off by a protease found only on the luminal side of the rough endoplasmic reticulum (Walkson and Blobel, 1977; Walter *et al.*, 1979).

The membrane of the rough endoplasmic reticulum is impervious to cytosolic proteins; only those proteins synthesized as precursors with signal sequences are able to pass through and reach the luminal side. Thus segregation of secretory and surface proteins from the cytoplasmic proteins of the cell occurs at a very early stage during their synthesis and is independent of the final size and shape of the mature proteins.

As the nascent polypeptide appears on the luminal side of the rough endoplasmic reticulum, preformed oligosaccharides composed of *N*-acetylglucosamine, mannose, and glucose (Tabas and Kornfeld, 1979) are transferred to certain asparagine residues (those occurring in the sequences Asn-X-Ser or Asn-X-Thr) (Neuberger *et al.*, 1972). This initial "core" glycosylation is but the first step in an elaborate program of reactions taking place both in the rough endoplasmic reticulum and later in the Golgi apparatus (Leblond and Bennett, 1977), in which sugars are added to and trimmed from the nascent protein (for review see Farquhar and Palade, 1981). It is also during this period that the secondary and tertiary structure of the mature protein molecule becomes established by disulfide linkages and association between polypeptide subunits.

Proteins destined to be secreted from the cell pass completely through the membrane of the rough endoplasmic reticulum and are released into the lumen; those that are to occupy a place on the surface of the cell remain anchored on the luminal surface of the membrane by a tract of 24–30 hydrophobic and nonpolar amino acids that are generally located near the C-terminus of the protein. These hydrophobic sequences, which are thought to form helical turns across the membrane, are usually followed by a short (2–50 amino acids) hydrophilic C-terminal tail that remains exposed on the proximal side of the membrane and that may act together with the hydrophobic sequences to halt the transfer of the final residues of the protein through the membrane (for review see Sabatini *et al.*, 1982).

In topological terms, the lumen of the rough endoplasmic reticulum is equivalent to the outside of the cell, and its luminal surface is equivalent to the cell's outer surface. Thus secretory proteins are released from the cell when vesicles pinched off from the Golgi apparatus fuse with the plasma membrane (Palade,

1975), and integral membrane proteins become displayed on the cell surface as the lipid bilayer of coated vesicles becomes incorporated into the plasma membrane.

II. Why It Would Be Useful to Express Proteins on the Surface of the Mammalian Cell

Although the biosynthetic scheme just discussed is undoubtedly correct in its major aspects, it is drawn with very broad brush strokes; there are large areas of detail that remain to be filled in, and there are huge regions that are entirely blank. For example:

1. Neither the significance nor the mechanism by which the intricate sequence of glycosylation reactions occurs in the rough endoplasmic reticulum and the Golgi apparatus is understood.
2. It is a puzzle to know how the signal-recognition protein reacts with the great diversity of amino acid sequences that constitute the signals of different integral membrane and secreted proteins and that block their translation. There are indications from the work of Blobel and colleagues (for review see Newmark, 1982) that an abundant species of 7-S RNA may be a part of the signal-recognition complex and may be involved either in its binding or blocking activity. However, the details of this involvement are entirely unknown.
3. Why should it be that the target sequences for glycosylation (ASn-X-Thr and Asn-X-Ser) should occur much less frequently in glycoproteins than in intracellular proteins?
4. How does intracellular sorting of glycoproteins occur? The surfaces of certain types of epithelial cells are polarized into two domains, the apical and basolateral, separated by an impermeable barrier of intercellular tight junctions; some integral membrane glycoproteins are inserted with equal efficiency into both domains, whereas other glycoproteins show a high degree of preference for one domain or the other. For example, in intact monolayers of polarized canine kidney cells, the glycoproteins of vesicular stomatitis virus are found at the basolateral surfaces, whereas the major glycoprotein of influenza virus—the hemagglutinin—is inserted preferentially into the apical surface of the cell (Rodriguez-Boulan and Sabatini, 1978; Rodriguez-Boulan and Prendergast, 1980; Roth *et al.*, 1979; Green *et al.*, 1981). Because the glycoproteins of simple, enveloped viruses are synthesized, processed, and transported using host-cell mechanisms, it seems likely that cellular-surface proteins will also be

sorted into different domains of the plasma membrane. The simplest explanation of sorting is that glycoproteins of different polarity are not present in the same vesicle during transport from the Golgi to the cell surface. Thus the events responsible for sorting surface molecules probably occur before their exit from the Golgi. However, the molecular basis of the phenomenon is not understood.

Undoubtedly, many of these questions can be answered by elegant biochemistry. However, to map precisely the various functional regions of membrane proteins and their precursors it also may be essential to have available genetic systems in which cloned genes coding for secretory and integral membrane proteins can be mutated at specific sites *in vitro* and then returned to mammalian cells and expressed with high efficiency. The use of recombinant DNA technoogy also opens up the possibility of constructing genes that code for chimeric glycoprotein molecules, to determine whether features important for directional transport reside in the cytoplasmic, membrane-spanning, or ''externally'' oriented portion of integral membrane proteins. Once these details are known, it may prove possible to specifically target molecules to specific domains or to individual cell organelles.

III. Hemagglutinin of Influenza Virus Is the Best-Characterized Integral Membrane Protein

For years, the hemagglutinin (HA) of influenza virus has been the object of intensive study. Initially, this work was undertaken in an effort to understand the mechanism by which the virus is so easily able to change its antigenic properties and thereby be a perennial cause of illness and death in the human population (for review see Schild, 1979). The major antigen of the virus particle is the HA glycoprotein, which is inserted into the lipid membrane that envelops the virion. Influenza viruses with the potential to cause new pandemics or epidemics in an immune human population have antigenically novel HAs (for review see Stuart-Harris, 1979). Two types of antigenic change have been recognized. Major shifts in the structure and antigenicity of HA define the beginning and end of each pandemic era, with the novel HA entering the human population from an influenza virus circulating or sequestered in an animal population other than man (Webster and Laver, 1975). Between these major shifts, which occur at intervals of ~10 yr, the HA undergoes antigenic drift—a series of stepwise alterations in amino acid sequence that represent the virus' response to the increased selective pressure that results from rising levels of antibody in the human population (see Laver and Air, 1980).

Apart from its importance to man as a mutable antigen, HA plays a major role in the penetration of the host cell by the virus. The HA is responsible not only for the initial attachment of the virus to receptors on the surface of the host cell (Hirst, 1942) but also for the fusion of viral cellular membranes that marks the onset of infection (Gething *et al.*, 1978; Garten *et al.*, 1981; Richardson *et al.*, 1980).

Initially, work on HA depended almost entirely on the techniques of classical protein chemistry (Waterfield *et al.*, 1979; Ward and Dopheide, 1980), but since the advent of recombinant DNA technology the genes encoding the HAs of many strains of influenza virus have been cloned and sequenced, and this has allowed the complete amino acid sequences of the proteins to be deduced (Verhoeyen *et al.*, 1980; Porter *et al.*, 1979; Gething *et al.*, 1980; Both and Sleigh, 1980; Min Jou *et al.*, 1980; Winter *et al.*, 1981; Hiti *et al.*, 1981). At the same time, techniques to produce monoclonal antibodies in hybridoma cells have become available, and such antibodies raised against influenza virus have allowed a detailed analysis of the antigenic structure of HA. Finally, in 1977, the glycoprotein spike of HA was crystallized (Wiley and Skebel, 1977), and since then its three-dimensional structure has been determined at 3-Å resolution using X-ray crystallography (Wiley *et al.*, 1981; Wilson *et al.*, 1981). The culmination of all this work has been a detailed description of the physical domains of the molecule, the location of its major antigenic sites, the points at which is glycosylated, its organization into trimeric structures, and its orientation with respect to the membrane. Hemagglutinin is the best characterized of all eukaryotic membrane proteins.

The HA gene of the Japan strain of influenza virus codes for a protein of 562 amino acids. The N-terminal signal sequence of 15 residues is cleaved from the protein during membrane translocation to yield a molecule called HA0 that consists of 547 amino acids (Gething *et al.*, 1980). After its passage through the Golgi and incorporation into virus particles at the surface of the infected cell, HAO is cleaved by an as yet unidentified cellular protease to produce the mature HA molecule, which consists of two polypeptide subunits HA1 and HA2 (324 and 222 amino acids, respectively) linked by a single disulfide bond (Waterfield and Scrace, 1981). This cleavage, which is essential for the formation of infectious virus (Lazarowitz and Choppin, 1975; Klenk *et al.*, 1975), results in the removal of a single arginine residue.

The crucial interactions between HA and membranes are mediated by the three separate hydrophobic regions of the molecule:

1. The N-terminal, hydrophobic signal draws the nascent polypeptide chain into contact with the intracellular membrane system. In the Japan HA, the signal

sequence consists of 15 amino acids, with a continuous hydrophobic stretch of 13 residues (Gething *et al.*, 1980). Comparison of the signal sequence of Japan HA with those of the HAs of other influenza viruses reveals that there is no conservation of length or specific amino acid sequence (Air, 1981). Furthermore, the signal sequences of other eukaryotic membrane or secretory proteins appear to be equally diverse. None of them is identical, and few of them show any striking degree of homology or conservation (Blobel *et al.*, 1979; Kreibich *et al.*, 1980). These findings suggest that it is either conformational features of the signals or simply their hydrophobicity that determines their interaction with the receptor protein or the membrane of the rough endoplasmic reticulum.

2. The C-terminal region anchors the completed molecule in the external membranes of both the virus and infected cell. Although there are a few examples of proteins that are attached to the membrane by N-terminal hydrophobic sequences (e.g., influenza virus neuraminidase (Fields *et al.*, 1981) and sucrose isomaltase (Frank *et al.*, 1978), the general rule is that "spike proteins" are anchored to membranes by 24–30 hydrophobic and nonpolar amino acids located near their C-termini (Wilson *et al.*, 1981). As in the case of the N-terminal signal sequences, there is little conservation, suggesting once again that hydrophobicity is the major and perhaps sole requisite for anchor function.

3. The N-terminus of HA2 appears to be involved in penetration of the cell membrane during infection. Usually, this hydrophobic region is buried in the middle of the HA molecule (Wilson *et al.*, 1981; Wiley *et al.*, 1981). However, it is believed to become exposed as a consequence of a conformational change that is triggered by exposure to the low pH of the lysosomes in which the virus is transported into the cell. The newly generated form of HA then forms a bridge between the envelope of the virus particle and the membrane of the lysosome (White *et al.*, 1982). The two lipid bilayers fuse, and the contents of the virus particle enter the cytoplasm of the cell.

As a consequence of a large quantity of work we now have a fascinating picture of the architecture of influenza virus HA. Furthermore, because the biosynthesis of HA involves host enzymes and processes during translation, membrane transport, glycosylation, and maturation (Choppin *et al.*, 1971; Compans and Choppin, 1975), it appears to be a profitable model for the study of the integral membrane proteins of eukaryotic cells. One approach to the further analysis of HA and of integral membrane proteins in general is to introduce alterations in their amino acid sequences and to analyze the subsequent effects on the biosynthesis, structure, and function of the molecules. It would be of particular interest to determine the effects of alterations in the hydrophobic sequences on the interactions of the protein with membranes, particularly those involving translocation. The first stage in this process is to synthesize HA from a cloned gene. The remainder of this chapter describes the construction and use of recombinant viral genomes consisting of both SV40 and influenza virus sequences that direct the expression of large amounts of HA at the surface of eukaryotic cells.

IV. The Gene Coding for Hemagglutinin Is of Simple Structure

In its natural state, the gene for HA occurs as one of the eight segments of negatively stranded RNA that constitute the influenza virus genome (Palese and Shulman, 1976; McGeoch *et al.*, 1976; Scholtissek *et al.*, 1976). Each of these RNA segments codes for only one viral protein (e.g., HA is encoded in segment 4), with the exception of segments 7 and 8, each of which codes for two proteins from overlapping reading frames (Lamb *et al.*, 1978; Inglis *et al.*, 1979; Lamb and Choppin, 1981). The Japan HA gene is 1773 nucleotides long and contains a continuous coding sequence of 1686 nucleotides. Termination codons occur frequently in two of the three reading frames so that only one frame is open for translation into protein. There are short untranslated regions at both the 3′ (43 nucleotides) and the 5′ (44 nucleotides) ends of the gene.

The messenger RNA for HA is an unspliced partial copy of the gene, lacking sequences complementary to the nucleotides at the 5′ and 3′ ends of the negatively stranded virion RNA. The mRNA carries poly(A) sequences at its 3′ end and heterogeneous extensions (including terminal cap structures) at its 5′ end. These are scavenged from the 5′ ends of nascent host-cell mRNAs and serve as primers for transcription of the viral HA gene (Krug *et al.*, 1979; Dhar *et al.*, 1981).

It is possible to clone the HA gene either from the genomic, negatively stranded RNA extracted from virus particles or from mRNA purified from infected cells (see Laver and Air, 1981). The clones used in our work (Gething *et al.*, 1980) were derived from the former source, using methods standardly employed to clone cellular mRNAs. The only extra step was the addition of a tract of polyadenylic acid to the 3′ end of the virion RNA to provide a binding site for the oligo(dT) used to prime reverse transcriptase. Double-stranded cDNA was inserted into the *Pst*I site of the plasmid pAT153 by dG:dC homopolymeric tailing. Following transfection of *Escherichia coli,* plasmids containing copies of the Japan HA gene were identified by *in situ* hybridization to colonies and to virion RNAs immobilized on DBM paper. The DNA sequence of the cloned inserts was carried out using the procedures of A. Maxam and W. Gilbert. Using these methods, a cDNA copy of the complete Japan HA gene was obtained (except for 11 nucleotides of the 3′ untranslated region). The exonuclease *Bal*31 was used to remove both of the sets of homopolymeric C:G and A:T tails used in the construction and also the 5′ untranslated region of the gene. *Hin*dIII linkers were added to the ends of the resected molecules, which were then recloned in pAT153. The resulting population of clones was scanned to identify an individual in which the nucleotide immediately following the *Hin*dIII linker was the A of the initiating AUG codon of the HA coding sequence. Cleavage of such a clone with *Hin*dIII and *Bam*HI yielded a fragment containing the entire coding sequence of HA in an orientation suitable for insertion into the various vectors described in the following section (Gething and Sambrook, 1981).

V. Vector Systems

Because the HA gene occurs naturally in the form of negatively stranded RNA, it contains none of the controlling elements (e.g., promoters and enhancers) that are required for efficient transcription of conventional DNA genes. Thus the vectors all were designed so that the HA coding region was placed under the control of either the SV40 early promoter–enhancer or the SV40 late promoters.

A. Vectors in Which the Hemagglutinin Coding Sequence Replaces the SV40 Early Genes

As mentioned previously, the cloned HA had been reconstructed so that the A residue immediately following the *Hin*dIII linker would correspond to the A of the initiating AUG of the protein (Fig. 1). This allowed the HA gene to be inserted into the early region of the SV40 genome (between the *Hin*dIII site at nucleotide 5171 and the *Bam*HI site at nucleotide 2533, see Tooze, 1980) so that its initiation codon was almost exactly in the position normally occupied by the initiation codon of large-T and small-t antigens. In addition, at the time that these experiments were undertaken, there were indications from other workers (Mulligan *et al.*, 1979; Hamer and Leder, 1979; Hamer *et al.*, 1979; Gruss and Khoury, 1981) that foreign genes inserted into SV40 recombinant vectors were expressed more efficiently when splice donor and acceptor sites were provided. We therefore built into the original vector a sequence of DNA containing the intron of the SV40 gene coding for small-t antigen. Finally, to ensure that transcripts of the HA gene would be terminated efficiently, the HA sequences were always inserted so that they lay upstream of a poly(A) addition signal (also

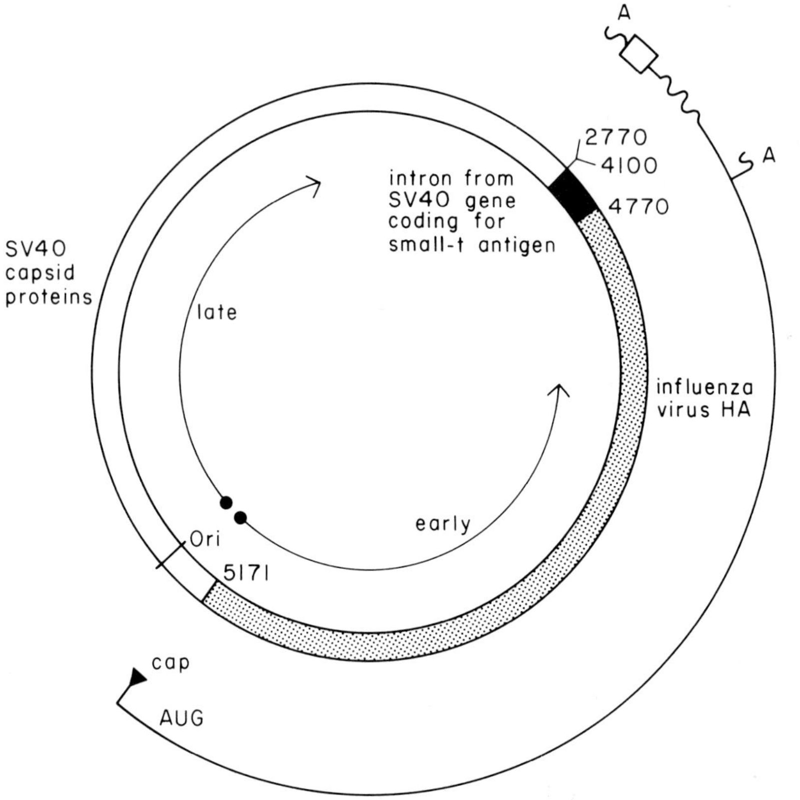

Fig. 1 Structure of the early-replacement vector, in which the hemagglutinin (HA) gene substitutes for the SV40 sequences coding for large-T and small-t antigens. The inner arc shows the direction of transcription of the recombinant genome, and the middle circle shows its structure. The numbers denote SV40 nucleotides numbered according to the BBB system (Tooze, 1980). Ori marks the position of the SV40 origin of DNA replication. The probable structure of the unspliced HA mRNA is shown in the outer arc. The two possible sites for poly(A) addition are indicated by A.

derived from SV40). It should be noted that the untranslated 3′ region of the HA gene contains the sequence AAUUAA, closely related to the consensus poly(A) addition signal AAUAAA that is believed to function during natural infections with influenza virus. However, it was not clear whether this signal would work efficiently in the form of double-stranded DNA, and a conventional poly(A) addition signal was therefore included in the construct.

The HA gene should therefore be transcribed under the control of the upstream SV40 promoter–enhancer to yield an mRNA that (1) initiates at the usual site (Ghosh *et al.*, 1981), (2) contains a 5′ untranslated region that apart from the last five nucleotides is identical to that of conventional SV40 early mRNAs, (3) carries a coding region specifying an HA molecule the sequence of which is identical to that of the natural form of the molecule, (4) has the opportunity to be spliced at the donor–acceptor sequences provided by the downstream intron, and (5) terminates at either the poly(A) addition site within the HA gene or at the SV40 site farther downstream.

This recombinant viral genome contained the origin of SV40 replication and an intact set of late genes. However, the SV40 early genes were replaced by the influenza virus HA gene, and the recombinant therefore could not replicate in simian cells unless functional T antigen was supplied. This was done by using as a permissive host the COS-1 line of SV40-transformed simian cells that carry an endogenously expressed copy of the SV40 T-antigen gene (Gluzman, 1981).

B. Vectors in Which the Hemagglutinin Coding Sequence Replaces the SV40 Late Genes

To place the HA coding sequences under the control of the SV40 late promoter(s), The HA gene was inserted into the late region of the SV40 genome between the HpaII site at nucleotide 346 and the *Bam*HI site at nucleotide 2533 (Tooze, 1980; see Fig. 2). The resulting recombinant viral genome contained the origin of SV40 replication and an intact set of SV40 early genes. The HA coding sequences were joined to the noncoding sequences of SV40 that would normally be transcribed into the untranslated 5′ region of SV40 late mRNAs. Late in the lytic infection of permissive simian cells, this recombinant genome should express an unspliced, hybrid mRNA consisting of untranslated SV40 sequences at its 5′ end and an intact set of HA coding sequences. Polyadenylation could occur either at the site within the 3′ untranslated sequences of the HA gene or at the normal site for late SV40 mRNAs at nucleotide 2674 (Tooze, 1980). Translation of this RNA should begin at the first AUG, located at the beginning of the HA coding sequences. Because the late-replacement vector contains a functional origin of DNA replication and an intact set of SV40 early genes, it will replicate its DNA efficiently in permissive simian cells. However, the late genes of SV40 have been deleted, and production of infectious virions containing the recombinant genome therefore requires that SV40 capsid proteins be supplied by a complementing helper virus such as *dl*1055, an early-deletion mutant of SV40 (Pipas *et al.*, 1980).

The late-replacement vector is similar to those used previously by many other workers to express cloned genes derived from a variety of different organisms and, more recently, to express cloned HA genes from other strains or influenza virus.

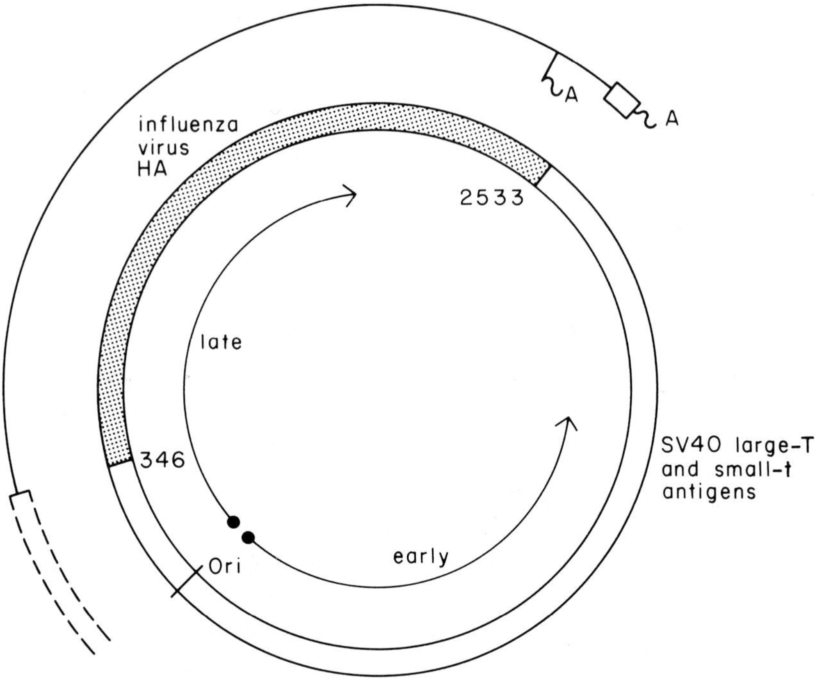

Fig. 2 Structure of the late-replacement vector in which the HA gene substitutes for the SV40 sequences coding for the capsid proteins. The inner arc shows the direction of transcription of the recombinant genome, and the middle circle shows its structure. The numbers denote SV40 nucleotides numbered according to the BBB system (Tooze, 1980). Ori marks the position of the SV40 origin of DNA replication. The probable structure of the unspliced HA mRNA is shown in the outer arc. The two possible sites for poly(A) addition are indicated by A.

C. Production of Virus Particles Carrying SV40–Influenza Recombinant Genomes

All of the intermediate constructions were carried out in bacterial plasmids of various sorts; the final recombinant viral genomes also were propagated in bacterial plasmids until their structures had been verified. They were then excised from the plasmids (by cleavage with *Bam*HI), purified by gel electrophoresis, and ligated at low concentrations of DNA to generate closed circular molecules, which were then introduced into the appropriate host cells using DEAE-dextran as a facilitator (McCutchen and Pagano, 1968). The early-replacement vectors were introduced into COS-1 cells in the absence of a helper viral DNA; the late-replacement vector was introduced into CV-1 cells together with an equal amount of the DNA of the SV40 deletion mutant *dl*1055. Only a small proportion of cultured mammalian cells are competent at any one time to take up DNA, so very few cells of the population are infected initially. However, during the next several days, each of these cells undergoes a lytic infection and produces over a million virus particles that are able to spread into neighboring cells and infect them with high efficiency. Usually, the lysate obtained from the first set of infected cells needs to be passed serially once or twice to obtain a high-titer virus stock. In the case of the early-replacement vector, this virus stock consists exclusively of recombinant genomes encapsidated in SV40 coat proteins; in the case of the late-replacement vector, it consists of approximately equal numbers of helper virus particles (*dl*1055) and recombinants. Both virus stocks can be used to infect permissive cells with high efficiency and to induce in them the lytic cycle of virus growth. During such productive infections the viral genomes are transported to the nucleus, where they are liberated from their capsids. The early promoter soon becomes active, and genes under its control are expressed. By 12 hr after infection, viral DNA replication is under way, and genes begin to be expressed from the late promoter with high efficiency. By 36–48 hr after infection, the newly synthesized viral genomes begin to be assembled into progeny virus particles—a process that continues for another 24 hr, when the cells detach from their substrate and die (see Tooze, 1980).

VI. Hemagglutinin Is Efficiently Expressed from Both the Early and Late SV40 Promoters

The simplest way to measure the quantity of HA synthesized in cells infected with the early- or the late-replacement vectors is by solid-phase radioimmune assay (Gething and Sambrook, 1981). In this text, extracts of the infected cells are allowed to react with HA-specific polyclonal antibody that has been adsorbed onto a polyvinyl chloride surface. Hemagglutinin is known to have several different epitopes, so it would be expected that a given molecule of HA might bind to the absorbed antibody through one antigenic site, leaving the others available for reaction. The number of these available sites (hence the number of HA molecules) can then be titrated with radiolabeled antibody.

When this test was applied to extracts made from cells infected with either the early- or the late-replacement vector, it was found that the amount of HA detected increased as the course of infection proceeded. By 62 hr, when the lytic infection was in its terminal phase, cells infected with the late-replacement vector contained ~6×10^8 molecules of HA per cell, whereas those infected with the early-replacement vector contained ~2.5×10^6 molecules of HA per cell. For comparison, a simian cell at a late stage during infection with influenza virus contains ~5×10^7 molecules of HA. These results show that the influenza virus HA gene is expressed with high efficiency when it is coupled to either the early or the late SV40 promoters.

Originally, the fact that COS-1 cells infected with the early-replacement vector synthesized 10-fold less HA than CV-1 cells infected with the late-replacement vector was thought to be simply a reflection of the relative strengths of the early and late SV40 promoters. However, we now know that other factors are involved that play a much more important role in determining the amount of HA synthesized by these recombinant viruses. Chief among these is the presence or absence from the early-replacement vector of the small DNA fragment that carries the intron of small-t antigen.

VII. Small-t Intron Leads to Genetic Instability of the Early-Replacement Vector

It is clear from the studies just described that splicing of mRNA is not required for expression of the HA gene from the late-replacement vector. To test whether the intron sequence built into the early-replacement vector is necessary for expression of HA, we constructed a derivative identical to the original early-replacement vector in every respect except that the it lacks the intron (Gething and Sambrook, 1982). A virus stock containing this new recombinant was raised and propagated in COS-1 cells in the absence of helper virus. When its ability to express functional HA was compared to that of the original intron-plus, recombinant virus, we found that the intron-minus virus stock was of a much higher titer, so that it could infect more cells of the population. Furthermore, each of these infected cells produced more HA than cells infected with the intron-plus virus (Table I).

Analysis of the recombinant DNA molecules isolated from cells infected with sequential passages of the intron-plus virus showed a rapid increase in the proportion of rearranged genomes. By the fourth and fifth passage the stock was essentially noninfectious and contained a preponderance of molecules shorter than those of the original construct. A number of the rearranged DNA molecules from the fourth-passage intron-plus stock were cloned into the bacterial plasmid Xf3, analyzed by restriction-enzyme digestion, and sequenced by the Max-

Table I Analysis of Hemagglutinin in Cell Extracts by Solid-Phase Radioimmunoassay

HA gene	Vector	HA (ngm/10^6 cells)	Average number of HA molecules produced per cell	Cells infected (%)	Number of HA molecules produced per infected cell
Wild type	Early replacement (intron +)	22	1.5×10^6	5	3×10^7
Wild type	Early replacement (intron −)	1,700	1.2×10^8	95	1.5×10^8
Wild type	Late replacement (intron −)	12,000	9×10^8	95	1×10^9
Anchor −	Late replacement (intron −)	12,000	9×10^8	95	1×10^9
Wild type	Influenza virus	700	4.5×10^7[c]	30	1.3×10^8[c]

[a]The figures given were obtained by radioimmunoassay of extracts of CV-1 or COS-1 cells late (48–62 hr) after infection with SV40–HA recombinants (Gething and Sambrook, 1981, 1982). The number of cells expressing HA was estimated by hemadsorption.

[b]Over 90% of the HA synthesized by the anchor-minus HA gene was secreted from the cells and accumulated in the supernatant during the course of the infection.

[c]These figures do not include any HA molecules budded from the cells in progeny HA molecules.

am–Gilbert procedure. The results showed that many forms of rearrangement had occurred, including (1) simple deletions, 50 bp to 1 kb in length, both within the HA gene and across both of the SV40-HA junctions; (2) deletion of a segment of the HA gene, together with an insertion of a similarly sized piece of SV40 DNA, derived from a distant region of the viral genome; and (3) deletion of the SV40–HA junction downstream of the HA coding region, together with an inversion of the sequence of SV40 DNA lying immediately downstream from the junction. These changes are reminiscent of the sorts of rearrangements that have been shown to occur in the SV40 genome when the virus is serially passaged many times in succession at high multiplicity. The genome of the intron-plus recombinant, however, seems to be particularly vulnerable to such rearrangements. In contrast, the recombinant genome lacking an intron is completely stable. No rearrangement of its genome could be detected by restriction analysis and Southern hybridization after six sequential passages at high multiplicity, and the intron-minus virus stock showed no reduction in infectivity with increasing passage number; it continued to direct the synthesis of large quantities of HA.

We therefore conclude that the provision of splice junctions is not required for efficient expression of influenza virus HA cloned within either the early- or the late-replacement vectors. The functional, unspliced mRNA for HA is apparently synthesized with great efficiency, no matter whether the HA gene is presented to the cell in the form of double-stranded DNA or as negatively stranded RNA. It might be tempting to argue that the presence of the intron in the early-replacement vector is actually detrimental to the expression of the HA gene. However, the rapid accumulation of rearrangements in the recombinant genome suggests that the deleterious effects of the intron are mediated more at the level of DNA than RNA. Why genomic instability should be correlated with the presence of the intron remains to be elucidated.

VIII. Hemagglutinin Synthesized by SV40–HA Recombinants Is Biologically Active

To establish that the HA expressed by the SV40–HA recombinants is authentic both in structure and biological activity, a series of assays were performed on cells infected with either the early- or the late-expression vectors. These included immunoprecipitation, cytoplasmic and cell-surface immunofluorescence, and erythrocyte binding. By all of these and other tests, the protein was indistinguishable from HA produced naturally in influenza virus-infected cells. These results have been described in detail previously and are discussed only briefly here:

1. Immunoprecipitation. Extracts of cells infected with the recombinant SV40–HA viruses contained a protein that was specifically precipitated by an HA antiserum and that was indistinguishable in size from authentic, glycosylated HA precipitated from extracts of cells infected with influenza virus. So much of this protein did the recombinant cells contain that it was not necessary to use immunoprecipitation to detect it; HA could be seen either as a band stained with Coomassie Blue after the extracts of infected cells had been analyzed by SDS-polyacrylamide gel electrophoresis or as a prominent radioactive species when extracts were prepared from infected cells that had been labelled with [^{35}S]methionine.
2. Cell-free translation of mRNA. Hemagglutinin-specific mRNAs isolated from authentic, influenza virus-infected cells and from cells infected with SV40–HA recombinants can be translated in lysates of rabbit reticulocytes into a protein of molecular weight 63,000, that is, the nonglycosylated form of HA. When the translation was carried out in the presence of dog pancreas membranes, the apparent molecular weight of both HA products increased to that of the *in vitro*, glycosylated form of HA (75,000). Thus the two forms of HA synthesized *in vitro* from recombinant mRNA are identical to the corresponding forms synthesized from authentic HA mRNA.
3. Immunofluorescence. Cells infected with the SV40–HA vectors displayed bright, cytoplasmic fluorescence, with the Golgi apparatus staining with particular intensity. The surface of the cells also stained specifically with a uniform, dimmer fluorescence. The distribution of fluorescence in cells infected with influenza virus was similar to that displayed by recombinant-infected cells, but of lower intensity.
4. Binding of erythrocytes. Cells infected with the early- or the late-replacement vector absorbed a dense carpet of erythrocytes onto their surfaces.

It is clear from these studies that both types of recombinant viruses express large quantities of HA, which appears normal in all respects; its molecular weight is indistinguishable from that of authentic HA, and it displays on the infected cell's surface in a glycosylated form that is both antigenically and biologically active. Thus a protein that is normally encoded by a negative-strand RNA genome can be expressed in copious amounts when double-stranded DNA copies of its coding sequences are harnessed to strong SV40 promoters. To our knowledge, influenza virus HA is the first integral membrane to be expressed in eukaryotic cells from recombinant DNA vectors. It is therefore possible to begin to think of using site-directed mutagenesis to dissect those parts of the molecule important in its structure, function, and biosynthesis and that have been inaccessible to genetic analysis until now.

IX. Removing the C-Terminal Hydrophobic Sequence Converts Hemagglutinin from an Integral Membrane Protein to a Secreted Protein

As discussed previously, HA is normally anchored in the plasma membrane of the infected cell or in the envelope of influenza virus by a tract of nonpolar amino acids that are coded by the distal sequences of the HA gene. These sequences were removed from the standard late-replacement vector to create a new vector that differs from its parent in only one significant way: it lacks the DNA sequences coding for the 38 amino acids normally found at the C-terminus of HA. Substituting for them is a stretch of 11 amino acids, largely polar in nature, that are encoded by the dG:dC homopolymer tail, a synthetic *Bam*HI linker, and a short sequence of SV40 DNA (Gething, 1983).

This "anchor-minus" mutant synthesizes as much HA as the standard late-replacement vector. However, instead of being entirely cell associated (in the Golgi or on the cell surface), the mutant HA is found largely in the medium. The synthesis and secretion of the anchor-minus HA is so efficient that it is the only labeled protein that can be detected in the medium after labeling the infected cells for 1 hr with [^{35}S]methionine. The most straightforward explanation of the mutant's properties is that the removal of the C-terminal hydrophobic sequences results in a loss of anchoring function, so the nascent polypeptide, instead of remaining attached to the luminal face of the rough endoplasmic reticulum, passes completely through the membrane. Once free in the lumen, the mutant form of HA is treated by the cell as if it were an authentic secretory protein, and it is discharged into the medium.

The secreted form of HA differs from the membrane-bound form in only one major respect—its rate of glycosylation. The complex oligosaccharides added to membrane and secretory proteins are composed of "core" and "terminal" regions that are added in the rough endoplasmic reticulum and the Golgi apparatus, respectively. The inner core region, linked to asparagine, contains a minimum of two *N*-acetylglucosamine and three mannose residues (Neuberger *et al.*, 1972). Although fewer in number than those originally added, all of these core sugars are donated as a part of a preformed group from an activated donor molecule—dolichol. As far as known, all secretory and integral membrane proteins share the same asparagine-linked oligosaccharide core. However, the final pattern of glycosylation differs greatly from protein to protein. Most of these differences are generated during the passage of the protein through the Golgi apparatus, as a consequence of the activity of a series of specific mannosidase and glycosyltransferase enzymes. Because in general all copies of a particular protein show the same pattern of glycosylation, there must be information contained in either the sequence or the folding of the polypeptide chain that determines what alterations are to be made to the original precursor oligosaccharide. It is therefore interesting that the anchor minus and the wild type, which differ so little in primary amino acid sequence, should show differences in the rates at which they become glycosylated. The core-glycosylated form of wild-type HA (M_r = 73,000), which can be detected after short periods of labeling (10 min), is chased completely into its final glycosylated form within a matter of minutes. In contrast, the population of anchor-minus molecules takes up to 2 hr to become completely glycosylated. Even though the rate of glycosylation of anchor-minus HA is comparatively slow, its pattern of labeling with radioactive sugars is very similar to that of wild-type HA, and it therefore appears that final composition of the oligosaccharides attached to the two proteins is not significantly different.

These results lead to the following conclusions:

1. The hydrophobic amino acid sequence at the C-terminus of HA is required to anchor the protein in the outer membrane of the cell. Removal of this sequence converts HA from an integral membrane protein into a secretory protein. It therefore follows that, signal sequences apart, there is no unique sequence of amino acids common to all secretory proteins that automatically confers on them the ability to be secreted.
2. The slow rate of glycosylation of the anchor-minus HA indicates that secretory proteins in general may be processed more slowly than integral membrane proteins. This difference may lie at the level of transport to the Golgi apparatus or passage through it.
3. The signals governing the pattern of glycosylation of HA cannot lie in the 38 amino acids that normally constitute the C-terminus of the molecule.

X. Prospects

Influenza virus HA, a protein that is normally encoded by a negative-strand RNA genome, is expressed in copious amounts when DNA copies of its coding sequence are harnessed to strong SV40 promoters and replicated to high copy number in cultured cells. The protein appears to be transported and processed in much the same manner as other cellular integral membrane proteins, and it is displayed on the cell's surface in a form that is antigenically and biologically indistinguishable from authentic HA coded by influenza virus itself. These initial findings can be extended in several directions. First, they open the way to use site-directed mutagenesis to dissect those parts of the molecule important to its structure, function, and biosynthesis that have been inaccessible to detailed genetic analysis. Second, they may provide a means to synthesize large quantities of mutant, nonfunctional forms of the molecule for structural analysis. Third, they offer the possibility of understanding how cells make decisions about molecular sorting. Why do some molecules emerge from one aspect of the cell's surface and some from another? Why do other proteins remain fixed in the membrane of the rough endoplasmic reticulum or in the Golgi apparatus rather than being transported to the cell surface or to some other intracellular location? Fourth, they open the possibility of using genetic engineering to attach DNA segments coding for hydrophobic signal and/or anchoring sequences to DNA sequences coding for proteins that normally occupy intracellular locations. Such address markers might allow these proteins to be displayed on the cell surface or secreted into the medium in a form that may be readily purified. Finally, there is a hope that this work may be extended into a general method to produce membrane antigens in quantities sufficient for use as vaccines.

References

Air, G. (1981). *Proc. Natl. Acad. Sci. USA* **78,** 7639–7643.

Blobel, G., and Dobberstein, G. (1975). *J. Cell Biol.* **67,** 852.

Blobel, G., and Sabatini, D. D. (1970). *J. Cell Biol.* **45,** 130–145.

Blobel, G., Walter, P., Chang, C. N., Goldman, B. M., Erickson, A. H., and Lingappa, V. R. (1979). *Symp. Soc. Exp. Biol.* **33,** 9–36.

Both, G. W., and Sleigh, M. J. (1980). *Nucleic Acids Res.* **8,** 2561–2575.

Choppin, P. W., Klenk, H., Compans, R. W., and Caliguiri, L. A. (1971). *In* "From Molecules to Man" (M. Pollard, ed.), pp. 127–158. Academic Press, New York.

Compans, R. W., and Choppin, P. W. (1975). *Compr. Virol.* **1,** 179–238.

Davis, B. D., and Tai, P.-C. (1980). *Nature (London)* **283,** 433–438.

Dhar, R., Chanock, R. M., and Lai, C. J. (1981). *Cell* **21,** 495–500.

Farquhar, M. G., and Palade, G. E. (1981). *J. Cell Biol.* **91,** 77S–103S.

Fields, S., Winter, G., and Brownlee, G. G. (1981). *Nature (London)* **290,** 213–217.

Frank, G., Brunner, H., Hauser, H., Wacker, G., Senenza, G., and Zuber, H. (1978). *FEBS Lett.* **96,** 183–188.

Garten, W., Bosch, F.-X., Linder, D., Rott, R., and Klenk, H.-D. (1981). *Virology* **105,** 205–222.

Gething, M.-J. (1983). *Nature (London),* in press.

Gething, M.-J., and Sambrook, J. (1982a). *In* "Eukaryotic Viral Vectors" (Y. Gluzman, ed.). Cold Spring Harbor Lab., Cold Spring Harbor, New York.

Gething, M.-J., and Sambrook, J. (1982b). *Nature (London)* **293,** 620–625.

Gething, M.-J., White, J. M., and Waterfield, M. D. (1978). *Proc. Natl. Acad. Sci. USA* **75,** 2735–2740.

Gething, M.-J., Bye, J., Skehel, J. J., and Waterfield, M. D. (1980). *Nature (London)* **287,** 301–306.

Ghosh, P., Lebowitz, P., Frisque, R. J., and Gluzman, Y. (1981). *Proc. Natl. Acad. Soc. USA* **78,** 100–104.

Gluzman, Y. (1981). *Cell* **23,** 175–182.

Green, R. F., Meiss, H. K., and Rodriguez-Boulan, E. (1981). *J. Cell Biol.* **89,** 230–239.

Gruss, P., and Khoury, G. (1981). *Proc. Natl. Acad. Sci. USA* **78,** 133–137.

Hamer, D. H., and Leder, P. (1979). *Cell* **18,** 1299–1302.

Hamer, D. H., Smith, K. D., Boyer, S. H., and Leder, P. (1979). *Cell* **17,** 725–735.

Hirst, G. K. (1942). *J. Exp. Med.* **75,** 49–64.

Hiti, A. L., Davis, A. R., and Nayak, D. P. (1981). *Virology* **111,** 113–124.

Inglis, S. C., Barrett, T., Brown, C. M., and Almond, J. W. (1979). *Proc. Natl. Acad. Sci. USA* **76,** 3790–3794.

Jackson, R. C., and Blobel, G. (1977). *Proc. Natl. Acad. Sci. USA* **74,** 5598–5602.

Klenk, H.-D., Rott, R., Orlich, M., and Blodorn, J. (1975). *Virology* **68,** 426–439.

Kreibich, G., Freienstein, C. M., Pereyra, B. N., Ulrich, B. L., and Sabatini, D. D. (1978). *J. Cell Biol.* **77,** 488–506.

Kreibich, G., Czako-Graham, L., Grebenan, R. C., and Sabatini, D. D. (1980). *Ann. N.Y. Acad. Sci.* **343,** 17–33.

Krug, R. M., Broni, B. A., and Bouloy, M. (1979). *Cell* **18,** 329–334.

Lamb, R. A., and Choppin, P. W. (1981). *Virology* **112,** 729–737.

Lamb, R. A., Etkind, P. R., and Choppin, P. W. (1978). *Virology* **91,** 60–78.

Laver, W. G., and Air, G. M. (eds.) (1980). "Structure and Variation in Influenza Virus." Elsevier–North Holland, New York.

Lazarowitz, S. G., and Choppin, P. W. (1975). *Virology* **68,** 440–454.

Leblond, C. P., and Bennett, G. (1977). *In* "International Cell Biology, 1976–1977" (B. R. Brinkley, and K. R. Porter, eds.), pp. 326–336. Rockefeller Univ. Press, New York.

McCutchan, J. H., and Pagano, J. S. (1968). *J. Natl. Cancer Inst. (US)* **41,** 351–357.

McGeoch, D. J., Fellner, P., and Newton, C. (1976). *Proc. Natl. Acad. Sci. USA* **73,** 3045–3049.

Meyer, D., and Dobberstein, B. (1980). *J. Cell Biol.* **87,** 498–502.

Meyer, D. I., Krause, E., and Dobberstein, B. (1982). *Nature (London)* **297,** 647–650.

Milstein, C., Brownlee, G. G., Harrison, T. M., and Mathews, M. B. (1972). *Nature New Biol.* **239,** 117–120.

Min Jou, W., Verhoeyen, M., Devos, R., Saman, E., Fang, R., Huylebroech, D., and Fiers, W. (1980). *Cell* **19,** 683–696.

Mulligan, R. C., Howard, H., and Berg, P. (1979). *Nature (London)* **277,** 108–114.

Neuberger, A., Gottschalk, A., Marshall, R. O., and Spiro, R. G. (1972). *In* "The Glycoproteins: Their Composition, Structure and Function" (A. Gottschalk, ed.), pp. 450–490. Elsevier, Amsterdam.

Newmark, P. (1982). *Nature (London)* **297,** 624.

Palade, G. E. (1975). *Science (Washington, D.C.)* **189,** 347–358.

Palese, P., and Schulman, J. L. (1976). *Proc. Natl. Acad. Sci. USA* **73,** 2142–2146.

Pipas, J. M., Adler, S. P., Peden, K. W. C., and Nathans, D. (1980). *Cold Spring Harbor Symp. Quant. Biol.* **44,** 285–291.

Porter, A. G., Barber, C., Carey, N. H., Hallewell, R. A., Threlfall, G., and Emtage, J. S. (1979). *Nature (London)* **282,** 471–477.

Richardson, C. D., Scheid, A., and Choppin, P. W. (1980). *Virology* **105,** 205–222.

Rodriguez-Boulan, E., and Prendergast, M. (1980). *Cell* **20,** 45–54.

Rodriguez-Boulan, E., and Sabatini, D. D. (1978). *Proc. Natl. Acad. Sci. USA* **75,** 5071–5075.

Roth, M. G., Fitzpatrick, J. G., and Compans, R. W. (1979). *Proc. Natl. Acad. Sci. USA* **76,** 6430–6434.

Sabatini, D. D., Kreibich, G., Morimoto, T., and Adesnik, M. (1982). *J. Cell Biol.* **92,** 1–22.

Schild, G. C. (ed.) (1979). *Br. Med. Bull.,* Vol. 35.

Scholtissek, C., Harms, E., Rohde, W., Orlich, M., and Rott, R. (1976). *Virology* **74,** 332–344.

Stuart-Harris, C. H. (1979). *Br. Med. Bull.* **35,** 3–8.

Tabas, I., and Kornfeld, S. (1979). *J. Biol. Chem.* **254,** 11655–11663.

Tooze, J. (ed.) (1980). ''Molecular Biology of Tumor Viruses,'' Part 2, 2nd ed. Cold Spring Harbor Lab., Cold Spring Harbor, New York.

Verhoeyen, M., Fang, R., Min Jou, W., Devos, R., Huylebroeck, D., Saman, E., and Fiers, W. (1980). *Nature (London)* **286,** 771–776.

Walter, P., and Blobel, G. (1982). *J. Cell Biol.* **91,** 551.

Walter, P., Jackson, R. C., Marcus, M. M., Lingappa, V. R., and Blobel, G. (1979). *Proc. Natl. Acad. Sci. USA* **76,** 1795–1799.

Walter, P., Ibrahimi, I., and Blobel, G. (1982). *J. Cell Biol.* **91,** 545.

Ward, C. W., and Dopheide, T. A. A. (1980). *Virology* **103,** 37–53.

Waterfield, M. D., and Scrace, G. (1981). *Nature (London)* **289,** 422–424.

Waterfield, M. D., Espelie, K., Elder, K., and Skehel, J. J. (1979). *Br. Med. Bull.* **35,** 57–65.

Webster, R. G., and Laver, W. G. (1975). *In* ''The Influenza Viruses and Influenza'' (E. D. Kilbourne, ed.), pp. 209–314. Academic Press, New York.

White, J. M., Helenius, A., and Gething, M.-J. (1982). *Nature (London),* in press.

Wiley, D. C., and Skehel, J. J. (1977). *J. Mol. Biol.* **112,** 343–347.

Wiley, D. C., Wilson, I. A., and Skehel, J. J. (1981). *Nature (London)* **289,** 373–378.

Wilson, I. A., Skehel, J. J., and Wiley, D. C. (1981). *Nature (London)* **289,** 366–373.

Winter, G., Fields, S., and Brownlee, G. G. (1981). *Nature (London)* **292,** 72–75.

CHAPTER 13

Expression of Human Interferon-γ in Heterologous Systems

RIK DERYNCK
RONALD A. HITZEMAN
PATRICK W. GRAY
DAVID V. GOEDDEL

Department of Molecular Biology
Genentech, Inc.
South San Francisco, California

I. Introduction

On induction with various mitogens, mammalian T lymphocytes from peripheral blood lymphocytes, tonsils, or other sources are stimulated to synthesize and secrete "immune" interferon or interferon-γ (IFN-γ). This interferon is distinct from the leukocyte interferons (IFNs-α) or fibroblast interferon (IFN-β), not only by its cellular origin but also on the basis of neutralization studies with antibodies raised against different species (for review see Stewart, 1981). Its

Genentech, Inc. Contribution no. 134.

Experimental Manipulation
of Gene Expression

ISBN 0-12-372380-9

induction differs from IFN-α and -β in that neither viral infection nor double-stranded RNA is sufficient for inducing its synthesis. Interforons-α and -β are pH-2 stable and have a molecular weight of ~20,000 (Zoon *et al.*, 1980; Knight *et al.*, 1980; Allen and Fantes, 1980; Levy *et al.*, 1980; Stein *et al.*, 1980). On the other hand, IFN-γ is acid labile, and molecular weight values of 35,000–70,000 have been determined on the basis of gel filtration (Langford *et al.*, 1979; de Ley *et al.*, 1980; Yip *et al.*, 1981; Nathan *et al.*, 1981). Yip *et al.* (1982) have analyzed pure human IFN-γ by electrophoresis on SDS-polyacrylamide gels and have determined the presence of two species with values of 20 and 25 kdaltons. A similar low value for human IFN-γ has been reported by Hochkeppel and de Ley (1982).

Interferon-γ is also distinct in its biological activities from IFNs-α and -β. Differences in virus and cell specificity for antiviral action have been reported (Rubin and Gupta, 1980). A much higher activity against reovirus and vaccinia virus has been found for IFN-γ than for other interferons. The antiproliferative activity against transformed cell lines seems to be more potent than for IFNs-α and -β (Rubin and Gupta, 1980). Also, there are some indications that IFN-γ might play a primarily immunomodulatory role (Sonnenfeld *et al.*, 1978).

II. Structure of the Human Interferon-γ cDNA

We have isolated a cloned cDNA of human IFN-γ, synthesized using mRNA from induced peripheral blood lymphocytes (Gray *et al.*, 1982). Nucleotide-sequence analysis of the full-size cDNA insert of 1200 bp revealed the presence of a reading frame coding for 166 amino acids (Fig. 1). The N-terminal 20 amino acids constitute the putative signal sequence, cleaved off during the secretion from the cell. The mature protein is thus 146 amino acids long with a molecular weight of ~17,000 in its unmodified form.

Only a very low degree of amino acid sequence homology is observed with the IFNs-α (Goeddel *et al.*, 1981; Weissmann, 1981). This similarity is present in a small region preceding and containing the cleavage site of the signal peptide from the mature protein (Gray *et al.*, 1982) and also at amino acids 7 to 17 in IFN-γ, compared to amino acids 130–136 in IFN-α2 (Gray and Goeddel, 1982). Two potential N-glycosylation sites are present at amino acids 28–30 and 100–102.

The difference between the calculated molecular weight of 17,000 and the much higher values determined by gel filtration column chromatography is unlikely to be due entirely to glycosylation or other posttranslational modification, but it might be due to dimerization of IFN-γ molecules under physiological

S1 S10 S20

met lys tyr thr ser tyr ile leu ala phe gln leu cys ile val leu gly ser leu gly

1 10 20

CYS TYR CYS GLN ASP PRO TYR VAL LYS GLU ALA GLU ASN LEU LYS LYS TYR PHE ASN ALA

30 40

GLY HIS SER ASP VAL ALA ASP ASN GLY THR LEU PHE LEU GLY ILE LEU LYS ASN TRP LYS

50 60

GLU GLU SER ASP ARG LYS ILE MET GLN SER GLN ILE VAL SER PHE TYR PHE LYS LEU PHE

70 80

LYS ASN PHE LYS ASP ASP GLN SER ILE GLN LYS SER VAL GLU THR ILE LYS GLU ASP MET

90 100

ASN VAL LYS PHE PHE ASN SER ASN LYS LYS LYS ARG ASP ASP PHE GLU LYS LEU THR ASN

110 120

TYR SER VAL THR ASP LEU ASN VAL GLN ARG LYS ALA ILE HIS GLU LEU ILE GLN VAL MET

130 140

ALA GLU LEU SER PRO ALA ALA LYS THR GLY LYS ARG LYS ARG SER GLN MET LEU PHE ARG

146

GLY ARG ARG ALA SER GLN

Fig. 1 Deduced amino acid sequence of the coding region of IFN-γ cDNA. The putative signal sequence is represented by the residues labeled S1 to S20. The numbers above the residues refer to the amino acid position.

conditions. On the other hand, the difference between the 17-kdalton value and the slightly higher values obtained from electrophoretic analysis on polyacrylamide gels (Yip *et al.*, 1982; Hochkeppel and de Ley, 1982) could be due to glycosylation or other modifications.

Characterization of the IFN-γ cDNA allowed the detection and elucidation of the structure of the corresponding chromosomal gene. The IFN-γ gene contains three intervening sequences (Gray and Goeddel, 1982), in striking contrast to the lack of intervening sequences in the IFNs-α (Nagata *et al.*, 1980; Lawn *et al.*, 1981a,b; Weissmann, 1981) and IFN-β (Houghton *et al.*, 1981; Lawn *et al.*, 1981c; De Grave *et al.*, 1981; Ohno and Taniguchi, 1981). The gene for human IFN-γ has been localized on chromosome 12 (Naylor *et al.*, 1982), whereas the human IFNs-α and -β are situated on chromosome 9 (Owerbach *et al.*, 1981).

III. Heterologous Expression in Escherichia coli

The coding sequence of human IFN-γ has been expressed in several heterologous systems (i.e., in mammalian cells, in the bacterial *E. coli* system, and in the yeast *Saccharomyces cerevisiae*). For heterologous expression in mammalian

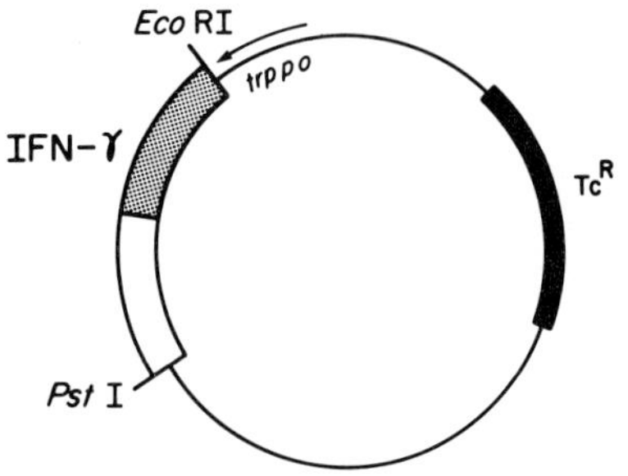

Fig. 2 Schematic representation of a plasmid (pIFN-γ*trp*48) coding for the direct synthesis of mature IFN-γ in *E. coli*. The thick segment represents the region coding for tetracycline resistance (Tc^R) in *E. coli*. The *trp* promoter operator region (*trp* p o) is shown with an arrow, indicating the direction of transcription. The boxed region is the coding sequence for mature IFN-γ (▦) and the 3′ untranslated sequence of the cDNA (▭).

cells, the complete coding sequence of human IFN-γ was placed under the control of the SV40 late promoter, itself located on a plasmid. The presence of the SV40 origin of replication on the plasmid allows replication of this vector in the monkey cell line COS-7 (Gluzman, 1981). In this system transient expression takes place, resulting in the secretion of human IFN-γ into the cell culture medium in levels up to 1500 units/ml over a period of 4 days following transfection (Gray *et al.*, 1982; H. M. Shepard, C. C. Simonsen, A. D. Levinson, and D. V. Goeddel, unpublished).

Bacterial expression of mature human IFN-γ has also been achieved in *E. coli* (Gray *et al.*, 1982). For this purpose we used the *E. coli trp* promoter and the *trp* leader ribosome-binding site (Yanofsky *et al.*, 1981), which had been incorporated in a high-copy-number plasmid originally derived from pBR322 (Fig. 2). Using this system, high-expression levels of several other foreign genes, including other human interferons (Goeddel *et al.*, 1980a,b; Yelverton *et al.*, 1981), have been obtained. Synthetic DNA was used to introduce a start codon, preceded by an *Eco*RI site, in front of the coding sequence of mature IFN-γ. Expression under control of the *trp* promoter results in the synthesis of $\sim 5 \times 10^8$ units of IFN-γ per liter of bacterial culture.

The availability of this high-level expression system makes it possible to study the effect of specific changes in the protein on its biological activity. This is illustrated here with a few examples. We originally expressed in *E. coli* the coding sequence for mature IFN-γ, coding for a glutamine at amino acid position 140 (Gray *et al.*, 1982). However, the presence of a codon for arginine at the same position in several other allelic IFN-γ cDNA clones was revealed by sequence analysis (Derynck *et al.*, 1982). We consequently made a bacterial IFN-γ cDNA expression plasmid, identical to the original one, except that the Gln codon was present instead of the Arg codon at position 140. When the expression levels of the two IFN-γ cDNAs were compared, a twofold lower level was detected for the IFN-γ cDNA with the Arg codon than for the original cDNA containing the Gln codon. This is apparently not due to a difference in specific activity of the bacterial IFN-γ, but it correlates well with a lower intensity of the stained IFN-γ protein band after gel electrophoresis of total bacterial extracts (N. Lin and M. Mumford, unpublished).

The presence within the coding sequence of the IFN-γ cDNA of several restriction-endonuclease sites allows simple manipulations for mapping the functional regions of the protein with respect to certain of its biological properties. A variety of deletions can be made using several restriction enzymes. One such deletion removed amino acids 98 to 141 but conserved the reading frame of the mRNA. This resulted in complete absence of IFN-γ antiviral activity in the *E. coli* extracts as measured by cytopathic-effect inhibition assay. Alternatively, a smaller deletion was introduced, removing amino acids 136 to 141. Threefold higher activity compared to the parent unmodified protein was detected in the bacterial extracts. This might be due either to a difference in specific activity or to an altered bacterial expression level. The apparent irrelevance of this C-terminal segment for antiviral activity was further investigated by a bacterial construction resulting in the replacement of the C-terminal amino acids 136 to 146 by the unrelated sequence Thr-His-Gln-Ser-Asn-Lys-Tyr-Leu. Also, here a higher interferon activity compared to the parent plasmid was present, showing that the antiviral activity of the molecule, as determined by the cytopathic effect of encephalomyocarditis (EMC) virus on human WISH cells, was not affected.

As illustrated here, it is clear that the introduction of specific changes in the coding sequence such as substitutions, deletions, and replacement of segments can lead to a better understanding of the structure–function relationship of human IFN-γ. This approach may have the potential for use in the construction of IFN-γ derivatives with a higher specific activity or with altered properties.

IV. Expression in the Yeast Saccharomyces cerevisiae

The coding sequence for mature human IFN-γ has also been expressed in a third heterologous system, the yeast *Saccharomyces cerevisiae*. Shuttle vectors, which can autonomously replicate both in *E. coli* and in yeast, are used for this purpose.

The bacterial replication origin and gene for β-lactamase, both derived from pBR322 (Bolivar *et al.*, 1977), are conserved in these plasmids. These can consequently be maintained in *E. coli* and allow proper selection of the transformed bacterial clones owing to the presence of the β-lactamase gene resulting in ampicillin resistance. The newly constructed hybrid yeast plasmids can thus be characterized easily and prepared from this bacterial source. These plasmids can also be maintained in yeast by the presence of a functional yeast origin of replication. The presence of the yeast chromosomal origin of replication *ars1* (Tschumper and Carbon, 1980) on the plasmid results in a stability of 25–35% of the recombinant plasmids. A much higher stability of about 95% is apparent when the replication origin of the 2-μm plasmid (Hartley and Donelson, 1980) is incorporated into these plasmids. Proper selection of the transformed yeast clone is achieved by the presence of the yeast *TRP1* gene (Tschumper and Carbon, 1980) as a marker on the chimeric plasmids. Transformation of a *trp*$^-$ yeast strain with plasmids containing this selection marker results in tryptophan prototrophy.

In principle, expression of foreign genes in yeast can be achieved by insertion of the coding sequence downstream and under the control of any yeast promoter for RNA polymerase II. We have described the use of the promoter for the abundant glycolytic enzyme alcohol dehydrogenase I (Bennetzen and Hall, 1982) for the expression of human leukocyte interferon D (Hitzeman *et al.*, 1981). Up to 3×10^6 antiviral units/ml of culture, accounting for about 0.4% of total cell protein, were obtained.

The glycolytic enzyme phosphoglycerate kinase (PGK) is one of the most abundant proteins in yeast, accounting for 4–10% of total protein (Scopes, 1973). The nucleotide sequence of the gene and its flanking regions has been elucidated. Also, the transcription-initiation site has been determined (Hitzeman *et al.*, 1982). For the expression of IFN-γ we used the PGK promoter, which is contained within the 300 bp of DNA upstream from the coding sequence. The 10-bp segment immediately preceding the PGK-translational start was replaced by a synthetic combined *Xba*I–*Eco*RI recognition site (5′ dTCTAGAATTC 3′). The introduction of these sites allows the versatile insertion of a coding sequence, as for mature IFN-γ. For expression in *E. coli,* an *Eco*RI recognition site, followed by the ATG initiation codon, was present in front of the coding sequence for mature IFN-γ (Gray *et al.*, 1982). In this way, the IFN-γ sequence could be inserted via the *Eco*RI site under the control of the PGK promoter.

The introduced gene should be followed by the sequences required for transcription termination and polyadenylation of the mRNA in yeast. A restriction fragment downstream of the coding sequences for either the *TRP1* (Tschumper and Carbon, 1980), the PGK (Hitzeman *et al.*, 1982), or the 2-μm plasmid "Able" gene (Hartley and Donelson, 1980) has been placed behind the IFN-γ sequence for this purpose.

The mature IFN-γ sequence in yeast was expressed using several plasmids, combining the needed structural elements just described. Also, variable lengths of the 3′ untranslated end of IFN-γ cDNA have been retained in these plasmids. Two plasmid constructions are shown in Fig. 3. Plasmid pPGK-γ4Δ3 contains the yeast *ars1* origin of replication (Tschumper and Carbon, 1980). The coding sequence of IFN-γ is followed by only the 50 bp of the 3′ untranslated end

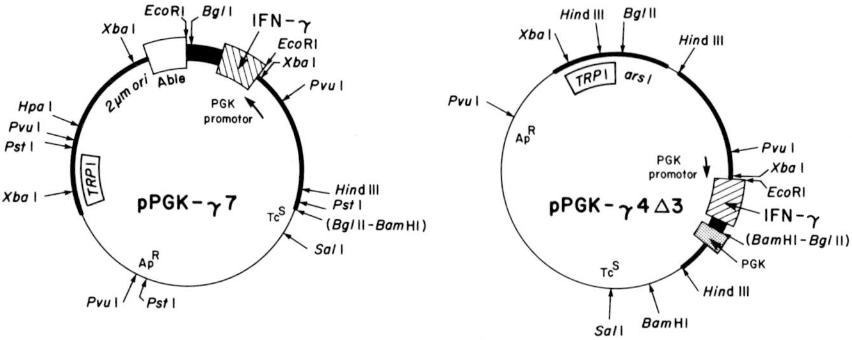

Fig. 3 Schematic representation of plasmids for expression of IFN-γ in yeast (pPGK-γ7, ~9200 bp; pPGK-γ4Δ3, ~8000 bp). The thin segments show the pBR322 sequences, and the yeast DNA is represented by thick segments. The boxed regions are the total or partial coding sequences for IFN-γ (▨), PGK (▩), and the "Able" protein (▭). The 3′ untranslated end of IFN-γ is shown as a black, boxed region. The arrow indicates the direction of transcription from the PGK promoter. Ap^R, Ampicillin resistance; Tc^S, tetracycline sensitivity; *ori*, origin of replication.

proximal to the coding sequence. The PGK-terminator fragment (Hitzeman *et al.*, 1982) is inserted behind this cDNA sequence. Plasmid pPGK-γ7 contains the 2-μm origin of replication (Hartley and Donelson, 1980). The 287 proximal base pairs of the 3′ untranslated sequence of the IFN-γ cDNA are retained and followed by the "Able" terminator fragment of the 2-μm plasmid. Both plasmids contain the *TRP1* gene as selection marker.

Yeast strain *S. cerevisiae* 20B-12 (α*trp pep4-3*) (Jones, 1976) was transformed with these IFN-γ expression plasmids, and selection for the transformed yeast took place using tryptophan-depleted medium. The presence of the *TRP1* gene on the IFN-γ cDNA expression plasmids enabled the transformed yeast to form colonies on agar lacking this amino acid. After disruption of the yeast cells with glass beads, an antiviral interferon activity of 8×10^5 and 5×10^6 units/liter of culture at an A_{660} of 4 was found with the plasmids pPGK-γ4Δ3 and -γ7, respectively. However, the level of IFN-γ activity is clearly dependent upon the extraction procedure. Spheroplasting of the yeast cells with zymolyase (Kuo and Yamamoto, 1975) followed by a hypotonic shock resulted in the release from the cells of a three- to fivefold higher activity than when the yeast was mechanically disrupted with glass beads. Because of the lability of IFN-γ, it was not possible to determine the absolute quantity of IFN-γ using very strong disruption conditions. After [^{35}S]methionine labeling, we were not able to detect unambiguously an IFN-γ–specific band after SDS-polyacrylamide gel electrophoresis of a total cell lysate.

The difference of IFN-γ expression level between the yeast transformed with either pPGK-γ4Δ3 or -γ7 is probably largely due to the difference in replication origin. The presence of the *ars1* origin in pPGK-γ4Δ3 results in a 25–30% stability, whereas a 95% stability is obtained with the 2-μm origin in pPGK-γ7. This higher stability is also reflected in the growth of transformed yeast under selective conditions of Trp-deficiency. The yeast 20B-12/pPGK-γ7 grows much faster to saturation than 20B-12/pPGK-γ4Δ3 and thereby closely follows the growth curve of the untransformed yeast in the same but Trp-supplemented medium (data not shown).

The presence of a foreign gene (e.g., that for human IFN-γ) under the control of the endogenous yeast PGK promoter can be used as a model system for the study of transcription in yeast. Polyadenylated RNA was therefore isolated from the yeast, transformed with the IFN-γ expression plasmids. The mRNA was electrophoresed on formaldehyde–agarose gels (Dobner *et al.*, 1981) and, after transfer to a nitrocellulose filter (Thomas, 1980), analyzed by Northern hybridization with ^{32}P-labeled restriction fragments specific for either the IFN-γ coding sequence or the PGK or "Able" terminator region. One distinct IFN-γ mRNA was visible after hybridization with the ^{32}P-labeled IFN-γ cDNA probe (Fig. 4). As expected, the higher stability of pPGK-γ7, when compared with other *ars1*-containing IFN-γ plasmids like pPGK-γ4Δ3, resulted in a clearly higher intensity of the yeast IFN-γ mRNA band.

The length of the IFN-γ mRNA from yeast 20B-12/pPGK-γ4Δ3 is ~850 bases, as estimated from the migration in formaldehyde–agarose gels. Using the restriction fragment containing the transcription-termination region as hybridization probe, a band of exactly the same size was detected after autoradiography (Fig. 4). This result clearly suggests that in the case of plasmid pPGK-γ4Δ3, transcription termination takes place in the PGK-terminator region. This termination has been localized at 86 to 93 nucleotides behind the stop codon in the case of the PGK mRNA (Hitzeman *et al.*, 1982). In the case of yeast 20B-12/pPGK-γ7, the yeast IFN-γ mRNA has a length of 720–750 bases. However, this mRNA does not hybridize with a probe specific for the terminator egion (Fig. 4), indicating that the termination of transcription takes place in the 287-bp segment of the 3′ untranslated region of IFN-λ cDNA, which is present in pPGK-λ7. This was also observed for mRNA derived from yeast containing other IFN-γ expression plasmids that had retained at least a 287-base region of the 3′ untranslated end of the IFN-γ cDNA closest to the stop codon. We consequently constructed another yeast IFN-γ expression plasmid, which contained the total 3′ untranslated region of the IFN-γ cDNA, but not followed by a region specific for transcription termination. The IFN-γ mRNA derived from this plasmid had the same length as in the case of plasmid pPGK-γ7. This supports the finding that transcriptional termination takes place in the 287-bp segment of the IFN-γ cDNA 3′ untranslated region. In contrast, the presence of only the 50 proximal base pairs in pPGK-γ4Δ3 results in a read-through of transcription, stopping in the PGK-terminator region.

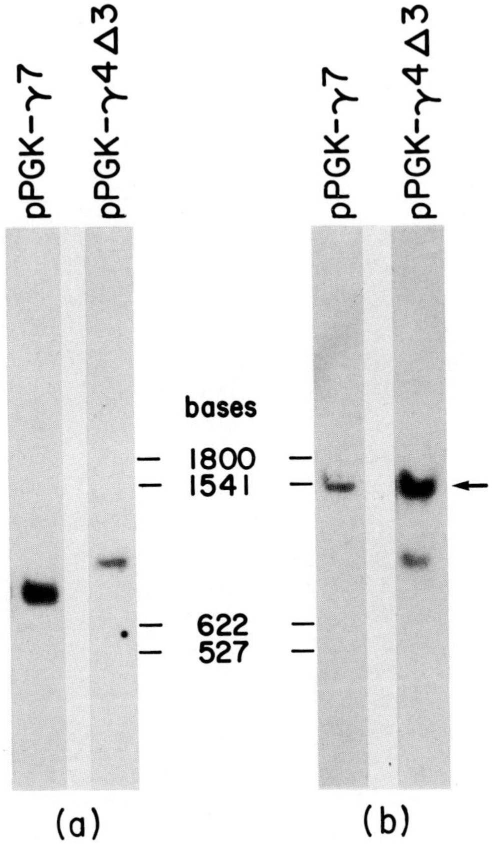

Fig. 4 Northern analysis of polyadenylated RNA from yeast 20B-12 containing the indicated plasmids after electrophoresis on formaldehyde–1.2% agarose gels. (a) Hybridization with the IFN-γ cDNA probe. (b) Hybridization with the PGK-terminator probe. The arrow indicates endogenous PGK mRNA, transcribed from the chromosomal gene. The indicated length markers are yeast 18-S ribosomal RNA (1800 bases; Warner, 1982), *E. coli* 16-S ribosomal RNA (1540 bases, Brosius *et al.*, 1978), and the denatured 620- and 520-bp *Hpa*II fragments of pBR322 (Sutcliffe, 1978).

Zaret and Sherman (1982) have studied the transcriptional termination of the yeast *CYC1* locus and of a mutation in the terminator region. Their results suggest that polyadenylation might be coupled to transcription termination in yeast. From these results and from the comparison of the 3′ untranslated region of various yeast genes, they proposed a consensus signal sequence for transcription termination. This sequence [TAG . . . TA(T)GT . . . TTT . . .] or a variant of it precedes the polyadenylation site at a variable distance in many but not all yeast genes. Very strikingly, a similar sequence (TAG . . . TAAGT . . . TTT) is present in the segment of the 3′ untranslated end of IFN-γ cDNA that causes the transcription termination. It is therefore possible that this sequence triggers the termination of transcription from the IFN-γ cDNA and subsequent polyadenylation, as observed in the case of pPGK-γ7. It is then obvious that removal of the segment containing this signal sequence results in a read-through of transcription until the next terminator sequence is encountered.

It is evident that many more studies are needed to understand the processes involved in gene expression. This knowledge is certainly needed to obtain high-level expression of artificially introduced foreign genes. The expression of the coding sequence of human mature IFN-γ can serve as a model system to investigate the structural elements and different aspects of transcription and translation in yeast.

In summary, we have expressed the cDNA of human IFN-γ in several heterologous systems. This has allowed high-level production of this lymphokine, enabling further studies of its biological and biochemical properties and leading to the exploration of its potential clinical use. Specific manipulations of the coding sequence of the IFN-γ cDNA (e.g., deletions or insertions), coupled with expression in *E. coli,* can lead to some understanding of the structure–function relationship of this intriguing molecule. Also, the introduction of this IFN-γ cDNA under the control of a heterologous promoter can be used as a model system for the study of gene regulation, as illustrated here in yeast using the PGK promoter.

V. Conclusion

Human interferon-γ (IFN-γ) is distinct from the other interferons, IFNs-α and -β, not only in its biochemical properties, but also in its biological activities. We have cloned and characterized the cDNA of human IFN-γ (Gray *et al.,* 1982). This cDNA codes for 166 amino acids, of which the N-terminal 20 residues are considered as the signal sequence, cleaved off during secretion from induced human lymphocytes. The IFN-γ cDNA has been expressed in several heterologous systems: the bacterium *Escherichia coli,* the yeast *Saccharomyces cerevisiae,* and monkey cells. Expression in *E. coli* results in the synthesis of IFN-γ as the major protein produced by *E. coli.* Introduction of specific changes in the IFN-γ coding sequence can lead to some understanding of the structure–function relationship of this molecule. This approach has led to the finding that the C-terminal 10 amino acids are not needed for the antiviral activity.

The IFN-γ cDNA has also been introduced into shuttle vectors, which are able to replicate in both *E. coli* and yeast. Expression of the coding sequence for mature IFN-γ has been obtained in yeast using the endogenous promoter for phosphoglycerate kinase. Analysis of the IFN-γ-specific mRNA from the transformed yeast shows that the presence of a 287-bp segment of the 3′ untranslated IFN-γ cDNA region closest to the stop codon acts as a terminator of transcription in yeast.

References

Allen, G., and Fantes, K. H. (1980). *Nature (Lodon)* **287,** 408–411.

Bennetzen, J. L., and Hall, B. D. (1982). *J. Biol. Chem.* **257,** 3018–3025.

Bolivar, F., Rodriguez, R. L., Greene, P. G., Betlach, M. C., Heyneker, H. L., Boyer, H. B., Crosa, J. H., and Falkow, S. (1977). *Gene* **2,** 95–113.

Brosius, J., Palmer, M. L., Kennedy, P. J., and Noller, N. F. (1978). *Proc. Natl. Acad. Sci. USA* **75,** 4801–4805.

De Grave, W., Derynck, R., Tavernier, J., Hageman, G., and Fiers, W. (1981). *Gene* **14,** 137–143.

de Ley, M., Van Damme, J., Claeys, H., Weening, H., Heine, J. W., Billiau, A., Vermylen, C, and De Somer, P. (1980). *Eur. J. Immunol.* **10,** 877–883.

Derynck, R., Leung, D. W., Gray, P. W., and Goeddel, D. V. (1982). *Nucleic Acids Res.* **10,** 3605–3615.

Dobner, P. R., Kawasaki, E. S., Yu, L. Y., and Bancroft, F. C. (1981). *Proc. Natl. Acad. Sci. USA* **78,** 2230–2234.

Gluzman, Y. (1981). *Cell* **23,** 175–182.

Goeddel, D. V., Yelverton, E., Ullrich, A., Heyneker, H. L., Miozzari, G., Holmes, W., Seeburg, P. H., Dull, T., May, L., Stebbing, N., Crea, R., Maeda, S., McCandliss, R., Sloma, A., Tabor, J. M., Gross, M., Familetti, P. C., and Pestka, S. (1980a). *Nature (London)* **287,** 411–416.

Goeddel, D. V., Shepard, H. M., Yelverton, E., Leung, D. W., Crea, R., Sloma, A., and Pestka, S. (1980b). *Nucleic Acids Res.* **8,** 4057–4074.

Goeddel, D. V., Leung, D. W., Dull, T. J., Gross, M., Lawn, R. M., McCandliss, R., Seeburg, P. H., Ullrich, A., Yelverton, E., and Gray, P. W. (1981). *Nature (London)* **290,** 20–26.

Gray, P. W., and Goeddel, D. V. (1982). *Nature (London)* **298,** 859–863.

Gray, P. W., Leung, D. W., Pennica, D., Yelverton, E., Najarian, R., Simonsen, C. C., Derynck, R., Sherwood, P. J., Wallace, D. M., Berger, S. L., Levinson, A. D., and Goeddel, D. V. (1982). *Nature (London)* **295,** 503–508.

Hartley, J. L., and Donelson, J. E. (1980). *Nature (London)* **286,** 860–865.

Hitzeman, R. A., Hagie, F. E., Levine, H. L., Goeddel, D. V., Ammerer, G., and Hall, B. D. (1981). *Nature (London)* **293,** 717–722.

Hitzeman, R. A., Hagie, F. E., Hayflick, J. S., Chen, C. Y., Seeburg, P. H., and Derynck, R. (1982). *Nucleic Acids Res.* **10,** 7791–7808.

Hochkeppel, H. K., and de Ley, M. (1982). *Nature (London)* **296,** 238–239.

Houghton, M., Jackson, I. J., Porter, A. G., Doel, S. M., Catlin, G. H., Barber, C., and Carey, N. H. (1981). *Nucleic Acids Res.* **9,** 247–266.

Jones, E. (1970). *Genetics* **85,** 23–33.

Knight, E., Jr., Hunkapiller, M. W., Korant, B. D., Hardy, R. W. F., and Hood, L. E. (1980). *Science (Washington, D.C.)* **207,** 525–526.

Kuo, S.-C., and Yamamoto, S. (1975). *Methods Cell Biol.* **11,** 169–183.

Langford, M. P., Georgiades, J. A., Stanton, G. J., Dianzani, F., and Johnson, H. M. (1979). *Infect. Immun.* **26,** 36–41.

Lawn, R. M., Gross, M., Houck, C. M., Franke, A. E., Gray, P. W., and Goeddel, D. V. (1981a). *Proc. Natl. Acad. Sci. USA* **78,** 5435–5439.

Lawn, R. M., Adelman, J., Dull, T. J., Gross, M., Goeddel, D. V., and Ullrich, A. (1981b). *Science (Washington, D.C.)* **212,** 1159–1162.

Lawn, R. M., Adelman, J., Franke, A. E., Houck, C. M., Gross, M., Najarian, R., and Goeddel, D. V. (1981c). *Nucleic Acids Res.* **9,** 1045–1052.

Levy, W. P., Shively, J., Rubinstein, M., Del Valle, U., and Pestka, S. (1980). *Proc. Natl. Acad. Sci. USA* **77,** 5102–5104.

Nagata, S., Mantei, N., and Weissmann, C. (1980). *Nature (London)* **287,** 401–408.

Nathan, I., Groopman, J. E., Quan, S. G., Gersch, N., and Golde, D. W. (1981). *Nature (London)* **292,** 842–844.

Naylor, S., Sakaguchi, A. Y., Shows, T. B., Law, M. L., Goeddel, D. V., and Gray, P. W. (1982). *J. Exp. Med.,* in press.

Ohno, S., and Taniguchi, T. (1981). *Proc. Natl. Acad. Sci. USA* **78,** 5305–5309.

Owerbach, D., Rutter, W. J., Shows, T. B., Gray, P. W., Goeddel, D. V., and Lawn, R. M. (1981). *Proc. Natl. Acad. Sci. USA* **78,** 3123–3127.

Rubin, B. Y., and Gupta, S. L. (1980). *Proc. Natl. Acad. Sci. USA* **77,** 5928–5932.

Scopes, P. K. (1973). *In* "The Enzymes" (P. D. Boyer, ed.), Vol. 8, pp. 335–351. Academic Press, New York.

Sonnenfeld, G., Mandel, A. D., and Merigan, T. C. (1978). *Cell. Immunol.* **40,** 285–293.

Stein, S., Kenny, C., Friesen, H.-J., Shively, J., Del Valle, U., and Pestka, S. (1980). *Proc. Natl. Acad. Sci. USA* **77,** 5716–5719.

Stewart, W. E., II (1981). "The Interferon System" 2nd ed. Springer-Verlag, Berlin.

Sutcliffe, J. G. (1978). *Nucleic Acids Res.* **5,** 2721–2728.

Thomas, P. S. (1980). *Proc. Natl. Acad. Sci. USA* **77,** 5201–5205.

Tschumper, G., and Carbon, J. (1980). *Gene* **10,** 157–166.

Warner, J. R. (1982). *In* "The Molecular Biology of the Yeast Saccharomyces" (J. N. Strathern, E. W. Jones, and J. R. Broach, eds.), Vol. 2. Cold Spring Harbor Lab., Cold Spring Harbor, New York.

Weissmann, C. (1981). *Interferon* **3,** 101–134.

Yanofsky, C., Platt, T., Crawford, I. P., Nichols, B. P., Christie, G. E., Horowitz, H., Van Cleemput, M., and Wu, A. M. (1981). *Nucleic Acids Res.* **9,** 6647–6668.

Yelverton, E., Leung, D. W., Weck, P. W., and Goeddel, D. V. (1981). *Nucleic Acids Res.* **9,** 731–741.

Yip, Y. K., Pang, R. H. L., Urban, C., and Vilcek, J. (1981). *Proc. Natl. Acad. Sci. USA* **78,** 1601–1605.

Yip, Y. K., Barrowclough, B. S., Urban, C., and Vilcek, J. (1982). *Proc. Natl. Acad. Sci. USA* **79,** 1820–1824.

Zaret, K. S., and Sherman, F. (1982). *Cell* **28,** 563–573.

Zoon, K. C., Smith, M. E., Bridgen, P. J., Anfinsen, C. B., Hunkapiller, M. W., and Hood, L. E. (1980). *Science (Washington, D.C.)* **207,** 527–528.

CHAPTER 14

Commercial Production of Recombinant DNA-Derived Products

J. PAUL BURNETT

Molecular and Cell Biology Research
Lilly Research Laboratories
Indianapolis, Indiana

I. Introduction

The purpose of this chapter is to discuss some general concepts about applications of recombinant DNA technology in the pharmaceutical industry and to consider how some of the information contained in other chapters in this book may play an important role in pharmaceutical applications of this technology.

Some of the various steps necessary for production of a protein product via molecular cloning are diagrammed in Fig. 1. Once a DNA fragment coding for a particular protein is obtained, derived either from DNA synthesis or natural sources, the steps necessary to obtain expression of this gene in the chosen host cell can begin. As one begins to design the so-called expression vehicle, it is obviously extremely important to consider the fermentation and product-purifica-

Experimental Manipulation
of Gene Expression

ISBN 0-12-372380-9

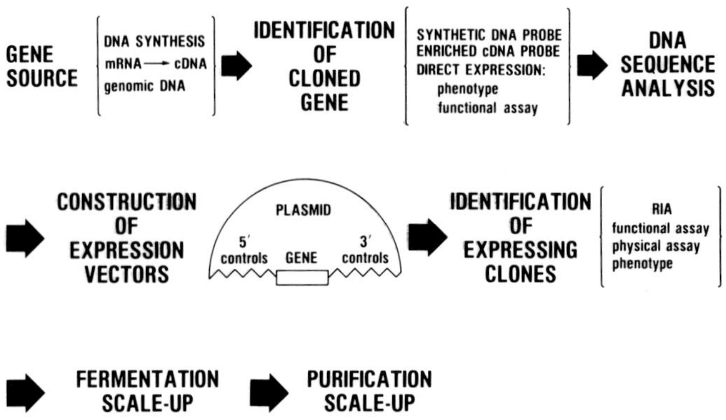

Fig. 1 Pathway to production of proteins by recombinant DNA technology.

tion scale-up steps that must be undertaken to produce a protein on large scale. The opportunity exists at the stage of expression-vehicle construction to design in features that will make these final production steps easier. Of course, what generally happens is that a cyclic path is followed between vehicle construction, fermentation scale-up, and isolation scale-up until the process is optimized.

Some of the types of product areas to which this technology might be applied in the pharmaceutical industry are listed in Table I. The first five categories represent areas in which proteins might have direct therapeutic application, whereas that of antibiotics represents a case in which the actual product is a result of secondary metabolism catalyzed by enzymatic proteins.

In Table II are listed a few of the proteins that are either known to have therapeutic utility (e.g., insulin or growth hormone) as well as some that can reasonably be expected to be useful therapeutically. Genes for a number of these polypeptides have already been cloned or synthesized. Insulin serves as a useful example for further discussion.

Table I Pharmaceutical Applications of Recombinant DNA

Hormones and growth factors
Analgesics
Plasma proteins
Enzymes
Immunology—vaccines, cytokines
Antibiotics

Table II Human Polypeptides Potentially Attractive for Biosynthesis

Polypeptide	Amino acid residues	Molecular weight
Insulin	51	5,734
Proinsulin	82	9,396
Growth hormone	191	22,005
Calcitonin	32	3,421
Glucagon	29	3,483
Corticotropin (ACTH)	39	4,567
Prolactin	198	22,900
Placental lactogen	192	22,400
Parathyroid hormone	84	9,562
Nerve growth factor	118	13,000
Epidermal growth factor		6,100
Insulin-like growth factors (IGF-1 and IGF-2)	70; 67	7,649; 7,471
Thymopoietin	49	—

II. Production of Biosynthetic Human Insulin

There are two possible pathways that could be used to produce insulin via fermentation (Fig. 2). Because insulin consists of two chains (A and B), the individual chains could be independently produced and coupled together through disulfide bridges to form insulin. Alternatively, one could produce proinsulin, the natural precursor of insulin, and enzymatically remove the connecting peptide to produce the final product. In collaboration with Genentech, Inc., both routes have been accomplished. Plasmids were constructed by David Goeddel and colleagues that, when transformed into *Escherichia coli,* led to the expression of either A chain, B chain, or proinsulin fused to an *E. coli* polypeptide. These plasmids have the general structure shown in Fig. 3. Into the plasmid pBR322 was inserted DNA coding for regulatory sequences and insulin polypeptides; the insulin gene segment was either a synthetic oligonucleotide coding for the A or B chain (which was synthesized by K. Itakura) or a proinsulin gene (partly synthetic and partly derived from cDNA). The regulatory sequences and coding region for the *E. coli* portion of the chimeric protein in the plasmid shown was derived from the tryptophan operon. This is actually the second generation of plasmids constructed. The first plasmids, the construction of which has been published (Goeddel *et al.*, 1979), used *E. coli* sequences from the *lac* operon to control expression of the foreign DNA. The overall result of switching to the *trp* system was a marked improvement in yield of insulin polypeptide. This result

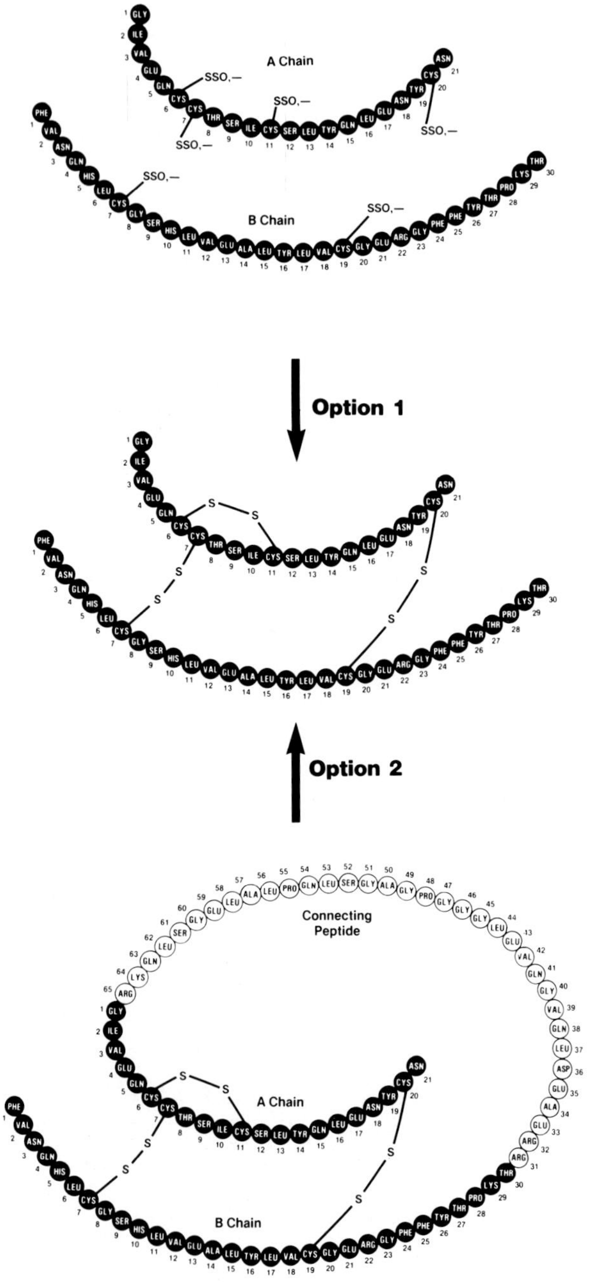

Fig. 2 Methods for producing human insulin by combination of A and B chains (Option 1) or from proinsulin (Option 2).

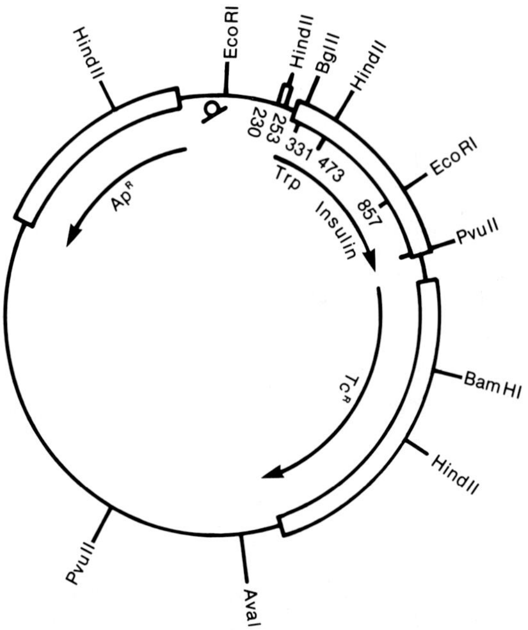

Fig. 3 Structure of insulin-coding plasmids. Ap^R, Ampicillin resistance; Tc^R, tetracycline resistance; *trp,* tryptophan operon 5′ regulatory sequences.

came about for several reasons. First, the *trp* promoter is known to belong to a class of strong promoters in *E. coli*—a class that also includes the P_L and lipoprotein (lpp) promoters described in Chapters 1 and 2. From studies with both insulin and growth hormone, we know that the tryptophan regulatory elements lead to at least 10-fold greater expression of polypeptide when compared to the *lac* system. Using the *lpp* regulatory region described by Masui *et al.* (Chapter 2), we have obtained expression of growth hormone at about the same level as that which we observed with the *trp* system. For these comparisons, the natural promoter and ribosome-binding-site sequences were used. Clearly, from work originating in a number of laboratories (Gold *et al.*, 1981), we also know that the coding sequence itself influences the overall level of expression of the final product, possibly through effects on mRNA secondary structure.

Figure 4 illustrates one other advantage of the *trp* system over the first-generation *lac* system. For an equivalent number of molecules of chimeric protein produced, the *trp* system results in a far greater percentage of protein as the desired product because the chimeric protein in the case of *trp* contains a much smaller contribution of *E. coli* polypeptide. It might be questioned why we did not just decrease the size of the *lac*-coding fragment used in order to achieve the same end. We did undertake several experiments in which we removed segments

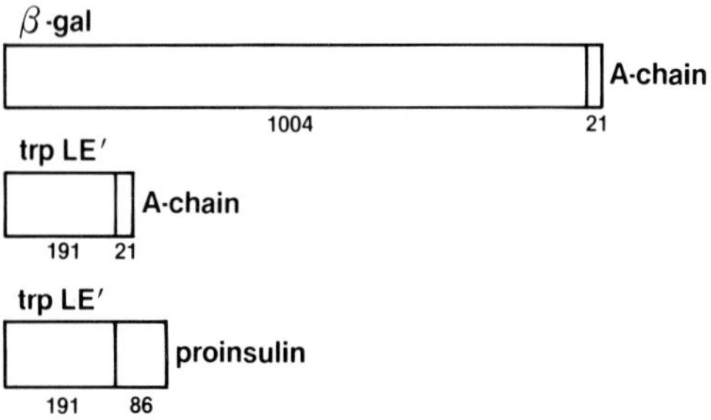

Fig. 4 Relative molecular size of various human insulin chimeric polypeptides. The numbers indicate the number of amino acids in each fragment.

of DNA coding for β-galactosidase polypeptide. In our hands, however, these plasmids resulted in even less final product, as measured in molecules per cell. This may have resulted from increased protein turnover owing to proteolysis. One of the features of both the large *lac* chimeric protein and the smaller *trp* chimeric protein is that both are highly insoluble. Both accumulate intracellularly as insoluble aggregates (Fig. 5), and this may reasonably be expected to decrease the rate at which they are degraded. Of course, this points out the importance of knowing more about the *E. coli* proteases and how they recognize proteins for degradation. We know, for instance, from the work of Fred Goldberg (Swamy and Goldberg, 1982) that *E. coli* has at least two proteases that efficiently degrade insulin polypeptides. We believe this partly explains why we are not able to recover significant yields of insulin polypeptides when these are produced by direct expression rather than as chimeric proteins. In contrast, human growth hormone seems to be remarkably stable in *E. coli.*

There are other possible solutions to the stability problem, if one exists for a particular protein. These include use of another host, such as yeast [as described by Derynck *et al.* (Chapter 13)] or *Bacillus,* or development of systems for export of the protein so that it is removed from accessibility to proteases. This could take the form of export from the cell entirely, such as can be obtained with yeast or *Bacillus,* or by export into the periplasmic space or outer membrane for *E. coli,* using vectors of the type described by Masui *et al.* (Chapter 2). There are pros and cons to be considered relative to secretion of protein products. The pros would include possible improvements in recovery and purification in addition to a possible reduction in degradation by proteases. Also, the potential physical limitations of intracellular space for accumulation of product could be overcome, thereby enhancing yield. The cons, however, include the necessity for recovery of protein from relatively dilute solutions, compared to the high concentration one can obtain from simple centrifugation of cells containing intracellular product. Here one must balance the relative difficulty of recovery from solution

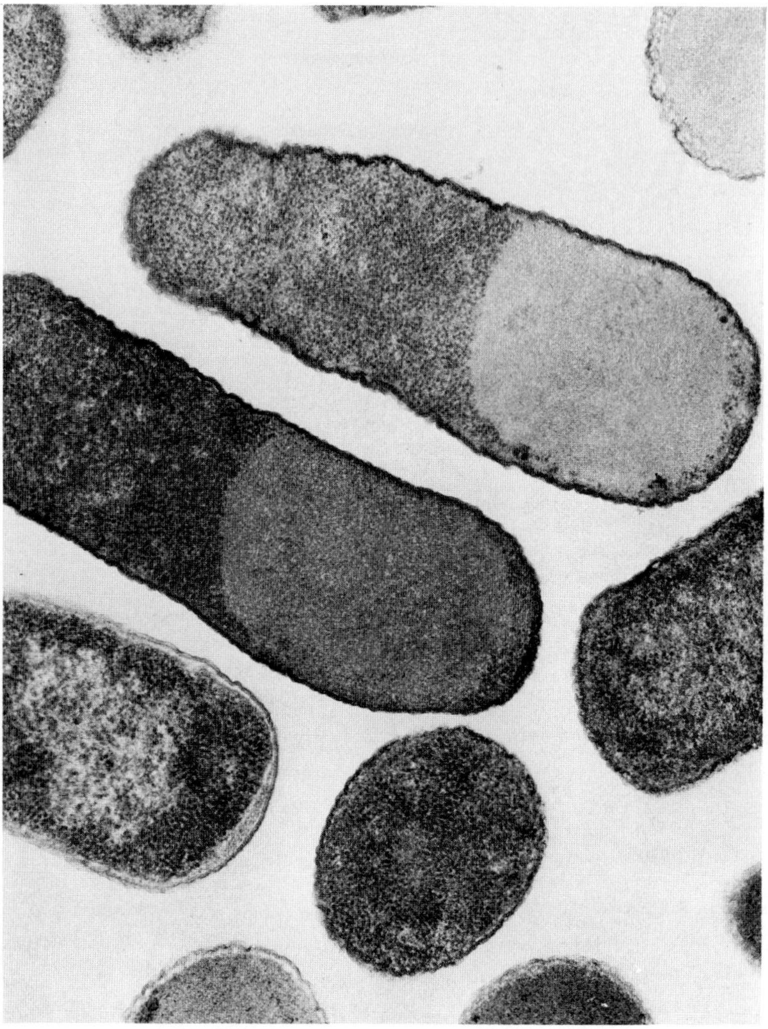

Fig. 5 Electron micrograph (thin section, magnification 44,500×) of *Escherichia coli* containing human insulin chimeric polypeptide.

versus that of cell breakage for recovery of intracellular product. Another problem that may be anticipated for certain proteins, following secretion from the cell, is denaturation at air–liquid interfaces. In commercial fermenters of large scale, air is forced into the fermenter at very high rates, and mechanical agitation is also very high. For easily denatured proteins, this would be a problem. The

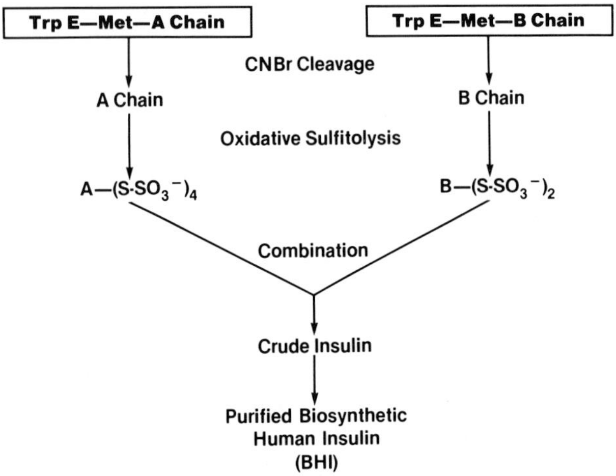

Fig. 6 Preparation of human insulin from A- and B-chain chimeric proteins.

size and amino acid sequence of the protein being produced and its resulting tertiary structure will determine whether it will be best to use a secretion or intracellular-accumulation system.

In the case of insulin, because it is produced as a chimeric protein, the first step in the recovery process is cleavage to release the insulin chain (Fig. 6). The individual chains are purified and then combined to yield the final product. Alternatively, we can produce insulin via proinsulin (Fig. 7). The construction of expression vehicles coding for proinsulin results in an improved process, because

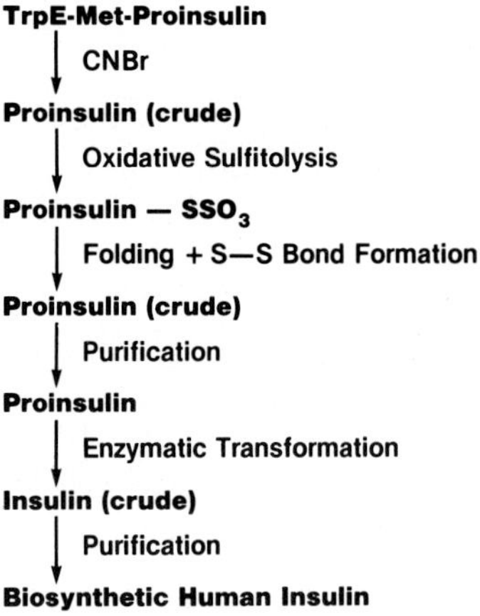

Fig. 7 Preparation of human insulin from proinsulin chimeric protein.

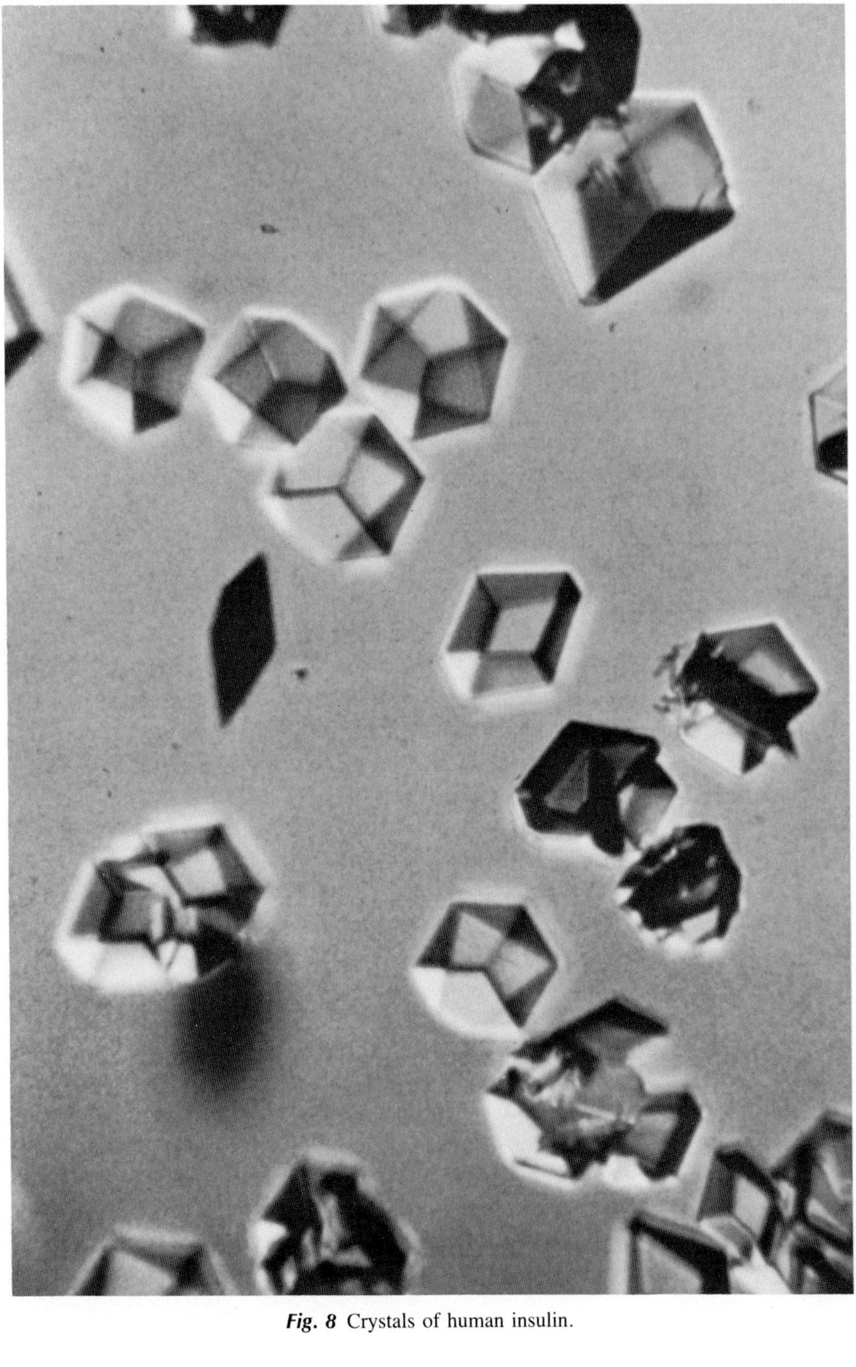

Fig. 8 Crystals of human insulin.

Table III Evaluation of Biosynthetic Human Insulin[a]

Test	Results
USP rabbit hypoglycemia assay (144 rabbits)	28.0 ± 2.2 units/mgm
Insulin radioimmunoassay	106 ± 10% of PHI standard
Insulin radioreceptorassay	96 ± 3% of PHI standard
Amino acid composition	Excellent
Gel electrophoresis	Excellent
UV and CD spectra	Identical to pork insulin standard
HPLC	Same retention time as PHI
Zinc crystallization	Excellent
Limulus assay for bacterial endotoxin	<0.1 ngm/mgm
USP rabbit pyrogen test	Nonpyrogenic
BP proteolytic activity assay	Satisfactory
Proinsulin radioimmunoassay	11.3 ppm
C-Peptide radioimmunoassay	<1 ppm
Escherichia coli peptide radioimmunoassay	<4 ppm

[a]BHI; derived from human proinsulin.

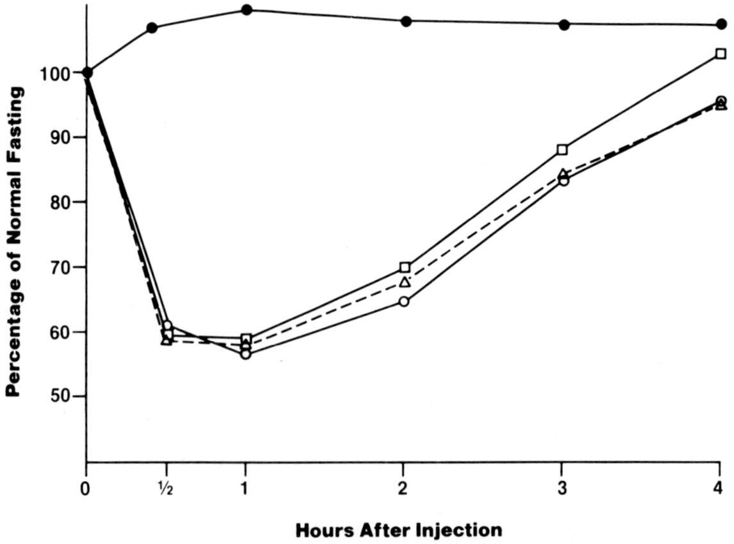

Fig. 9 Rabbit hypoglycemia test (0.2 unit/kgm) with human (□, pancreatic, $n = 12$; △, biosynthetic, $n = 24$) and porcine (○ , 27 μl/mgm, $n = 36$) insulin. Rabbits were injected with insulin at time 0, and blood glucose (ordinate; 72 ± *1* mgm %) was monitored over a 4-hr period. ●, Saline ($n = 24$). Average standard error of mean = ±20%.

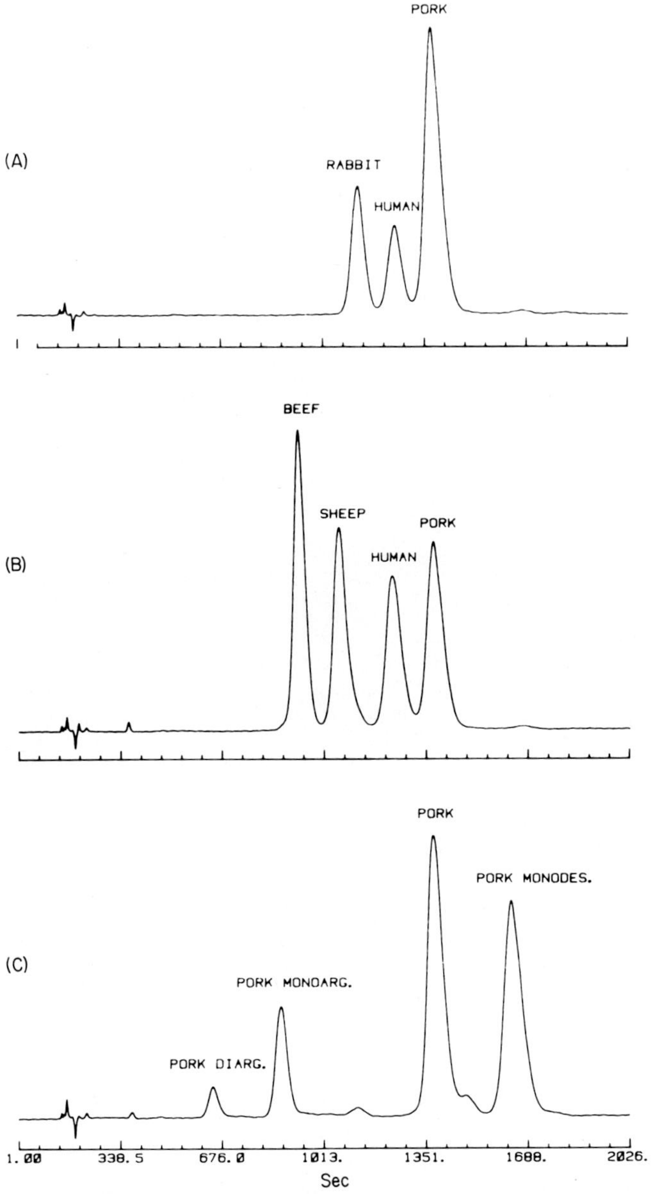

Fig. 10 High-performance liquid chromatography (HPLC) profiles of various species of insulin. PORK DIARG., PORK MONOARG., Arginine derivatives of insulin prepared by incomplete cleavage of proinsulin via trypsin and carboxypeptidase; PORK MONODES., monodesamido derivative of pork insulin.

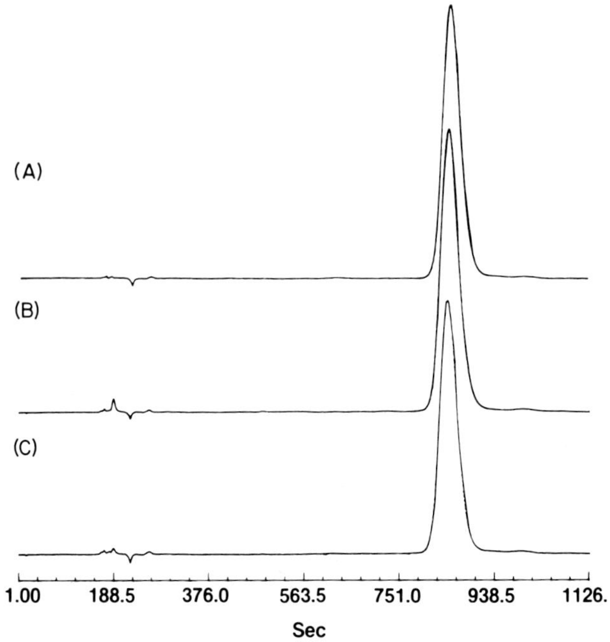

Fig. 11 HPLC profiles of human insulin. (A) Biosynthetic human insulin produced in *E. coli* via recombinant DNA technology. (B) Human insulin extracted from human pancreas glands. (C) A mixture of biosynthetic and pancreatic insulin.

only a single fermentation is required. The final product in either case is crystalline insulin (Fig. 8).

The characteristics of the final product are shown in Table III. There are two important features to be considered for a recombinant DNA-derived product: identity with the natural or designed product and purity. In addition to biological properties such as hypoglycemic effect (Fig. 9), radioimmunoassay, and radioreceptor assay, we have relied on the demonstration of correct amino acid composition and sequence as well as on analytical biochemical procedures, especially high-performance liquid chromatography (HPLC). Newer methods of HPLC allow one to separate insulin differing by only a single amino acid (Fig. 10). In addition, HPLC allows one to detect impurities present in trace amounts. For instance, with low column loads insulin would appear as shown in Fig. 11. At high column load (Fig. 12), trace impurities can be detected, isolated, and characterized. We know the ones shown here are carbamyl or formyl derivatives, as well as a trace of insulin dimers. One finds similar trace derivatives in glandular insulin isolated by similar protein-purification procedures.

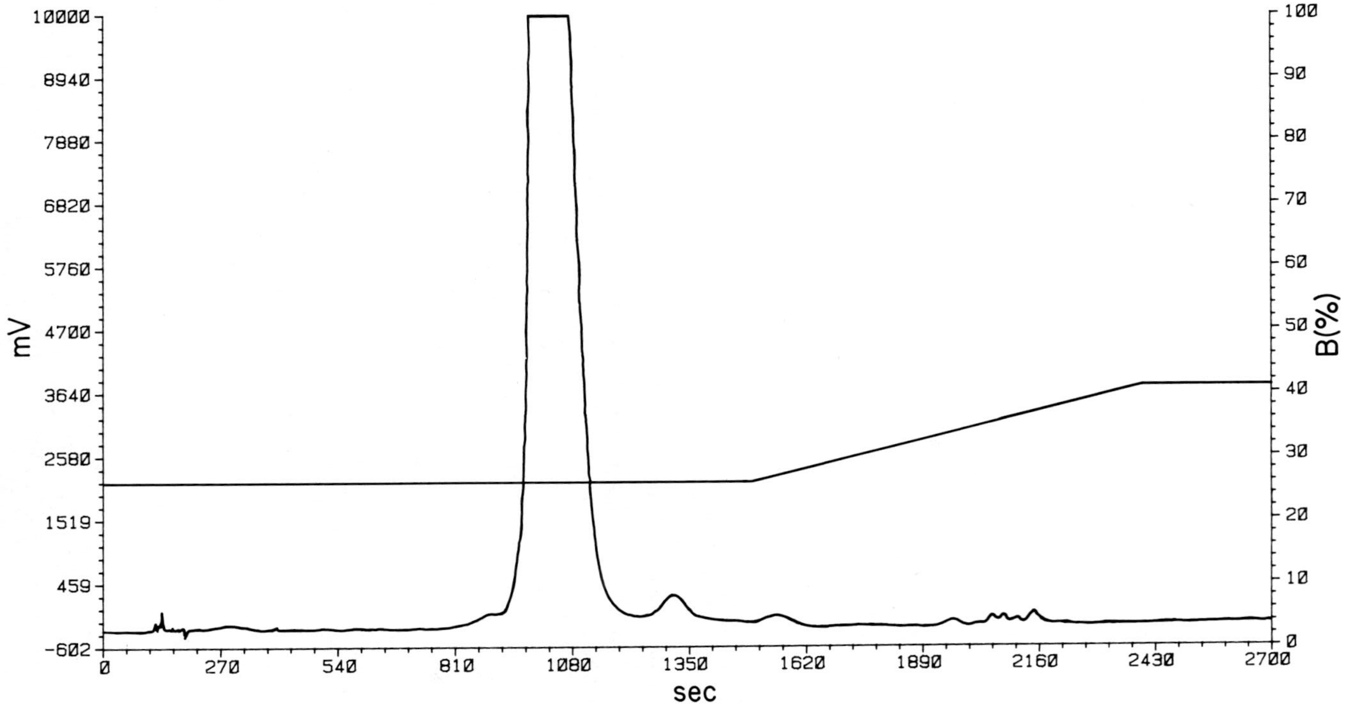

Fig. 12 HPLC profile of biosynthetic human insulin employing a very high sample concentration to expose trace impurities. Zorbax column eluted with sodium phosphate containing CH_3CN.

Table IV Lymphocyte-Derived Cytokines

Mediators affecting macrophages
Migration inhibitory factor (MIF)
Macrophage activating factor (MAF)
Chemotactic factor (CF)
Mediators affecting lymphocytes
Allogenic effect factor (AEF)
Mitogenic factor (MF)
Factors enhancing antibody formation
Factors suppressing antibody formation
T-cell replacing factor (TRF)
Chemotactic factors for basophils (BCFs) and eosinophils (ECFs)
Mediators affecting other cells
Cytotoxic factor (lymphotoxin)
Collagen-producing factor
Osteoclast-activating factor
Interferon (IF)

III. Other Pharmaceutical Applications of Recombinant DNA

I should like to close by briefly mentioning some of the other areas in the pharmaceutical industry in which recombinant DNA technology may be applied (Table I). The first of these is immunology. Here, certainly, new and improved vaccines are being made. However, potentially the most exciting area is that of the cytokines. This general class of proteins, which includes lymphokines such as the interferons, represent a group of biological messengers. Table IV lists some of these molecules that are known to be produced by lymphocytes. There are many cytokines that have only been identified on the basis of biological activity. Molecular cloning may afford us the possibility of producing these molecules in amounts sufficient to establish their biological roles and therapeutic potentials. Some of the diseases that might be likely targets for cytokine therapy, based on the fact that these molecules may play a role in inflammation, a helper function in some cases, or a supressor function, are shown in Table V.

The other area for application of recombinant DNA about which the pharmaceutical industry can be particularly enthusiastic is that of antibiotics. A typical antibiotic biosynthetic pathway is illustrated by that for tylosin (Fig. 13). These various biosynthetic steps were elucidated in our laboratories by Richard Baltz, using blocked mutants in the producing organism, *Streptomyces fradiae*.

Table V Clinical Application of Cytokines[a]

Factors affecting inflammation	Helper factors	Suppressor factors
Patients with overwhelming infection	Tumor patients	Allergy
Postsurgical immune suppression	Aging	Autoimmune disease, systemic lupus erythromatosus, arthritis
Burn patients	Diabetics	Multiple sclerosis, thyroiditis, myasthenia gravis, etc.
Postsurgical peritonitis	Dialysis patients	Transplantation
Tumor patients	Immunodeficiency disease	
Aging	Allergy	
Immunodeficiency disease	Natural killer-cell activity	
Dialysis patients	Burn patients	
Thyroiditis	Trauma	
Multiple sclerosis	Postsurgical immune suppression	
Allergy	Chronic diseases—hepatitis, parasitic disease	
	Hodgkins disease	

[a]Afflictions for which each of the three roles of cytokines may be of use are listed.

There are three important ways one could use recombinant DNA in conjunction with this type of information. First, yield improvement could be achieved by gene-dosage effects. Second, adding genes coding for particular enzymes to a producing organism might allow one to carry out *in vivo* modifications of the natural antibiotic; many of the most useful antibiotics are chemical derivatives of natural products. Third, combining genes from various antibiotic biosynthetic pathways in a single organism may lead to entirely new antibiotic structures.

One of the crucial steps necessary for any of these approaches is the development of a good vector system for use in producing organisms. Because *Streptomyces* accounts for approximately 80% of useful antibiotics, we have concentrated our vector development on this genus. Figure 14 shows the structure of a cryptic 20-kb plasmid isolated at Lilly Research Laboratories (Richardson *et al.*, 1982) from an antibiotic-producing strain. Because the plasmid had no identifiable genetic marker, Jeff Fayerman from our laboratories added one coding for thiostrepton resistance and subsequently removed a large amount of nonessential DNA to create pFJ105 (Fig. 15). This plasmid efficiently transforms *Streptomyces*. In order to make it even more useful, Fayerman next added pBR322 DNA and a DNA fragment coding for neomycin resistance, another useful marker in *Streptomyces* (Fig. 16). Plasmid pFJ123 now had two useful markers together with the ability to be more easily manipulated in *E. coli,* with subsequent return to *Streptomyces* for observation of expression of *Streptomyces* DNA sequences.

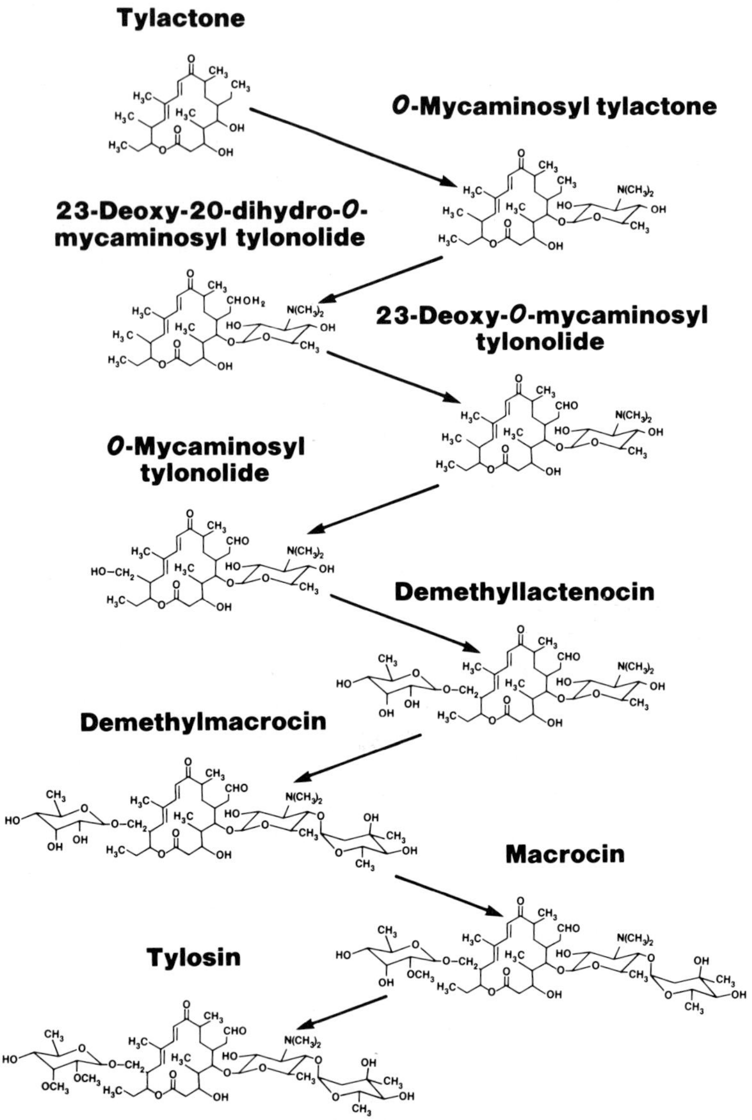

Fig. 13 Biosynthetic pathway for production of tylosin in *Streptomyces fradiae*.

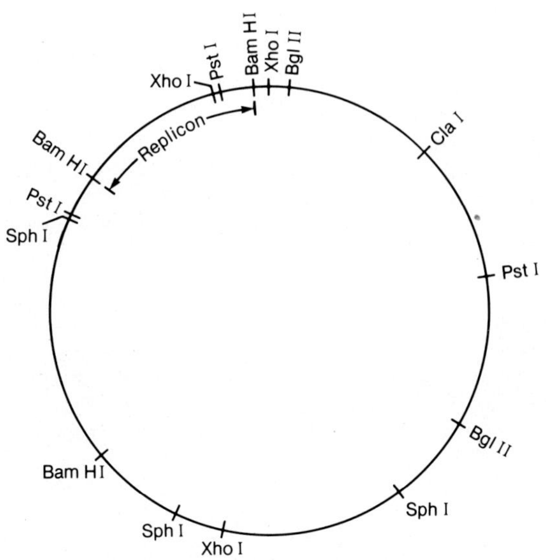

Fig. 14 Plasmid pFJ103 restriction map.

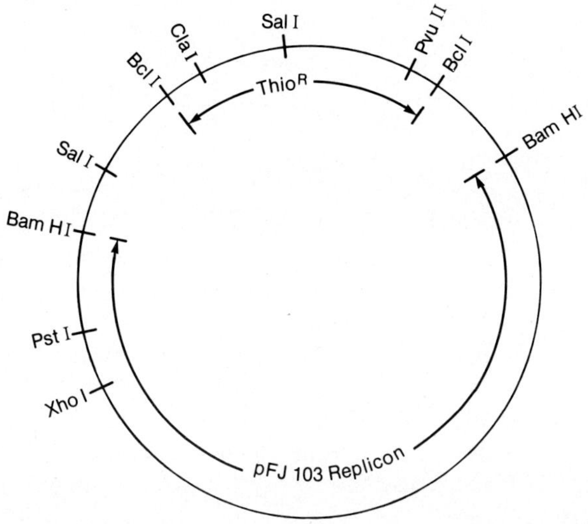

Fig. 15 Plasmid pFJ105 restriction map. $Thio^R$, Thiostrepton resistance.

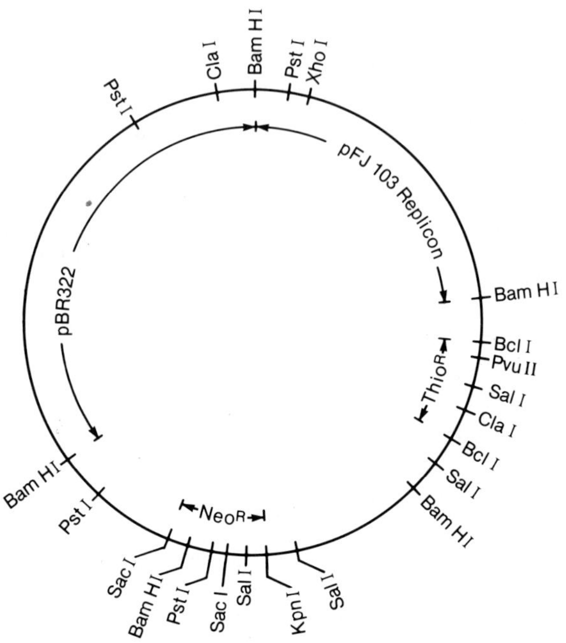

Fig. 16 Plasmid pFJ123 restriction map. $Thio^R$, Thiostrepton resistance; Neo^R, neomycin resistance; pBR322, DNA sequences derived from *E. coli* plasmid pBR322.

IV. Conclusion

From these few examples of potential applications of recombinant DNA in the pharmaceutical industry, it is readily apparent that the overall impact of this technology for our industry is enormous. It will allow the synthesis of many proteins heretofore difficult to obtain. In addition, however, it is important to keep in mind that the greatest benefit to the pharmaceutical industry may well be eventually derived from the fundamental understanding of biological systems that will come about from the use of these and other techniques of molecular biology. This knowledge can lead to the rational design of new therapeutically useful molecules. Many of these may be small chemical "drugs" synthesized by more traditional chemistry.

References

Goeddel, D. V., Kleid, D. G., Bolivar, F., Heyneken, H. L., Yansura, D. G., Crea, R., Hirose, T., Kraszewski, A., Itakura, K., and Riggs, A. D. (1979). *Proc. Natl. Acad. Sci USA* **76,** 106–110.

Gold, L., Pribnow, D., Schneider, T., Shinedling, S., Singer, B. S., and Stormo, G. (1981). *Annu. Rev. Microbiol.* **35,** 365–403.

Richardson, M. A., Mabe, J. A., Beerman, N. E., Nakatsukasa, W. M., and Fayerman, J. T. (1982). *Gene* **20,** 451–457.

Swamy, K. H. S., and Goldberg, F. (1982). *J. Bacteriol.* **149,** 1027–1033.

APPENDIX 1

Two-Dimensional DNA Electrophoretic Methods Utilizing *in Situ* Enzymatic Digestions

THOMAS YEE
MASAYORI INOUYE

Department of Biochemistry
State University of New York at Stony Brook
Stony Brook, New York

Experimental Manipulation
of Gene Expression

ISBN 0-12-372380-9

I. Introduction

In a gel matrix, the diffusion of DNA restriction fragments is retarded to a far greater extent than the diffusion of proteins of comparable or smaller molecular weight. This differential retardation stems from the rigid linear nature of double-stranded DNA. In agarose or acrylamide gels, DNA restriction fragments that are large relative to the pores of the gel matrix can become virtually immobilized, remaining susceptible to digestion by enzymes that can diffuse relatively freely. This is the basis for a variety of two-dimensional electrophoretic techniques that can be applied to DNA (Yee and Inouye, 1982). A gel strip containing a one-dimensional separation of DNA digested by some restriction enzyme is immersed in a solution of an appropriate second enzyme, which diffuses in and redigests the DNA fragments, which can then be separated in a second dimension.

In agarose gels, a proper choice of restriction enzymes for the first and second dimensions can resolve virtually an entire bacterial genome on a single gel. Two-dimensional digestions by isochizomers differing in methylation sensitivity can be used to probe an entire bacterial genome for methylation. Bacterial DNA digested by a 4-base restriction enzyme can be denatured, renatured, separated on an acrylamide gel, then redigested by S1 nuclease to reveal repetitive sequences including tandem and inverted repeats as well as other arrangements of duplicated DNA. DNA from two bacterial strains can be denatured and renatured together, in which case S1 nuclease can often reveal sites of mismatch heteroduplexing between the two genomes. In short, therefore, two-dimensional electrophoretic separations with *in situ* enzymatic redigestions make possible the enzymatic probing of total bacterial chromosomes to the level of individual restriction fragments (Yee and Inouye, 1982).

II. Experimental Procedures

A. Enzymes

Restriction enzymes were purchased from Bethesda Research Laboratories or from New England BioLabs. Digestions were conducted in the buffers recommended by the suppliers. S1 nuclease was purchased from Bethesda Research Laboratories or from Miles Laboratories. In this appendix, one restriction enzyme unit is defined as the amount of enzyme completely digesting 1 μgm of bacteriophage λ DNA in 1 hr in a 50-μl reaction volume, whereas one S1 nuclease unit is the amount acid solubilizing 1 μgm of denatured calf thymus DNA in 1 min at 37°C. It is assumed that the various 1- and 30-min S1 nuclease units and the 15-min and 1-hr restriction-enzyme units employed by different suppliers may be related by simple multiplication factors corresponding to the differences in assay times.

B. Two-Dimensional Agarose Electrophoresis

Total chromosomal restriction digests were electrophoresed using a 2 V/cm voltage gradient in vertical slab gels 170 mm square by 2.54 mm thick using 0.7% (w/v) agarose prepared in E buffer [40 m*M* Tris-acetate (pH 8), 20 m*M* sodium acetate, 2 m*M* EDTA]. Vertical slab gels were preferred over horizontal ones for their higher resolution, and also because thin gels are more easily poured, which conserved enzyme in the soak steps. The slots used were at least 10 mm in width. Both xylene cyanol and bromophenol blue were employed as tracking dyes. After electrophoresis, the central 7 mm of each lane was cut out, avoiding the retarded edges of the lanes. The cuts were centered between the well edges and the xylene cyanol; occasional lateral deviations in the running of the bromophenol blue were ignored. The strips were not stained, because ethidium bromide inhibits restriction enzymes; however, the trimmed edges were stained to determine the position of the chromosomal bands. The upper portion of the gel strip containing no DNA and the lower portions less than 2 kb were trimmed off.

The gel strips were presoaked for at least 3 hr in a large excess of the reaction buffer appropriate for the second-dimension enzyme, with a buffer change after 90 min.

To the tubes containing the strips was then added 1.5 strip volumes of reaction buffer containing 15 units/ml of restriction enzyme. The strips were soaked for 18–24 hr at 4°C, with occasional gentle agitation. After the soak period, the tubes are incubated for 4–6 hr at the temperature appropriate for the enzyme. Both the presoak and the enzyme soak should contain 100 μgm/ml of bovine serum albumin (Sigma A-4378). The albumin should be free of nucleases, but excessively expensive grades are not necessary.

After incubation, the gel strips were soaked at least 2 hr in a large excess of 0.2× E buffer containing a small amount of bromophenol blue.

The second-dimension agarose gels were prepared using 0.7% agarose prepared in 2× E buffer. The strips were laid down and sealed in place with 0.7% agarose prepared in 0.2× E buffer. A small amount of phage λ DNA digested with *Hin*dIII was optionally included as a molecular-weight marker. The second-dimension run was begun immediately after the sealing agarose solidified, using 2× E buffer in the upper and lower chambers. The 10-fold ionic strength difference between the strips and the second-dimension gel results in stacking and compression of the first-dimension bands into tight spots in the second dimension.

C. Two-Dimensional Polyacrylamide Electrophoresis for Study of Methylation

Electrophoresis of 40 μgm of total chromosomal DNA digested with *Hpa*II was conducted at a voltage gradient of 18 V/cm on vertical slab gels made using 5% (w/v) acrylamide prepared in TBE buffer [50 m*M* Tris-borate (pH 8.3), 1 m*M* EDTA]. The slots employed were 7 mm in width, but they can be up to 16 mm in width without excessive loss of resolution. Enzyme diffusion rates are lower in acrylamide than in agarose. We originally used 1.27-mm-thick acrylamide gels and soak times identical with the times employed in two-dimensional agarose electrophoresis. If 2.54-mm-thick acrylamide gels are desired, we recommend increasing all soak times by 50%.

Because only negligible amounts of edge retardation occur in acrylamide electrophoresis, we cut out the entire width of the lanes. The upper portions of the gel strips containing no DNA bands and the lower portions with bands of material of less than 100 bp were trimmed off.

Except for increased soak times as noted, the strips were treated exactly as described for two-dimensional agarose electrophoresis, except that postsoaking was in 0.2× TBE buffer. Acrylamide has a tendency to stick to test-tube walls, lowering soak efficiency; the soaking solutions were forced between the strips and the test-tube walls so that the strips floated freely and were bathed on all sides. The enzyme soak was with 50 units/ml of *Msp*I. Because acrylamide swells significantly during the soaks and incubations, the second-dimension gels should be poured between spacers that are 25% thicker than the first-dimension spacers.

D. Two-Dimensional S1 Nuclease Heteroduplex Mapping

Total chromosomal DNA restriction digests from two strains of bacteria presumed to differ by a chromosomal rearrangement were heat denatured in hybridization buffer consisting of 50% formamide, 100 m*M* Tris-HCl (pH 8.5), and 10 m*M* EDTA and were allowed to renature at room temperature. Three hybridization reactions were conducted: two homoduplex reactions and one heteroduplex reaction in which equal amounts of DNA from each bacterial strain were mixed together prior to heat denaturation. Each reaction volume of 60 μl contained 80 μgm of restriction digest. Hybridization was allowed to proceed to a C_0t of

200–400 (up to 24 hr). Excessively long hybridizations give poor results, possibly because the formamide solution drifts towards acidity with increasing time. Another possibility is that heteroduplexes may be unstable, and excessive renaturation times may result in their loss.

Four-base-pair-sequence-recognition restriction enzymes were used for the chromosomal digests. Even small amounts of partial digestion severely degraded the two-dimensional heteroduplex analysis, so we routinely digested to at least a fivefold excess. In addition, certain 4-base enzymes were incapable of complete digestion, even after prolonged reactions with large amounts of enzyme; this was presumably due to partial methylation of their recognition sites. *Taq*I, for example, was unsuitable for use with any strain of *Escherichia coli* that we tested, whereas *Msp*I was marginally suitable with certain strains, but gave excellent results with others. DNA from stationary-phase cultures appeared to be more difficult to cut to completion than DNA from cells harvested from earlier stages. DNA that had been badly sheared in isolation gave higher backgrounds than DNA carefully isolated with minimum shear.

The hybridization mixtures were loaded directly into 16-mm-wide wells of a 170 mm square by 2.54 mm thick vertical slab gel made with 5% (w/v) acrylamide, as described in Section II,C. After the gel run, the entire 16 mm width of each strip was chopped out with a long knife. The strips were trimmed off at the bottom to eliminate low-molecular-weight bands, but the tops were retained.

Presoaking and enzyme soaks were conducted in a buffer consisting of 300 m*M* NaCl, 30 m*M* sodium acetate (pH 4.6), 4.5 m*M* $ZnCl_2$, and 100 μgm/ml BSA (Shenk *et al.*, 1975). S1 nuclease was used at a concentration of 70 units/ml. Enzyme digestions were conducted at 37°C for 4 hr and were stopped by soaking the strips for one hr in 2× TBE, with shaking. The strips were then soaked in 0.2× TBE and run in a second-dimension 2× TBE gel as described in Section II,C.

III. Examples

A. Two-Dimensional Separation of the Entire *Escherichia coli* Genome

Figures 1–5 illustrate different applications of the various two-dimensional electrophoretic methods described here. Figure 1 shows an agarose two-dimensional double restriction enzyme separation of *E. coli* DNA, in which *Bam*HI was employed in the first dimension and *Bst*EII in the second dimension. Nearly

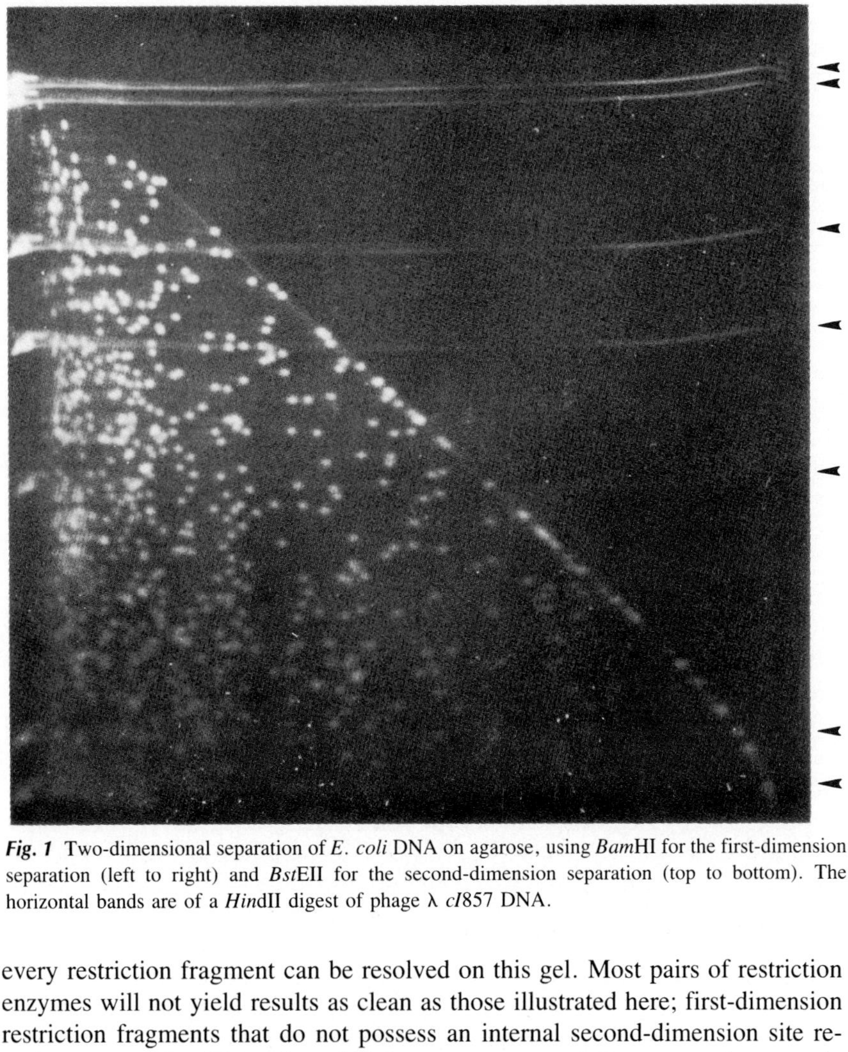

Fig. 1 Two-dimensional separation of *E. coli* DNA on agarose, using *Bam*HI for the first-dimension separation (left to right) and *Bst*EII for the second-dimension separation (top to bottom). The horizontal bands are of a *Hin*dII digest of phage λ *cI*857 DNA.

every restriction fragment can be resolved on this gel. Most pairs of restriction enzymes will not yield results as clean as those illustrated here; first-dimension restriction fragments that do not possess an internal second-dimension site remain uncut on the diagonal, and if there are many such fragments, they remain unresolved. To achieve nearly complete resolution as shown here, it is necessary to use a first-dimension enzyme that cuts the genome into a minimum number of separable pieces and a second-dimension enzyme that cuts frequently enough such that few first-dimension fragments remain on the diagonal, but not so frequently as to generate an excessive number of fragments. The *Bam*HI and *Bst*EII employed here cut *E. coli* DNA to average molecular sizes of 10.7 and 3.8 kb, respectively (Yee and Inouye, 1982). For other species of bacteria, different pairs of enzymes prove to be optimal.

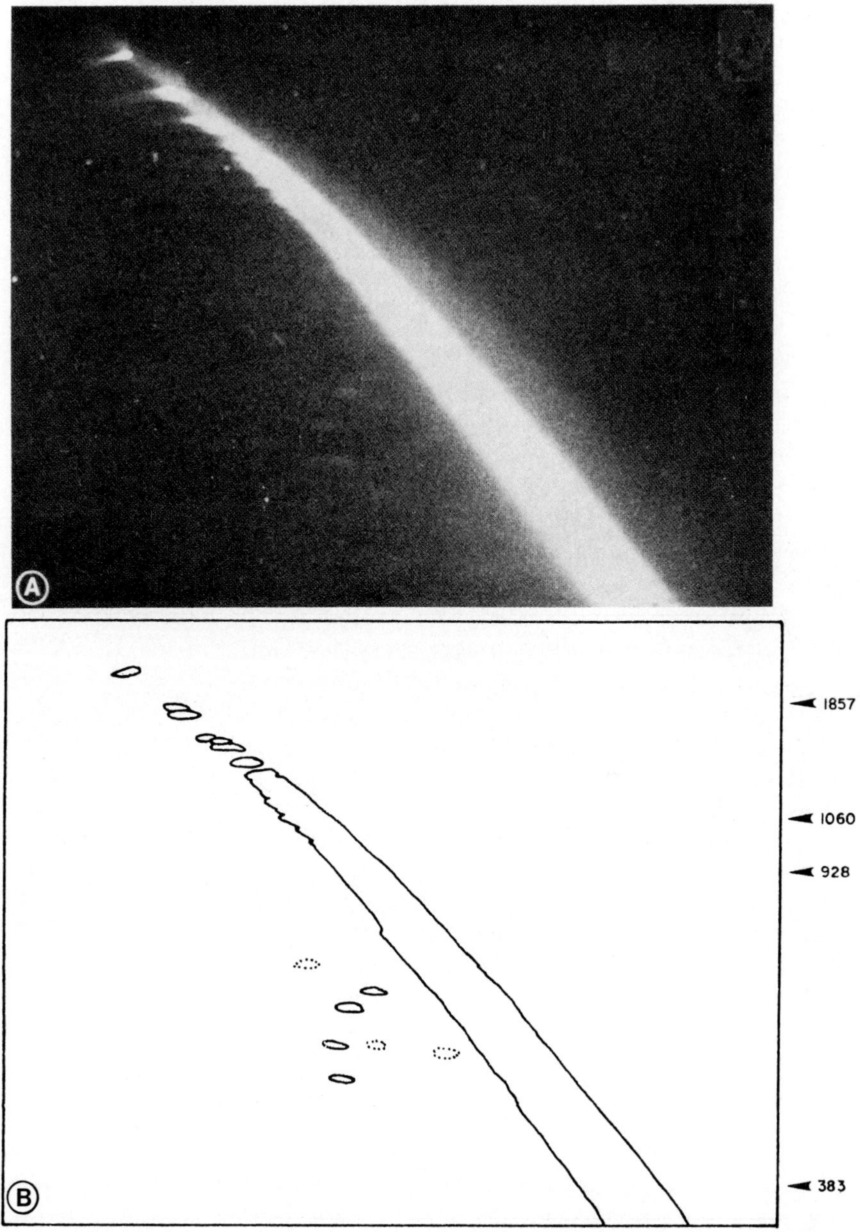

Fig. 2 (A) Methylation pattern of *Myxococcus xanthus* DNA extracted from developmental cells (22 hr growth on starvation agar). The first-dimension digest was with *Hpa*II, and the second-dimension digest was with *Msp*I. (B) Tracing of (A). The molecular-weight markings on the right are from a *Bst*NI digest of pBR322 DNA.

B. Two-Dimensional Separation of Methylated DNA Fragments

Figure 2 illustrates a methylation study that we conducted on DNA from *Myxococcus xanthus* (Yee and Inouye, 1982). *Hpa*II-digested DNA from developmental cells was subjected to acrylamide electrophoresis, redigested with *Msp*I, and run in a second dimension. The amount of DNA employed, 40 μgm, was such that if any *Hpa*II sites were completely methylated at the internal cytosine of its recognition sequence (CCGG), recutting by *Msp*I should have resulted in bright spots below the diagonal. Instead, we observe in Fig. 2A faint spots, some of which are at the limit of detectability. Figure 2B is a tracing of Fig. 2A in which we have indicated by dotted outlines the positions of several additional spots that may not be distinct in the halftone reproduction, but the existence of which we have confirmed in independent experiments. The interesting conclusion we reached was that, during development, *M. xanthus* DNA becomes methylated at highly specific *Hpa*II sites, but that only a fraction of the population of DNA molecules become methylated at any particular site. Developmental cells of *M. xanthus* have various fates: some are destined to become spores, whereas by far the greater proportion remain vegetative and eventually lyse. The heterogeneity that we observed in DNA methylation may reflect this heterogeneity in the population of developmental cells.

C. Two-Dimensional S1 Nuclease Heteroduplex Mapping

Figures 3 and 4 illustrate an early version of two-dimensional S1 nuclease heteroduplex mapping applied to a model system. This early version employed 6-base-recognition restriction enzymes and agarose electrophoresis rather than 4-base-recognition restriction enzymes and acrylamide electrophoresis. Although we have abandoned the use of 6-base enzymes because of insurmountable background problems, the model system illustrates quite well the basic principles involved. Plasmids pAM008 and pAM009, obtained from Pamela Green (State University of New York at Stony Brook), are 10.2-kb plasmids differing only in the orientation of a 257-bp insert. As indicated in Fig. 3, we linearized the plasmids with *Sal*I, mixed the two, denatured, and renatured. The renatured mixture was subjected to agarose electrophoresis, and the gel strips were then digested with S1 nuclease and run in a second dimension. Three major spots were predicted: a 10.2-kb spot resulting from homoduplex annealing and 5.9- and 4.0-kb spots resulting from S1 digestion of the heteroduplex. Additional minor spots can result, for instance, from further annealing of the single-strand regions of the heteroduplexes, annealing via the 257-bp inserts, and so on, but

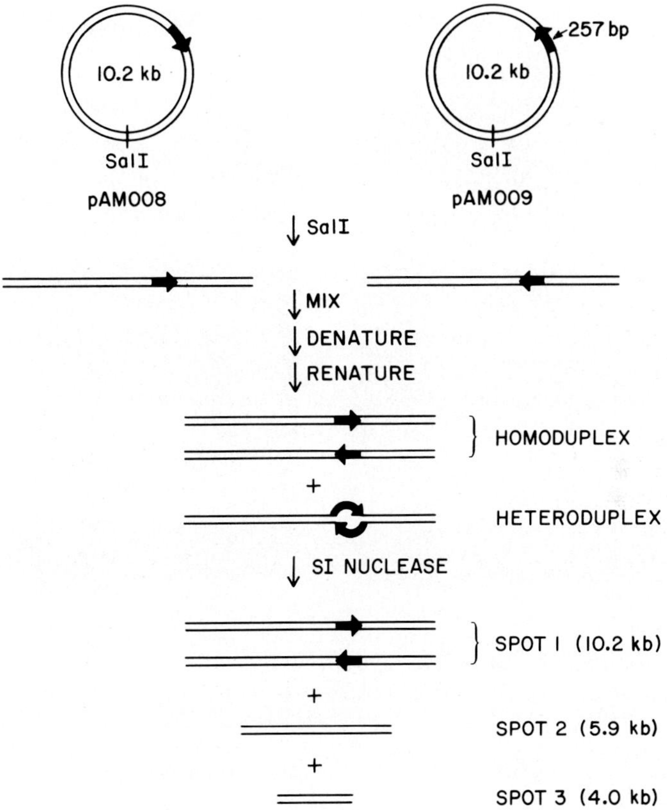

Fig. 3 Model system for detection of a phase variation. Spots 1, 2, and 3 correspond to those indicated in Fig. 4.

these are not illustrated. Figure 4A shows the plasmid mixture subjected to two-dimensional S1 nuclease heteroduplex mapping, and Fig. 4B shows the result of mixing the plasmid with *E. coli* total chromosomal DNA prior to denaturation. The 10.2-kb spot in this case is obscured by the vast excess of total chromosomal DNA, but the 5.9- and 4.0-kb spots are clearly visible beneath the diagonal. In Fig. 4A (but not Fig. 4B) are visible faint secondary spots resulting from the more complex annealings to which we just alluded.

Figure 5 illustrates the present acrylamide method of two-dimensional S1 nuclease heteroduplex mapping applied to a second, more realistic model system. We obtained from Michael Silverman (University of California, San Diego) two derivatives of *E. coli* in which the *Salmonella* phase-variation region has been inserted into the *E. coli* genome via λ lysogeny. The *hin* gene within the insert, necessary for switching, was inactivated by an insertion of a fragment of

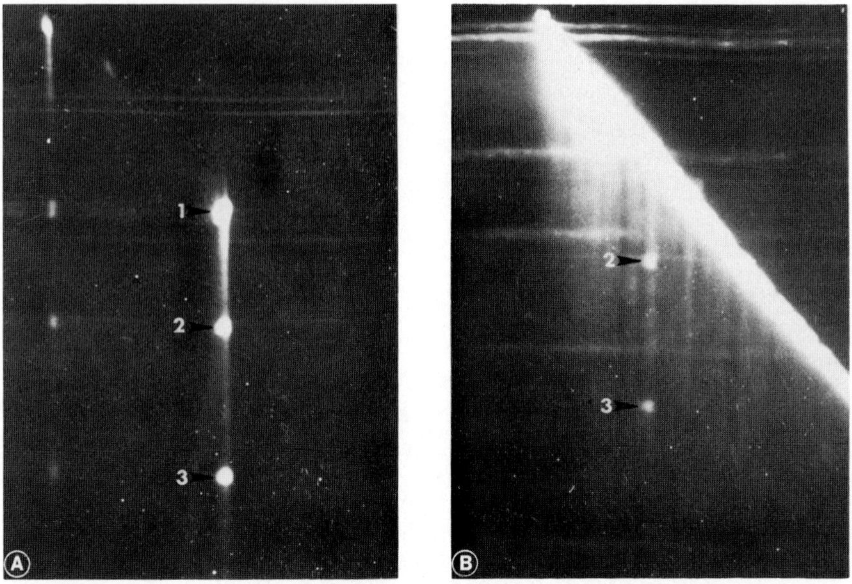

Fig. 4 Two-dimensional S1 electrophoresis model system using 6-base restriction enzymes and agarose electrophoresis. Plasmids pAM008 and pAM009 (both cut with *Sal*I) and *E. coli* total chromosomal DNA (cut with *Eco*RI) were mixed together, heat denatured, and renatured. (A) Plasmids pAM008 and pAM009, mixed together without any *E. coli* total chromosomal DNA. Spot 1 consists of pAM008 and pAM009 homoduplexes not susceptible to S1. Spots 2 and 3 represent the long and short arms of the pAM008 × pAM009 heteroduplexes after digestion by S1. (B) Full model system consisting of pAM008, pAM009, and *E. coli* DNA. The total amount of DNA loaded on the gel was 3.5 μgm.

Tn5. The two derivatives differ from each other only in that in one the phase-variation region has been fixed in an ''on'' orientation, expressing the H2 gene, whereas in the other the phase-variation region has been fixed into the ''off'' orientation. Figure 5 shows a heteroduplex mixture of equal amounts of DNA from the two strains. Below the bright diagonal streak are visible many spots of varying intensity that result from the presence within the genome of various repetitive sequences. The bright spot indicated by the arrow, however, is unique to the heteroduplex mixture. The restriction enzyme employed for this heteroduplex analysis was *Hha*I. When *Hin*fI was employed, no unique heteroduplex spots were seen. The fragments generated by *Hin*fI may have been too small, or the heteroduplex mismatches may have been too near the ends of large fragments such that insufficient size change resulted from the S1 digestion to drop the spots off the diagonal. In general, it is not possible to predict in advance which 4-base restriction enzyme will generate appropriately sized and oriented restriction frag-

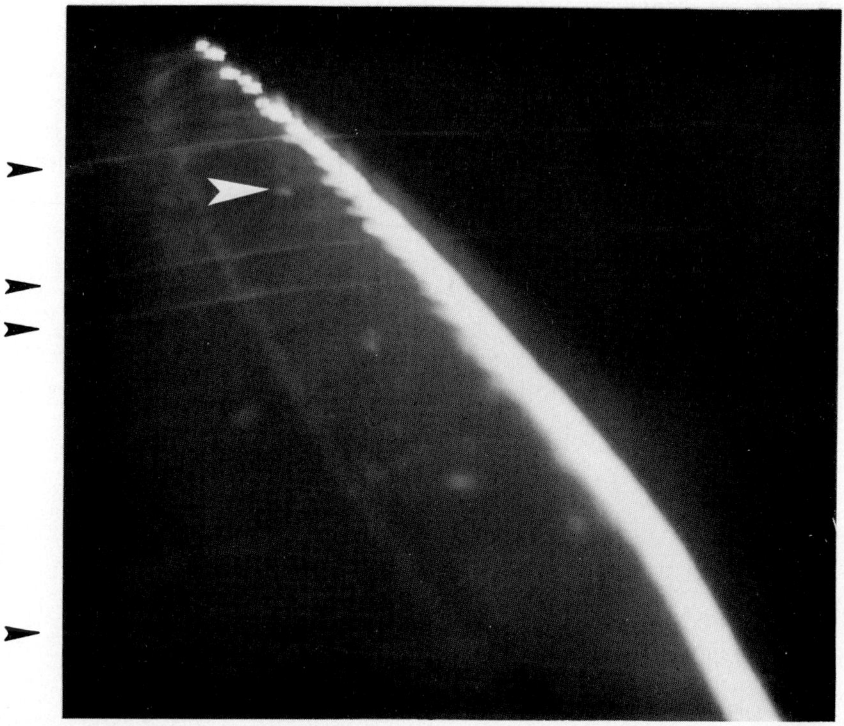

Fig. 5 Two-dimensional S1 electrophoresis model system using 4-base restriction enzymes and acrylamide electrophoresis. *Hha*I digests of two *E. coli* strains lysogenic for λ *fla378* carrying the *Salmonella* phase-variation region fixed in the ''on'' and ''off'' orientations were mixed together, heat denatured, renatured, and subjected to two-dimensional S1 heteroduplex analysis as described in the text. The white arrow points to a spot that appears only when ''on'' and ''off'' DNA are mixed together; it does not appear when either DNA is homoduplexed with itself. The horizontal bands are of a *Bst*NI digest of pBR322 DNA.

ments for visualizing an arbitrary heteroduplex arrangement. Therefore, several must be tried at random.

IV. Conclusion

The various two-dimensional electrophoretic methods discussed here are potentially very powerful tools for the analysis of structural features of total bacteria chromosomes. Using two-dimensional agarose electrophoresis, for example, we definitively settled a controversy concerning the genome size of *Myxococcus xanthus* by the simple expedient of counting all the spots (Yee and Inouye, 1982). As mentioned previously, we observed methylation changes associated with this organism's developmental cycle. Using two-dimensional S1 nuclease heteroduplex mapping, we have observed DNA rearrangements associated with phase variation in this organism (T. Yee and M. Inouye, unpublished observations).

The basic principle common to these methods of two-dimensional electrophoresis is the virtual immobilization of DNA within a gel matrix during enzymatic digestion. Doubtless other variants of two-dimensional electrophoresis are possible employing this same principle.

It remains to be seen whether further improvements in sensitivity and lowered backgrounds are possible that will allow scaling up these procedures for the analysis of small eukaryotic genomes.

References

Shenk, T. E., Rhodes, C., Rigby, P. W., and Berg, P. (1975). *Proc. Natl. Acad. Sci. USA* **72,** 989–993.

Yee, T., and Inouye, M. (1982). *J. Mol. Biol.* **154,** 181–196.

APPENDIX 2

Site-Specific Mutagenesis Using Synthetic Oligodeoxyribonucleotides as Mutagens

GEORGE P. VLASUK
SUMIKO INOUYE

Department of Biochemistry
State Univeristy of New York at Stony Brook
Stony Brook, New York

Experimental Manipulation
of Gene Expression

ISBN 0-12-372380-9

I. Introduction

The ability to create specific *in vitro* mutations within a known DNA sequence has proven to be a valuable tool in a variety of genetic studies. The most versatile and specific form of site-specific mutagenesis is oligodeoxyribonucleotide-directed mutagenesis, first developed to alter the DNA sequence of the single-stranded DNA phage ϕX174 (Razin *et al.*, 1978; Hutchison *et al.*, 1978; Gillam and Smith, 1979a,b). This method involves the use of a short, synthetic oligodeoxyribonucleotide, which acts as a primer of *in vitro* DNA synthesis on a complementary single-stranded circular DNA template. Following transformation of competent cells with the resulting heteroduplex containing the incorporated oligodeoxyribonucleotide carrying the appropriate mutation, *in vivo* semiconservative replication resolves the heteroduplex into homoduplexes derived from mutant and parental strands. The versatility of this method has been greatly enhanced by use of double-stranded recombinant plasmids as the source of the single-stranded templates and by use of improved colony-hybridization techniques employing the synthetic oligodeoxyribonucleotide mutagen as the specific ^{32}P probe, allowing the isolation of site-directed mutants without the use of phenotypic screening (Wallace *et al.*, 1980).

Mutagenesis through synthetic oligodeoxyribonucleotides has been used to introduce deletions, insertions, transitions, and transversions at specific sites within a given DNA sequence (Gillam and Smith, 1979a,b; Gillam *et al.*, 1979, 1980; Kudo *et al.*, 1981; Montell *et al.*, 1982; Simons *et al.*, 1982). In addition, the use of this method in the creation of mutant proteins with specific amino acid changes has allowed precise investigations of structure–function relationships (Inouye *et al.*, 1982; Charles *et al.*, 1982; Dalbadie-McFarland *et al.*, 1982). In this regard, we have used oligodeoxyribonucleotide-directed mutagenesis to create specific mutations within the signal-peptide region of the major outer-membrane lipoprotein of *Escherichia coli* to elucidate the function of this N-terminal extension in the secretion of this lipo-protein across the cytoplasmic membrane (Inouye *et al.*, 1982, 1983; Vlasuk *et al.*, 1983). These experiments require that an inducible system be used to express the mutant lipoprotein genes, because it is expected that many of the mutations are lethal to the cell. Therefore, by using the wild-type lipoprotein gene inserted into an inducible double-stranded plasmid, we were able to mutagenize and analyze the biological effects of the mutation using the same plasmid vector, thereby eliminating extensive manipulation of the mutagenized gene.

The methods described here detail the rapid construction and isolation of mutants using synthetic oligodeoxyribonucleotides and double-stranded DNA plasmids. In addition, an example of this method in the construction of a lipoprotein mutant is also given.

II. Experimental Procedures

A. Purification of Synthetic Oligodeoxyribonucleotides

Following synthesis and complete removal of all protecting groups, the synthetic oligodeoxyribonucleotide was purified by polyacrylamide gel electrophoresis. Prior to electrophoresis, however, the DNA was partially purified using one of two methods: thin-layer chromatography (TLC; developed by the Biologicals Co.) or DEAE chromatography on a minicolumn. These steps remove many contaminants (e.g., monodeoxyribonucleotides, remaining protecting groups, etc.) that may interfere with gel purification.

1. THIN-LAYER CHROMATOGRAPHY

The dry DNA was dissolved in 300–500 μl of 20% ethanol prior to application on the TLC plate. The sample (100 μl/TLC plate) was applied with a micropipet to the TLC plate (silica gel 60 F_{254}, 20 × 20 cm with plastic backing; E. Merck, Darmstadt), 1.5 cm from the bottom of the plate in a straight line 14 cm long. The sample should be applied in one application, even if the application must be continued over portions still wet with sample. The sample was allowed to air dry (do *not* use a blower). Standards (if available) may be applied next to the sample in a line approximately 5–6 cm long.

The TLC plate was developed in a tightly sealed chromatography tank for 12–16 hr at room temperature in a solvent composed of *n*-propanal–NH_4OH–water 55:35:10 (v/v/v). This solvent is suitable for resolving oligodeoxyribonucleotide sequences 7–16 units long. After development, the plate was removed and air dried. The DNA was visualized on the dried TLC plate using shortwave ultraviolet light. In general, the desired product should be the slowest running band, which is usually the most intense band apart from the origin. The migration of oligodeoxyribonucleotides in this system is dependent mostly on the guanosine content and length (i.e., increasing guanosine content results in slower migration). Therefore, the use of standard oligodeoxyribonucleotides of known length and guanosine content should be run concomitantly with the sample to allow unambiguous identification. Separation of synthetic oligodeoxyribonucleotides containing mixed bases is more difficult in this system; therefore, care should be taken.

After the appropriate band was located on the TLC plate, it was scraped off and collected in a sterile, siliconized 1.5-ml Eppendorf tube. The DNA was eluted by filling the tube (1 ml) with 20% ethanol, vortexing, and incubating at room temperature for 15 min. The tube was vortexed again and centrifuged for 3 min. The supernatant containing the oligodeoxyribonucleotide was collected in a siliconized glass test tube and the polymer support washed again as before. The combined supernatants from the two washes were frozen and lyophilized overnight. Generally, the oligodeoxyribonucleotide was more than 85% pure after this stage. The dried TLC-purified oligodeoxyribonucleotide was used directly for polyacrylamide gel purification (see Section II,A,3).

2. MINI-DEAE CHROMATOGRAPHY

The dry, deprotected DNA was dissolved in 100 μl of 0.1× TE buffer [1× TE = 10 m*M* Tris-HCl (pH 7.5), 1 m*M* EDTA] and loaded onto a mini-DE52 cellulose column (1.0 cm high in a siliconized Pasteur pipet supported on siliconized glass wool) that was equilibrated with 0.2 *M* triethylammonium bicarbonate, pH 8.0 (TEAB; made from distilled triethylamine). The DNA was washed with 3–4 ml of 0.2 *M* TEAB prior to elution with 1 ml of 2 *M* TEAB into a siliconized glass test tube. The eluted DNA was quantified spectrophotometrically. The yield should be greater than 90%. Following quantification, the eluted DNA must be diluted 10 times with sterile, distilled water prior to freezing and lyophilization. The degree of purity obtained from this step is much less than the TLC method described previously. However, owing to the time involved (which takes a day less than the TLC method), it may be preferable.

3. POLYACRYLAMIDE GEL PURIFICATION

Following lyophilization, the partially purified synthetic DNA (100–200 μgm) was dissolved in 24 μl of 7 *M* urea and applied in a well 3.1 cm wide and 1.5 cm deep on a 0.075 × 14 × 17 cm polyacrylamide gel containing 20% polyacrylamide (40:1.3 = acrylamide : methylene bisacrylamide), 7 *M* urea, 100 m*M* Tris-borate (pH 8.3), 2 m*M* EDTA, and 0.1% (w/v) ammonium persulfate. A dye mixture containing 0.025% (w/v) bromophenol blue, 0.025% (w/v) xylene cyanol FF, and 7 *M* urea was loaded in a well adjacent to the sample-containing well. The gel was prerun at 400 V (constant voltage) for 20 min prior to the loading of the sample. Electrophoresis of the sample was conducted at the same voltage for 2–3 hr. In this type of gel, the bromophenol blue migrates as a single-stranded oligomer of 5–6 bases and xylene cyanol FF as an oligomer of 20 bases.

Following electrophoresis, the DNA was visualized by placing the gel on a 20 × 20 cm TLC plate containing fluorescent indicator covered with plastic wrap. The gel was also covered with plastic wrap and exposed to shortwave ultraviolet light. The exposure should only be long enough to visualize and mark the appropriately sized band, which appears dark on a bright green background. The band was cut out, placed in a siliconized glass homogenizer, and crushed with a siliconized glass pestle in 1 ml of elution buffer containing 0.2 *M* TEAB and 1 m*M* EDTA. The gel slurry was transferred to a sterile, siliconized 1.5-ml Eppendorf tube and the homogenizer rinsed with 0.5 ml of the elution buffer. The wash was added to the gel slurry and the mixture incubated overnight at 37°C. Following the incubation, the gel fragments were pelleted by centrifugation and the supernatant removed. The pellet was washed with 0.25 ml of elution buffer and the combined supernatants applied to a mini-DEAE column as described previously. The DNA was eluted with 1 ml of 2 *M* TEAB, quantified, and lyophilized. Generally, the yield of full-length oligomer using this gel elution procedure is 60–70%. The lyophilized DNA was resuspended in sterile 0.1× TE buffer to a concentration of 20 pmol/μl.

B. Preparation of Oligodeoxyribonucleotide Primer

First, 10 μci of [γ-^{32}P]ATP (3000 ci/mmol; synthesized according to Walseth and Johnson, 1979) were dried down in a siliconized 1.5-ml Eppendorf tube. Second, 800 pmol of oligomer was phosphorylated in this tube in 100 μl of 66 m*M* Tris-HCl (pH 7.5), 10 m*M* $MgCl_2$, 10 m*M* β-mercaptoethanol, 2400 pmol of cold ATP (three times the oligomer content), and 12 units of T4 polynucleotide kinase (New England Nuclear, 8 units/μl). The reaction mixture was incubated at 37°C for 60 min, followed by chilling on ice. The mixture was loaded on a 10-ml Sephadex G-50 superfine column (in a 10-ml disposable plastic pipet supported on siliconized glass wool) that had been equilibrated with 50 m*M* TEAB buffer. The elution of the phosphorylated oligomer with 50 m*M* TEAB was monitored with a hand-held Geiger counter. The first peak of radioactivity containing the phosphorylated oligomer was collected in a siliconized glass test tube. A sample was counted using liquid scintillation and the purity checked by PEI–TLC in a 0.75 *M* potassium phosphate buffer (pH 3.5) system. In this system, the phosphorylated oligomer does not migrate appreciably from the origin, whereas unreacted [γ-^{32}P]ATP and ^{32}Pi, which migrate much faster, can be clearly separated. The phosphorylated oligomer was frozen at −70°C for 20 min and lyophilized. The dried primer was resuspended in sterile 0.1× TE to a concentration of 10 pmol/μl and stored at −20°C until use.

C. Preparation of Oligodeoxyribonucleotide Hybridization Probe

First, 1000 μci of [γ-^{32}P]ATP were dried down in a siliconized 1.5-ml Eppendorf tube. Second, 100 pmol of oligomer were phosphorylated in this tube in 30 μl of 66 m*M* Tris-HC1 (pH 7.5), 10 m*M* $MgCl_2$, 10 m*M* β-mercaptoethanol, and 8 units of T4 polynucleotide kinase. The reaction mixture was incubated at 37°C for 60 min, followed by purification and characterization of the labeled oligomer as described previously for the preparation of the primer. A yield of ~1 × 10^9 cpm was generally obtained. The labeled probe *not* lyophilized was stored at −20°C after adding EDTA to a final concentration of 1 m*M*.

D. Preparation of Open Circular and Single-Stranded DNA

First, 5 μgm of closed circular plasmid DNA (pKEN125; Nakamura *et al.*, 1982) were nicked in one strand by 200 units of restriction endonuclease *Eco*RI (Bethesda Research Laboratories; 10 units/μl) in 0.5 ml of 50 m*M* Tris-HC1 (pH 7.5), 10 m*M* NaCl, 10 m*M* $MgCl_2$, 1 m*M* dithiothreitol, 0.1 mg/ml bovine serum albumin, and 0.15 mgm/ml ethidium bromide for 90 min at 37°C in the dark. The reaction mixture was brought to 10 m*M* EDTA and incubated at 60°C for 5 min. Then, 10 μl of sample were removed and checked by agarose-gel electrophoresis on 0.7% agarose in 40 m*M* Tris-acetate (pH 8.0) and 2 m*M* EDTA. These conditions generally produced 75–80% open circular DNA and 15–20% linear DNA. A restriction site close to the area of mutagenesis should be used to assure that exonuclease III treatment produces an appropriately sized single-stranded region. The remaining DNA was extracted twice with phenol saturated with TE buffer (first extraction 1 volume, second ½ volume) and three times with diethyl ether (2 volumes). The open circular DNA was precipitated by adding 1/10 volume of 3 *M* $NaOCOCH_3$, 1/40 volume of 1 *M* $MgCl_2$, and 2½ volumes of 100% ethanol, followed by freezing at −70°C for 10 min and centrifugation. The DNA was precipitated again in ½ volume of 0.3 *M* $NaOCOCH_3$ and 2½ volumes of 100% ethanol and dried by lyophilization.

The dried open circular DNA was resuspended in 40 μl of exonuclease III buffer [66 m*M* Tris-HC1 (pH 8.0), 0.66 m*M* $MgCl_2$, and 1 m*M* β-mercaptoethanol]. Digestion with 50 units of exonuclease III (BioLabs; 1500 units/μl) was conducted for 90 min at 37°C. The reaction was terminated by bringing the reaction mixture to 10 m*M* EDTA, followed by incubation at 60°C for 5 min. The partially single-stranded DNA was extracted with phenol and ether, recovered by ethanol precipitation, and dried as before. The DNA should be kept dry at −20°C until use.

E. Annealing of Synthetic Oligodeoxyribonucleotide Primer and Heteroduplex Formation

The dried exonuclease III–treated DNA was resuspended in 50 μl of polymerase/ligase buffer [200 m*M* NaCl, 13 m*M* Tris-HCl (pH 7.5), 9 m*M* $MgCl_2$, and 20 m*M* β-mercaptoethanol]. This mixture plus 90 pmol of oligodeoxyribonucleotide primer was heated to 100°C for 3 min, followed by rapid cooling in ice and incubation at 4°C for 2 hr. To the mixture were added deoxyribonucleoside triphosphates (333 μ*M* each), ATP (833 μ*M*), 5 units of the Klenow fragment of DNA polymerase I (Boehringer–Mannheim; 5 units/μl), and 10 units of T4 DNA ligase (Bethesda Research Laboratories; 2.5 units/μl), for a final reaction volume of 90 μl. The reaction was allowed to proceed overnight at 12.5°C. Generally, half of the reaction mixture was used to transform competent *E. coli* strain JA221 (lpp^-/F′ $lacI^q$ $lacZ^+$ $lacY^+$ $proA^+$ $pro B^+$). Approximately 1.6×10^4 Amp^R transformants were obtained.

F. Selection of Transformants

Initial selection of transformants carrying the appropriate mutation was accomplished using a modified colony-hybridization procedure described by Holland *et al.* (1979). Transformants (500–1000 per plate) spread on L-broth agar plates supplemented with 50 μgm/ml ampicillin (LA-50) were transferred to sterile 3 MM Whatman chromatography paper disks (8 cm in diameter, appropriately marked to allow the proper orientation with the replica plate) by firmly placing the paper disk on the transformants, followed by a gentle peeling off of the colonies from the plate. The transferred colonies were placed face up on a fresh LA-50 plate and allowed to grow overnight at 37°C, as was the original plate on which the colonies regrew, forming the replica plate. Colonies were fixed to the paper disk by using the following procedure:

1. Soaking for 7 min in 15 ml of 0.5 *N* NaOH
2. Neutralization with 20 ml of 0.5 *M* Tris-HCl (pH 7.5) for 3.5 min
3. Treatment with 25 ml of 1× SSC (0.15 *M* NaCl, 0.015 *M* sodium citrate, pH 7.2) for 3.5 min with gentle swirling
4. Vigorous swirling in 50 ml of 100% ethanol for 7 min

Treatment of individual disks was in glass petri dishes, and the liquid was removed after each step by aspiration. Each step was repeated twice for individual filters. Following the ethanol wash, the paper disks were air dried on a sheet of absorbant paper at room temperature for 0.5–1 hr, followed by baking at 80°C *in vacuo* for 2.5 hr. Paper disks were kept under desiccation until use. This method of paper-disk preparation was found to be optimal for reducing excessive spreading of colonies.

Hybridization with the ^{32}P-labeled oligodeoxyribonucleotide probe was done in a solution (3.0 ml/filter) containing 5× Denhardt's solution (Denhardt, 1966), 10% Dextran sulfate (Pharmacia), 0.5% SDS, 6× NET [1× NET = 150 m*M* NaCl, 15 m*M* Tris-HCl (pH 7.5), and 1 m*M* EDTA], and ~0.5 pmol of ^{32}P-labeled oligodeoxyribonucleotide (8×10^6 cpm/pmol). Hybridization times were 16–18 hr at 4°C below the theoretical temperature calculated according to the rules described by Suggs *et al.* (1981) for synthetic oligodeoxyribonucleotide hybridization.

Following hybridization, the filters were washed in 400–500 ml of 6× SSC for three 20-min washes at room temperature (25°C), followed by two 10-min washes in 100% ethanol and air drying. Positive colonies were visualized by autoradiography using Kodak XAR-5 film for 2-3 hr at room temperature with a Cronex intensifier screen (DuPont). This method of hybridization and washing was found to be optimal for distinguishing mutant from wild-type sequences using 16 different oligodeoxyribonucleotide mutagens carrying single-base changes, deletions, and insertion mutations.

Positive colonies in the area of radioactivity were selected from the replica plate, and plasmid DNA was extracted from 0.5-ml cultures as described by Birnboim and Doly (1979). Individual colonies resulting from the transformation of JA221 *lpp*$^-$/*lacI*^{q} with the plasmid preparation were streaked on LA-50 plates in an orderly array, followed by overnight incubation at 37°C. Thirty of these colonies were picked onto square 3 MM chromatography paper, grown overnight at 37°C on LA-50 plates, and submitted to colony hybridization as described previously. Several of the positive colonies resulting from the second transformation were selected; from these plasmid DNA was isolated from a 100-ml culture as described by Nakamura and Inouye (1981). The purified plasmid DNA was submitted to DNA sequencing as described by Inouye *et al.* (1982).

III. Example

We have used the method described here (Fig. 1) to isolate many site-specific mutations within the signal sequence of the lipoprotein precursor prolipoprotein. As an example, the construction and isolation of one of these mutants (I-5) will be discussed. This mutation, which is at the N-terminus of the prolipoprotein signal peptide, alters the charge of this region from +2 to +1. The charge of this region is thought to play an important role in the secretion of lipoprotein across the bacterial cytoplasmic membrane (Inouye and Halegoua, 1980).

The synthetic oligodeoxyribonucleotide-5 (5′ GCTACTAATCTGGTA 3′) has a 1-bp mismatch in the fifth codon for lysine (−16) in the prolipoprotein signal sequence (Fig. 2). Oligodeoxyribonucleotide-directed mutagenesis of plasmid pKEN125 [an isopropyl-β-D-thiogalactoside (IPTG) inducible plasmid carrying the wild-type lipoprotein gene (Nakamura *et al.*, 1982)] changed lysine (−16) to asparagine (Fig. 2), which altered the charge at the N-terminus of prolipoprotein from +2 to +1 (mutant I-5).

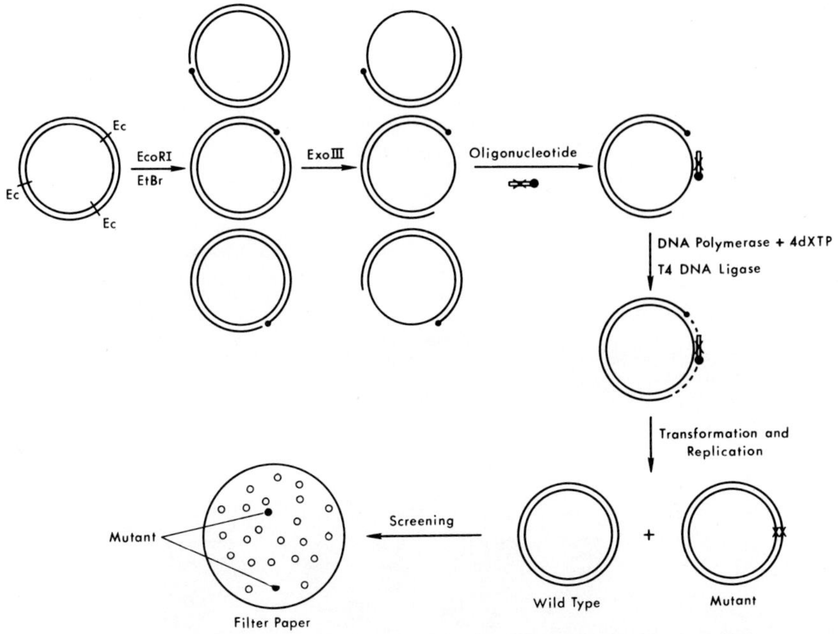

Fig. 1 Schematic diagram depicting the steps involved in oligodeoxyribonucleotide-directed mutagenesis of a closed circular double-stranded DNA plasmid. Closed circular plasmid containing three *Eco*RI restriction sites is nicked in one strand with *Eco*RI in the presence of ethidium bromide (EtBr). The open circular plasmid with exposed 5′ (●) and 3′ termini is made partially single stranded with the 3′-specific exonuclease III. The single-stranded region to be mutagenized is annealed with the synthetic oligodeoxyribonucleotide primer, which carries the mutation (X). Heteroduplex formation via *in vitro* DNA synthesis primed in part by the oligomeric primer is catalyzed by DNA polymerase and DNA ligase in the presence of the 4-deoxyribonucleoside triphosphates (dXTP). Transformation of competent cells with the heteroduplex mixture followed by *in vivo* replication resolves the heteroduplex into homoduplexes derived from mutant and parental strands. Identification of colonies carrying the appropriate mutation is accomplished through colony hybridization, using the synthetic oligodeoxyribonucleotide mutagen as the ^{32}P-labeled probe.

```
                  T
  5′ GCTACTAA  /    \  CTGGTA 3′        Oligodeoxyribonucleotide primer
TACTTTCGATGATT \    /  GACCAT. . . .    Complementary wild-type sequence
                  T

        −20              −15
H2N—MetLysAlaThr Lys LeuVal. . . .      Wild-type amino acid sequence
                  ↓
H2N—MetLysAlaThr Asn LeuVal. . . .      Mutant amino acid sequence
```

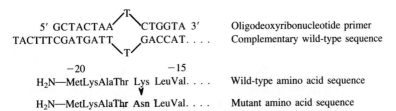

Fig. 2 Nucleotide sequences of the synthetic oligodeoxyribonucleotide mutagen and the complementary wild-type strand. The corresponding amino acid sequences encoded by the mutant and wild-type sequences are also shown. These amino acids are the first six in the prolipoprotein signal sequence.

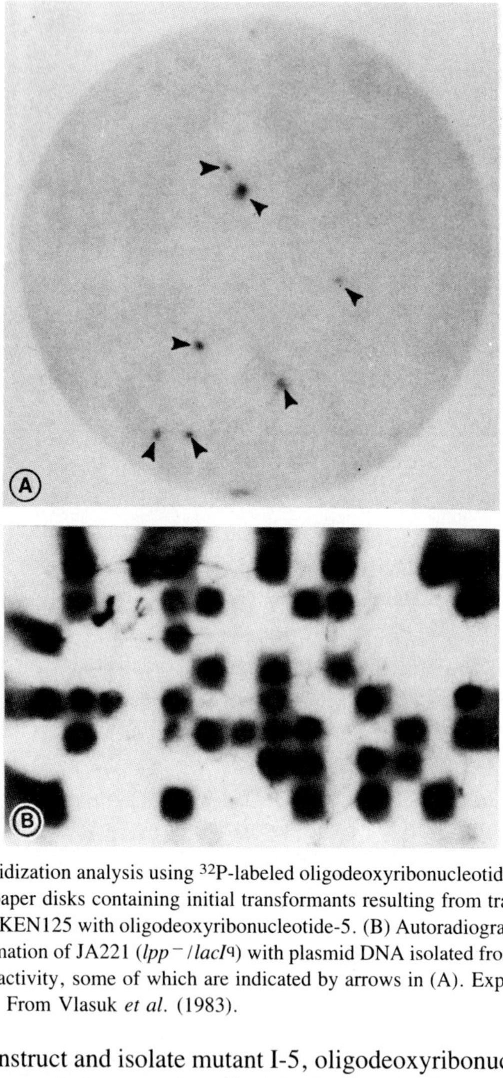

Fig. 3 Colony-hybridization analysis using ^{32}P-labeled oligodeoxyribonucleotide-5 as the probe. (A) Autoradiogram of paper disks containing initial transformants resulting from transformation with *in vitro* mutagenized pKEN125 with oligodeoxyribonucleotide-5. (B) Autoradiogram of colonies resulting from retransformation of JA221 (*lpp*$^{-}$/*lacI*^{q}) with plasmid DNA isolated from colonies withing a single area of radioactivity, some of which are indicated by arrows in (A). Experimental details are given in Section II. From Vlasuk *et al.* (1983).

To rapidly construct and isolate mutant I-5, oligodeoxyribonucleotide-directed mutagenesis was carried out as described in Section II. Selection of mutant I-5 from the high background of wild-type transformants that resulted from using the partially single-stranded template described previously was accomplished using a highly sensitive colony-hybridization procedure (Section II). Thousands of initial transformants were screened with this method using the ^{32}P-labeled oligodeoxyribonucleotide-5 as the probe. Figure 3A shows a typical autoradiograph of a filter containing ~800 transformants. The dark areas of radioactivity (marked by arrows), depicting colonies carrying the mutant plasmid, stand out clearly against the low background of radioactivity. We have found that the screening of

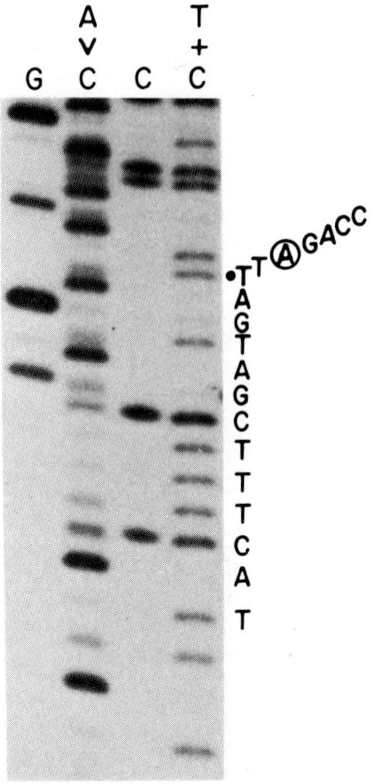

Fig. 4 DNA sequence analysis of the *lpp* gene in the region corresponding to the N-terminus of prolipoprotein for mutant I-5. Plasmid DNA from the mutant was digested with *Xba*I, labeled with [^{32}P]deoxyribonucleoside triphosphates and cleaved with *Hin*fI as described by Inouye *et al.* (1982). The cleavage products from the *Xba*I/*Hin*fI fragment were separated on a 20% polyacrylamide gel by electrophoresis and visualized by autoradiography. Only the regions from the initiation codon to leucine (−15) in the signal sequence are shown. The sequence at the side of the gel is complementary to the sense sequence ATGAA AGCTACTAA(T)CTGG. The circled base is the one changed from wild type using oligodeoxynucleotide-5. From Vlasuk *et al.* (1983).

mutants in which deletions or insertions were incorporated gives a much stronger signal over background. However, as can be seen in Fig. 3A, a 1-base change can easily be differentiated from the wild-type sequence. Colonies in the area of radioactivity from the first transformation (marked by arrows) were selected and plasmid DNA isolated using the mini-preparation method of Birnboim and Doly (1979). Retransformation of JA221 (*lpp*$^-$/F′ *lacI*^{q}) with this plasmid preparation followed by rescreening using colony hybridization assured the removal of any wild-type segregants (Fig. 3B). Several of the positive mutant colonies were chosen from this group of retransformants, from which plasmid DNA was purified on a large scale. The pure plasmid was sequenced using the chemical method of Maxam and Gilbert (1980) to demonstrate unambiguously the presence of the correct mutation in the expected position (Fig. 4).

IV. Conclusion

The method of oligodeoxyribonucleotide-directed mutagenesis described here has been used to successfully construct 15 site-specific mutations within the lipoprotein gene coding for the signal sequence (Inouye *et al.,* 1982, 1983; Vlasuk *et al.,* 1983). These mutations include 1-base changes, 3-base deletions, and 3-base insertions. It has proven to be a powerful and versatile tool in the study of protein secretion in prokaryotes, and it will surely be valuable in the rapid construction of many new mutant proteins, from which previously unavailable information can now be obtained.

References

Birnboim, H. C., and Doly, J. (1979). *Nucleic Acids Res.* **7,** 1513–1523.

Charles, A. D., Gautier, A. E., Edge, M. D., and Knowles, J. R. (1982). *J. Biol. Chem.* **257,** 7930–7932.

Dalbadie-McFarland, G., Cohen, L. W., Riggs, A. D., Morin, C., Itakura, K., and Richards, J. H. (1982). *Proc. Natl. Acad. Sci. USA* **79,** 6409–6413.

Denhardt, D. T. (1966). *Biochem. Biophys. Res. Commun.* **23,** 641.

Gillam, S., and Smith, M. (1979a). *Gene* **8,** 81–87.

Gillam, S., and Smith, M. (1979b). *Gene* **8,** 99–106.

Gillam, S., Astell, R. C., and Smith, M. (1980). *Gene* **12,** 129–137.

Gillam, S., Jahnke, D., Astell, C., Phillips, S., Hutchison, C. A., and Smith, M. (1979). *Nucleic Acids Res.* **6,** 2973–2985.

Holland, M. J., Holland, J. P., and Jackson, R. A. (1979). *Methods Enzymol.* **68,** 408–419.

Hutchison, C. A., Phillips, S., Edgell, M. H., Gillam, S., Jahnke, P., and Smith, M. (1978). *J. Biol. Chem.* **253,** 6551–6560.

Inouye, M., and Halegoua, S. (1980). *CRC Crit. Rev. Biochem.* **7,** 339–371.

Inouye, S., Soberon, X., Franceschini, T., Nakamura, K., Itakura, K., and Inouye, M. (1982). *Proc. Natl. Acad. Sci. USA* **79,** 3438–3441.

Inouye, S., Franceschini, T., Sato, M., Itakura, K., and Inouye, M. (1983). *EMBO J.* **2,** 87–91.

Kudo, I., Leinweber, M., and RajBhandary, U. L. (1981). *Proc. Natl. Acad. Sci. USA* **78,** 4753–4757.

Maxam, A. M., and Gilbert, W. (1980). *Methods Enzymol.* **65,** 499–559.

Montell, C., Fisher, E. F., Caruthers, M. H., and Berk, A. J. (1982). *Nature (London)* **295,** 380–384.

Nakamura, K., and Inouye, M. (1981). *Mol. Gen. Genet.* **183,** 107–114.

Nakamura, K., and Inouye, M. (1982). *EMBO J.* **1,** 771–775.

Nakamura, K., Masui, Y., and Inouye, M. (1982). *J. Mol. Appl. Genet.* **1,** 289–299.

Razin, A., Hirose, T., Itakura, K., and Riggs, A. D. (1978). *Proc. Natl. Acad. Sci. USA* **75,** 4268–4270.

Simons, G. F. M., Veeneman, G. H., Konings, R. N. H., Van Boom, J. H., and Schoenmakers, J. G. G. (1982). *Nucleic Acids Res.* **10,** 821–832.

Suggs, S. V., Hirose, T., Miyake, T., Kawashima, E. H., Johnson, M. J., Itakura, K., and Wallace,

R. B. (1981). *In* "Developmental Biology Using Purified Genes" (D. D. Brown, and C. F. Gox, eds.), pp. 683–693. Academic Press, New York.)

Vlasuk, G. P., Inouye, S., Ito, H., Itakura, K., and Inouye, M. (1983). *J. Biol. Chem.*, in press.

Wallace, R. B., Johnson, P. F., Tanaka, S., Schöld, M., Itakura, K., and Abelson, J. (1980). *Science (Washington, D.C.)* **209,** 1396–1400.

Index

C

D

E

I

L

R

X

Y